Climate Change and Agricultural Development

Two of the greatest current challenges are climate change (and variability) and food security. Feeding nine billion people by 2050 will require major efforts aimed at climate change adaptation and mitigation. One approach to agriculture has recently been captured by the widely adopted term of "Climate Smart Agriculture" (CSA). This book not only explains what this entails, but also presents practical on-the-ground studies of practices and innovations in agriculture across a broader spectrum, including agroecology and conservation agriculture, in less developed countries.

It is shown that CSA is not a completely new science and a number of its recommended technologies have been used for some time by local farmers all over the world. What is relevant and new is 'the approach' to exploit their adaptation and mitigation potential. However, a major limitation is the lack of evidence-based knowledge that is necessary for policy makers to prepare strategies for adaptation and mitigation. This book assembles knowledge of CSA, agroecology and conservation agriculture, and perspectives from different regions of the world, to build resilient food systems.

The first part analyzes the concept, opportunities and challenges, and provides a global perspective, drawing particularly on studies from Africa and Asia. The second part of the book showcases results from various studies linked to soil, water and crop management measures from India and other regions. The third section assesses the needs for an enabling policy environment and mainstreaming gender for up-scaling and/or out-scaling innovations.

Udaya Sekhar Nagothu is a Professor and Director (International Projects) at NIBIO (Norsk institutt for biookonomi /Norwegian Institute of Bioeconomy Research), Ås, Norway. He is the editor of *Food Security and Development: Country Case Studies* (Routledge, 2015).

Other books in the Earthscan Food and Agriculture Series

Climate Change and Agricultural Development

Improving resilience through climate smart agriculture, agroecology and conservation

Edited by Udaya Sekhar Nagothu

Routledge
Taylor & Francis Group

LONDON AND NEW YORK

earthscan
from Routledge

First published 2016 by Routledge

2 ParK Square, Milton ParK, AbinGdon, Oxfordshire OX14 4RN

52 Vanderbilt AVenue, New YorK, NY 10017

Routledge is an imprint of the Taylor & Francis Group, an informa business

First issued in paperbacK 2018

British Library Cataloguing-in-Publication Data
A catalogue record for this book is available from the British Library

Library of Congress Cataloging in Publication Data
Names: Nagothu, Udaya Sekhar, editor.
Title: Climate change and agricultural development : improving resilience through climate smart agriculture,
agroecology and conservation / edited by Udaya Sekhar Nagothu
Other titles: Earthscan food and agriculture.
Description: New York, NY : Routledge, 2016. |
Series: Earthscan food and agriculture series |
Includes bibliographical references and index.
Identifiers: LCCN 2016001148| ISBN 9781138922273 (hbk) |
ISBN 9781315685953 (ebk)
Subjects: LCSH: Crops and climate. | Climatic changes. | Agricultural systems. | Food supply.
Classification: LCC S600.7.C54 C56247 2016 | DDC
630.2/515--dc23
LC record available at http://lccn.loc.gov/2016001148

ISBN: 978-1-138-92227-3 (hbk)
ISBN: 978-1-138-36408-0 (pbk)

Typeset in Bembo
by HWA Text and Data Management, London

Contents

Figures

Tables

Contributors

Lakshmanan A. is professor of biotechnology at Tamil Nadu Agricultural University, Coimbatore, India.

Monday Ahonsi is a senior scientist and plant pathologist at the Biosciences Eastern and Central Africa Hub of the International Livestock Research Institute (BecA-ILRI Hub).

Rachel Bezner Kerr is associate professor in the Department of Development Sociology at Cornell University, USA, and the project director of the Malawi Farmer-to-Farmer Agroecology project.

Andrew Borell is crop physiologist at the University of Queensland, Australia, and Centre Leader of the Queensland Government's Hermitage Research Facility.

Molly E. Brown is associate professor at the University of Maryland, College Park, Maryland, USA.

Nicholas Clarke is research professor at the Norwegian Institute of Bioeconomy Research, Ås, Norway.

Laifolo Dakishoni is coordinator of the Soils, Food and Healthy Communities (SFHC) organization, and co-principal investigator of the Malawi Farmer-to-Farmer Agroecology project.

Johannes Deelstra is senior scientist at the Norwegian Institute of Bioeconomy Research, Ås, Norway.

Appolinaire Djikeng is director of Biosciences eastern and central Africa–International Livestock Research Institute (BecA-ILRI) Hub.

Dong Hongmin is senior research fellow at the Institute of Environment and Sustainable Development in Agriculture, Chinese Academy of Agriculture Sciences in Beijing.

Niek van Duivenbooden is a senior scientist specializing in food security and soil fertility in Africa, based at WUR-Alterra in Wageningen, the Netherlands.

Sita Ghimire is a senior scientist at Biosciences eastern and central Africa – International Livestock Research Institute (BecA-ILRI) Hub.

Gong Daozhi is associate director of the Key Lab for Dryland Agriculture, Ministry of Agriculture, People's Republic of China.

Inga Greipsland is researcher at the Norwegian Institute of Bio-economy Research, Ås, Norway.

Hao Weiping is division head of Dryland Agriculture Research, Institute of Environment and Sustainable Development, Chinese Academy of Agricultural Sciences, Beijing.

Thorsten Huber is a senior professional specializing in ecosystem-based adaptation to climate change working for GIZ.

Raj K. Jat is cropping system agronomist in Borlaug Institute of South Asia (BISA).

Robyn Johnston is a principal researcher at IWMI Southeast Asia office, based in Vientiane, Lao PDR.

Gurava Reddy K. is principal agricultural information officer at the Agricultural University in Andhra Pradesh, India.

Krishna Reddy K. is regional researcher at International Water Management Institute, Hyderabad, India.

Senthilraja K. is research associate in Agro Climate Research Centre, Tamil Nadu Agricultural University, Coimbatore, India.

Suresh Reddy K. is an agronomist and works as an independent consultant in Andhra Pradesh, India.

Ritika Khurana is a research consultant with CGIAR Research Program on Climate Change, Agriculture and Food Security (CCAFS), IWMI, New Delhi, India.

Solveig Kolberg is research scientist at the Norwegian Institute of Bioeconomy Research, Ås, Norway.

Guillaume Lacombe is a principal researcher at IWMI Southeast Asia office, based in Vientiane, Lao PDR.

Li Yingchun is researcher at the Institute of Environment and Sustainable Development in Agriculture, Chinese Academy of Agricultural Sciences

Li Yue is Division head of climate change research at IEDA, Chinese Academy of Agricultural Sciences.

Liu Buchun is division head, agrometeorological disaster reduction research at IEDA, Chinese Academy of Agricultural Sciences.

Isaac Luginaah is a professor in the Department of Geography, Western University, and a Canada Research Chair in Health Geography.

Esther Lupafya is AIDS director of Ekwendeni Hospital and co-principal investigator of the Malawi Farmer-to-Farmer Agroecology project.

Jat M. L. is senior cropping systems agronomist & CIMMYT-CCAFS South Asia Coordinator at International Maize and Wheat Improvement Centre (CIMMYT).

Arasu M. S. is engineer at the Water Resources Department of the Government of Tamil Nadu, India.

Ma Xin is senior research fellow at the Institute of Agricultural Environment and Sustainable Development in Agriculture, Chinese Academy of Agricultural Sciences.

Kaushik Majumdar is the director of the South Asia Program of International Plant Nutrition Institute, based in Gurgaon, India.

Matthew P. McCartney is a principal researcher and head of IWMIs Southeast Asia office, based in Vientiane, Lao PDR.

Mei Xurong is director general, Department of Research Management, Chinese Academy of Agricultural Sciences.

Irene Moed is junior researcher specializing in water-food security issues in CSA in East-Africa. She is based at WUR-Alterra in Burundi.

Manikandan N. is research associate in Agro Climate Research Centre, Tamil Nadu Agricultural University, Coimbatore, India.

Udaya Sekhar Nagothu is professor of development studies and coordinator, international projects at the Norwegian Institute of Bioeconomy Research, Ås, Norway.

Hanson Nyantakyi-Frimpong is post-doctoral student at the University of Toronto, Canada, and a collaborating researcher with the Malawi Farmer-to-Farmer Agroecology project.

Rengalakshmi R. is gender specialist at the M.S. Swaminathan Research Foundation, Chennai, India.

Sumathi S. is research associate in Agro Climate Research Centre, Tamil Nadu Agricultural University, Coimbatore, India.

Tek B. Sapkota is a climate change and agricultural system scientist working in Sustainable Intensification Programme of CIMMYT.

Lizzie Shumba is the field coordinator of the SFHC, Malawi and co-Principal Investigator of the Malawi Farmer-to-Farmer Agroecology project.

Clare Maeve Stirling is senior scientist with the Sustainable Intensification Programme and project leader of CIMMYT's Climate Change, Agriculture and Food Security (CCAFS) project.

Mehreteab Tesfai is research scientist at the Norwegian Institute of Bio-economy Research, Ås, Norway.

Geethalakshmi V. is professor at Tamil Nadu Agricultural University, Coimbatore, Tamil Nadu, India, and specializes in agricultural meteorology and climate change.

Xiong Wei is principal researcher at the Institute of Environment and Sustainable Development in Agriculture, Chinese Academy of Agricultural Sciences.

Preface

The Earth's climate is always in flux, but recent changes pose a major threat to humanity and this is going to exacerbate food insecurity in many countries. We already see how this is offsetting political instability and crisis, one result being a dramatic increase in migration. With 30 per cent more people expected by 2050 and increasing risk of flooding, heat stress and water scarcity, farmers will face a more challenging situation than before. Climate preparedness is an urgent need and requires new paradigms of agricultural development to minimize risks.

This book demonstrates how crucial it is to adapt to climate change through innovative and sustainable farming approaches. The book contextualizes the complexity of addressing climate change through a multidisciplinary lens. Taking a pluralistic approach, it acknowledges the ongoing discourses enmeshed in the constructs of divergent scientific views and ideologies. The disparate chapters draw experiences of scientists from various disciplines and regions in the world and provide evidence to show the importance of developing smallholder-friendly and sustainable agriculture systems including climate smart agriculture, agro-ecology and conservation agriculture. The emphasis on the relevance of gender, social inclusion, capacity building, policy and investments is deliberate and seen as a necessity to address climate change.

The main message is that any new paradigm of agricultural development we advocate must benefit farmers and the environment at the end of the day. I strongly believe that the timing of the book published after the Paris Climate Change Conference 2015 will contribute to the contemporary dialogue, theory and practice related to smallholder adaptation to climate change and food (in) security around the world.

1 Climate smart agriculture

Is this the new paradigm of agricultural development?

Udaya Sekhar Nagothu, Solveig Kolberg and Clare Maeve Stirling

Introduction

Climate change poses a major threat to humanity and is likely to exacerbate political and economic instability in many countries as tensions fuelled by food insecurity rise. By 2050, climate change is expected to impact negatively on more than half of all food crops in sub-Saharan Africa, and at least 22% of the area cultivated by the world's most important crops, most notably rice (Campbell *et al.*, 2011). This is against a backdrop of a rising global population that is predicted to reach 9 billion by mid-century, requiring a 70% increase in global food production (Miller *et al.*, 2010; FAO, 2013). Some would argue that adaptation strategies that focus on increasing food production are misled because there is already sufficient food produced to feed more than the projected 2050 global population. Instead, the focus should be on creating equitable and efficient food systems that are sustainable and provide secure access for all (Holt-Giménez and Altieri, 2012; IAASTD, 2008). This includes addressing losses and waste in the food chain that result in as much as one-third of global food production not being consumed. The ambiguity over whether to produce more food or not, has whipped up a serious debate amongst the different groups who view the issues of food insecurity through their own ideological lenses. The fundamental question is, *what kind of agricultural development is desirable and just to feed the world in the future?*

Paradigms of future agricultural development need to balance growth with environmental sustainability. According to proponents, climate smart agriculture (CSA) is one such approach that aims to address the challenges of food security and climate change by sustainably increasing agricultural productivity whilst adapting to climate change and reducing greenhouse gas (GHG) emissions (FAO, 2011). A differentiating aspect of CSA is the notion that agriculture has a huge potential to mitigate climate change by reducing GHG emissions and sequestering carbon in soils. Those in favour of CSA argue that if soils can be used to fix carbon dioxide from the atmosphere, then they can generate carbon credits that can be sold to polluters who want to offset their emissions. That said, there has been heated debate recently about the mitigation potential of certain agricultural practices (Neufeldt *et al.*, 2013) with the general consensus being that practices such as no-tillage have a limited ability to sequester carbon in the soil (Powlson *et. al.*, 2014). Environmentalists argue that carbon markets generated due to CSA will only serve the corporate sector and marginalise smallholder farmers further. Therefore, an important question is, whether the ongoing debate on climate adaption, mitigation and food security is helping smallholders and the environment in any way.

The sustainable development model that evolved from the environmental movement recognised the need to limit growth but managed to avoid the conflict that exists between continued growth and environmental sustainability by reframing the economic objective as 'development'. A more mainstream model is that of 'Green Growth' or 'Environmental Keynesianism' which not only insists on the compatibility of growth and the environment but also argues that protecting the environment can actually stimulate better growth. Blackwater (2012) believes that all these paradigms are fundamentally flawed and that 'Green Growth' depends on the consumer economy that in turn relies on continuous economic growth that is damaging to the environment. Environmentalists, on the other hand, propose to reduce consumer demand and halt the promotion of endless growth which Blackwater (2012) argues may be good for the environment but is economically untenable. There are other concepts such agro-ecology and food sovereignty that are gaining momentum and do not prefer to be associated with CSA. Their main criticism is that CSA does not have as developed a set of approaches as agro-ecology. The role of corporations, governments and scientists in deciding what is CSA is currently under political contestation. CSA needs to be attentive to ecological limits and social inequities as well as clarify the specific practices associated with the term and recognising the trade-offs and limitations (Harvey *et al.,* 2014; Neufeldt *et al.,* 2013). Some realignment of development paradigms will be required to generate new models of agricultural development that involve some form of socialised investment, with social equity and environmental sustainability as the main objectives, rather than a purely profit motive (Patel, 2009; Patel, 2012).

Is CSA then, the new Holy Grail of agricultural development, as some scientific groups claim? According to Neufeldt *et al.* (2013), CSA may have the potential to establish the climate change-agriculture nexus. Many in the scientific community consider CSA as a strategic tool to leverage support and investment from policy makers, donor agencies and the private sector (CCAFS, 2014). As a relatively new concept, CSA has become relatively popular in the international community (FAO, 2009a, FAO, 2009b), although it has been criticised for being too all-encompassing and lacking an adequate scientific evidence base. Many CSA interventions are highly location-specific and knowledge-intensive and it will require considerable effort to develop the knowledge and capacities needed to make CSA a reality and to develop local specific CSA measures that are socially and environmentally acceptable and sustainable (FAO, 2013).

As scientists, we need to ask ourselves, *what does CSA or for that matter any new agricultural paradigm really mean to a smallholder remotely located?* During a visit to a coastal village named Rang Dong in Nam Dinh province in January 2015, we asked a farmer '*What is your main concern right now?*' The farmer's response was unsurprising, '*the increasing salinity levels affecting my rice crop.*' Salinity levels are increasing each year in Vietnam due to seawater intrusion along the coast. Like millions of other farmers in the Red River Delta or the Mekong River Delta in Vietnam and other such vulnerable regions across the world, farmers are desperately looking for crop varieties and technologies adaptable to extreme weather and climate. Whilst adaptation to extreme weather is a priority of the smallholder, GHG mitigation will rarely be a deciding factor unless there are co-benefits such as increased profits. What, in an overarching sense, should scientists then be developing to make the new agricultural approaches more meaningful to millions of smallholders and the environment?

This book will attempt to address some of these thought-provoking questions and search for suitable answers that will contribute to the ongoing debate and knowledge development process at large. The multidisciplinary approach reflects the need to address different biophysical, social and institutional aspects of adaptation in order to develop the most suitable interventions to address the complexity of climate change. In this chapter, we present the conceptual framework and the need for more clarity in the concept, discuss the different components or key elements desired for ensuring successful implementation of CSA, highlight the opportunities, as well as gaps and limitations and finally provide some concluding remarks.

New paradigms of agricultural development

Climate Smart Agriculture

The FAO coined the term CSA in 2010 for the first time in the background document prepared for The Hague Conference on Food Security, Agriculture and Climate Change. Here, CSA was defined as: 'agriculture that sustainably increases productivity, resilience (adaptation), reduces/removes GHGs (mitigation), and enhances achievement of national food security and development goals' (FAO, 2011). The notion was that the agriculture sector needed to become climate-smart in order to reach the sustainable agriculture and rural development goals which, if achieved, would contribute to the Millennium Development Goals (MDGs) of reducing hunger and improving environmental management (ibid.). The FAO and the World Bank have principally led the further development and use of the concept, while the Consultative Group on International Agricultural Research (CGIAR) has taken the scientific lead (Scherr *et al.*, 2012). The concept has evolved since then from one that attempts to set a global agenda for investments in agriculture research and innovation to one that aims to benefit principally smallholders and vulnerable people in developing countries (Neufeldt *et al.*, 2013).

The CSA concept by definition envisions a transformation of agriculture systems to achieve short-and-long-term agricultural development goals that integrate the three dimensions of sustainable development (economic,

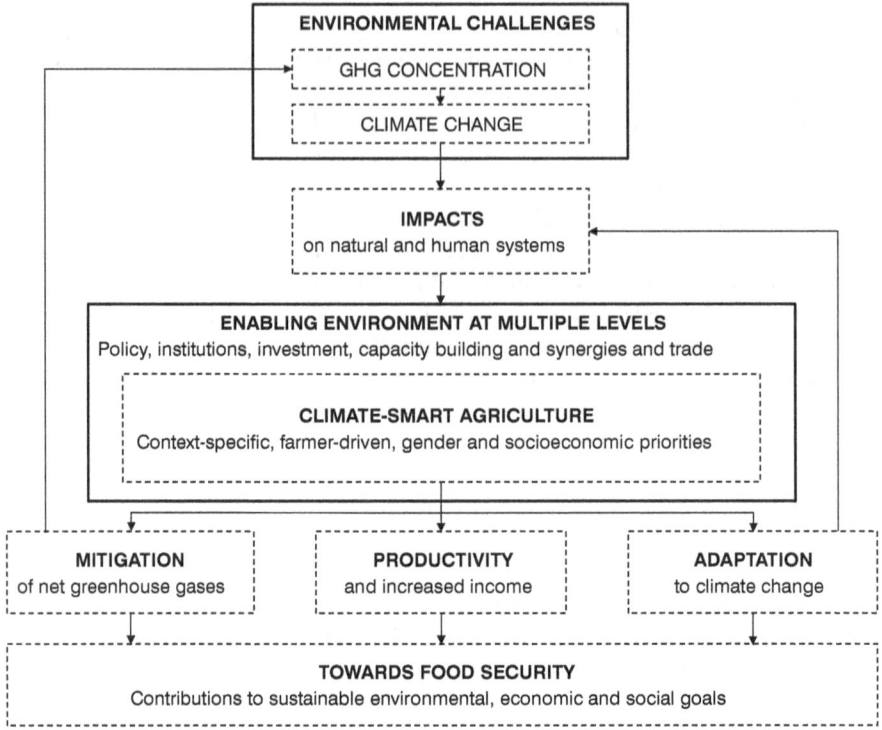

Figure 1.1 Framework for climate-smart agricultural landscapes

social and environmental) by simultaneously addressing food security and climate challenges (Branca *et al.,* 2011). The CSA perspective acknowledges agriculture's contribution to global GHG emissions as well as its vulnerability to climate change. It is now built on three main pillars similar to the above definition, includes income and separates adaptation from resilience (FAO, 2013) thus aiming at: i) Sustainably increasing agricultural productivity and incomes; ii) adapting and building resilience to climate change; and iii) reducing net greenhouse gases emissions, where possible. Figure 1.1 shows a proposed framework for CSA illustrating the main components and inter-linkages needed to achieve the desired outcomes.

Policy, institutions, synergies and trade-offs, investments and capacity building, besides technologies, make up a crucial part of the enabling environment that could help to integrate CSA into strategies and plans at regional, national, and local levels and across landscapes. At the same time, CSA needs to be locally specific, farmer-driven and socially equitable. Integration of these priorities at different levels is critical for the success of CSA. The relative importance of the three main CSA elements (adaptation, mitigation and productivity) depends on the local context and stakeholder preferences and need to identify potential synergies and trade-offs (Campbell *et al.*, 2014). The mitigation and adaptation components[1] are both considered climate-risk strategies, and the two components also address the efficiency of production, i.e. sustainable increase in agriculture yields and income. Mitigation here is primarily concerned with the potential of CSA technologies to reduce net GHGs in the atmosphere, and adaptation addresses climate change impacts[2]. It is not necessary that a particular technology should meet all the three criteria, and in most cases it may not.

Improving resource-use efficiency is at the core of the concept of a green economy[3]. Thus, CSA can be a central driver for a green economy, directly by increasing resource-use efficiency and resilience, and indirectly by developing related services and enterprises (FAO, 2012b). Given the complex interactions between adaptation, mitigation and productivity interventions, appropriate indicators and analytical methods are critical to establish a baseline that includes all relevant information, as well as for meaningful monitoring and evaluation of CSA initiatives (FAO, 2012a). CSA involves more than the adoption of individual agricultural practices such as conservation agriculture or alternate wetting and drying in rice systems, though such interventions may be critical components of a CSA strategy in specific locations and countries (ibid.). It not only includes proven sustainable land management techniques such as improved soil and water conservation, micro-watershed farming, crop diversification and crop rotation, agroforestry, improved grazing and pasture management and land restoration that can address CSA goals simultaneously (Scherr *et al.*, 2012), but also, inventive practices such as climate risk management that refers to various aspects of the risk management process using climate information to better cope with climate impacts on development and resource management problems (Selvaraju, 2010). Examples are early warning systems, risk insurance, capacity building and gender considerations added into plans, programs and policies. CSA, according to the proponents, is thus a holistic approach that promotes sustainable agricultural development, from farm to fork, all along the value chain, to guarantee a farmer better income and good working conditions. How it transforms itself when put into practice is yet to be seen.

Other relevant agricultural development concepts

A concept closely interlinked and relevant in the current discourse on CSA is sustainable intensification (SI). SI advocates the need to increase productivity whilst minimising pressure on the environment mainly through improved resource use efficiency. The theory of SI resonates very well with the majority of the policy makers' goals since it is consistent with the larger narrative of economic growth as the indispensable component of human progress (Garnett, 2015). Critics argue whether it is feasible to decouple growth from environmental impacts and question whether improvements in technology alone will be sufficient. It may need a change in human behaviour; the way we consume and manage our natural resources. Significant amounts of food produced is lost or wasted. This cannot be allowed to continue and drastic measures need to be taken to reduce the food wastage if we are serious about achieving the goals of environmental sustainability. CSA, SI and for that matter any other similar approaches to food security and climate change need to take a holistic view.

CSA and SI are both part of a multi-pronged approach towards global food security (Campbell *et al.*, 2014). Another concept that has gained popularity in recent years is 'Agroecology' which it is claimed is more suitable for smallholders compared to CSA and does not depend on fossil fuels thereby helping to reduce global warming (La Via Campesina, 2015). The 'landscape approach' can encompass CSA and involves the management of production systems and natural resources over an area sufficiently large enough to produce vital ecosystem services yet small enough that activities can be managed by people using the land (FAO, 2013). The Landscape approach includes three features: i) climate-smart practices at the field and farm scale; ii) diversity of land use across the landscape to provide resilience; and iii) land use interactions management at landscape scale to achieve social, economic and ecological impacts. This approach could be key to achieving the multiple goals of CSA (Scherr *et al.*, 2012) but to date there have been few studies of the processes required for implementation to achieve climate-smart landscapes.

On the one side, a diversity of CSA measures and mitigation options provide flexibility, but on the other side it also raises challenges in measuring, reporting and verification (Campbell *et al.*, 2011). Accordingly, CSA needs a clearer meaningful set of criteria that define and ensure an intervention is smart for the current and future climate. It should provide a clear direction for research and development, and acknowledge the possible synergies and trade-offs between the three postulated goals for the agriculture sector. In the following sections we will discuss these issues in more depth in terms of what a CSA-enabling environment should consist of and the cross-cutting issues that need to be addressed, together with the opportunities and barriers for successful implementation of CSA.

Measures needed to a balanced CSA approach

Global initiatives

At the global level, CCAFS has been the main flagbearer of CSA together with its CGIAR centres. Since 2010, CCAFS has been working at different levels to promote CSA through regional and local initiatives. A major step in CSA development was the establishment of the Global Alliance on Climate-Smart Agriculture, a voluntary alliance that came into existence at the UN Secretary General's Climate Summit in September 2014 (CCAFS, 2014). This strategic global initiative to scale climate smart agriculture has started to attract the attention of governments, international agencies and the private sector. Some, however, have criticised the Alliance for lacking transparency as well as sufficient environmental and social safeguards, and for not questioning the structural causes of climate change and hunger (CIDSE, 2015).

At the UN Secretary General's Climate Summit in September 2014, a few governments and international agencies made a verbal commitment to provide funding support for the implementation of CSA. Several countries are now in the process of preparing country profiles to map their status, implications and potential of climate smart interventions. Reliable country profiles can provide an entry point for agencies interested in investments in the agriculture sector (CCAFS, 2014). Besides, emerging regional (Africa) Alliances on Climate-Smart Agriculture (ACSA) provide a platform for shared learning and collaboration among partners in the respective regions. The global knowledge base on CSA has improved over the last 3 years with the addition of more evidence-based results gathered from different sources including farmers. For the first time more than 100 experts associated with the Knowledge Action Group of the Global Alliance on Climate-Smart Agriculture met in Montpellier in 2015. The Conference aimed to identify knowledge priorities for CSA and build partnerships to make these priorities possible. The coming years will show whether the Alliance will live up to its expectations and serve the interests of smallholders and the environment or remains another bureaucratic organization.

Enabling the right policy and institutional environment

Even though the technical side of CSA receives most attention, the policy and institutional aspects of CSA are undeniably crucial for successful adoption and contribution to resilience. Institutional and policy support will be required for major transformation of the agriculture sector with better aligned policy approaches across agricultural, environmental and financial boundaries (Behnassi *et al.*, 2014a). Institutions have a fundamental role in CSA implementation, especially when it comes to promoting social inclusiveness, developing a reliable knowledge base, infusing knowledge on climate and responses, enabling local-scale innovation, stimulating investments and providing insurance that can empower smallholders, women and poor resource-dependent societies to adopt and benefit from CSA (Meinzen-Dick *et al.*, 2012). Social inequalities will shape policy responses to global climate change. Appropriate policy and institutional support could help in promoting CSA systems. Mainstreaming climate change into agricultural policy is essential for promoting CSA, but the link between climate change and agriculture is still limited in most national development, food security and climate adaptation plans. In many developing countries, the design and operation of agricultural support could be substantially improved, and SI and CSA objectives could be integrated within this comprehensive policy context (Campbell *et al.*, 2014).

Achieving the multiple objectives of CSA requires a cross-sectorial approach (landscape level) that involves an alignment of sectorial policies and planning frameworks (FAO, 2013). Nevertheless, people's willingness to embrace climate change policies depends on their perception of the attached benefits and costs (Peters *et al.*, 1999). Adger *et al.* (2011) argues that limits to adaptation go beyond thresholds in terms of biological, technological or economic variables, but emerge from within society and depend on ethics, knowledge, attitudes to risk and culture. There are considerable gaps in knowledge about the complexity of policy processes and its link to adaptation to climate change. Thus, there is a need to develop further, an evidence base on how to effectively combine social safety net measures to mitigate vulnerability to climate change in different contexts (FAO, 2012a). In particular, local-level institutions and inter-institutional synergies play a crucial role in facilitating the adoption of CSA strategies. Therefore, it is important to identify the local institutions needed to support the transition to CSA and related adaptation and mitigation systems.

The need for local-specific and a farmer-driven approach

There are divergent views about CSA and the way we can move forward. Multiple views open doors for discussions and this shapes stakeholder arguments for future actions. In the current discussions, we are concerned with CSA and whether it should be primarily technology focussed or farmer driven. The role of technology and its potential to address the problems we face in agriculture today is without doubt important and undisputable. Technology has played a key role in the advancement of agriculture and food production during the past decades. At the same time, technology has not been able to solve the myriad of problems we face today, from climate and economic changes. The Green Revolution had both positive and negative impacts on food security and the livelihoods of smallholders and the environment (Pingali, 2012; Swaminathan, 2012; Patel, 2012). We need to learn from the experiences as we devise new strategies for the future (Mosley, 2003).

Rather than approaching from a technological perspective, looking at CSA and similar farming systems from the perspective of farmers and the environment, their needs, resources, practices and policies may offer a sustainable outcome. This involves going beyond food security, adaptation and mitigation dimensions and realising the local circumstances, social agenda and farmer needs. This implies the need for a strong sociological filter to sharpen the focus of the CSA lens not only on the technological, but also on the socio-ecological dimension of adaptation and mitigation. CSA is criticised for increasing smallholder participation in carbon markets and carbon mitigation, citing risks of land grabs and displacement of smallholders, supporting conventional agricultural practices, placing emphasis on technology development and transfer, and potentially excluding or marginalizing smallholder voices and priorities (Sharma and Suppan, 2011; ActionAid, 2012; Pearce, 2011; Naess, 2011). Looking from a farmer perspective, the success of CSA adoption will demand increased investments at the farm level and capacity building of smallholders.

Many effective indigenous as well as modern agricultural technologies already exist in climate vulnerable areas and several of the most effective interventions have been built on practices and technologies evolved over generations of accumulated farm experience in managing climatic risks (Campbell *et al.*, 2011). It makes sense to promote such locally existing technologies rather than introducing new ones that are unfamiliar to farmers. Even where agriculture-based practices and technologies could have the potential to increase food production and the adaptive capacity of the food production system, in addition to reducing emissions or increasing carbon storage, capturing such synergies may entail significant costs, particularly for smallholders in the short term (McCarthy *et al.*, 2011). Hence, proper care should be taken to consider the costs or risks and trade-offs they may face in the CSA development process. Extended transition times and some form of financing to compensate for low or negative returns to agriculture may be needed to realise the CSA-benefits of productivity or increased resilience during the transition and in the long term (FAO, 2012a).

A key feature of CSA is that it requires site-specific assessments to identify suitable agricultural production practices. The effects and timing for these interventions also depend on agro-ecological conditions and current and historical land-use patterns. Local engagement and farmer-led innovations are crucial in the co-development of new portfolios of practices, and information systems. Nevertheless, to sustainably increase yields, knowledge and insights gained from all current systems of agricultural production, including those based on organic principles, local indigenous knowledge, and innovative soil management interventions and plant-breeding technologies need to be harnessed and developed (Taneja *et al.*, 2014; Thornton and Lipper, 2014).

Research and development agencies play a significant role in identifying and promoting CSA strategies. They should strengthen rural societies, improve smallholder livelihoods and employment, and avoid adverse social and cultural effects on land tenure and forced migration (Campbell *et al.*, 2014). To link farmers with new information sources on climate change will be imperative and 'translating' the risks and potential margin of error that exist will help farmers make appropriate decisions (FAO, 2012a).

Knowledge base, investments and scaling out

Investments in climate change mitigation and adaptation are necessary. Fundamentally, growth and waste must be curtailed to avoid environmental disaster (Blackwater, 2012). What is required is a new mechanism for directing a significant proportion of funds or resources into such projects – one not based on earning profits. This is tricky, and it will not be easy to sell such a concept to the private sector. Such investments may have to come from the state agencies. Nevertheless, sustainable scaling up and out of CSA technologies will not be possible without support from both the state and private sources.

Factors that matter for scaling up and out of CSA initiatives involve developing a strong knowledge base (formal and informal knowledge), filling the knowledge gaps and provision of adequate investments to scale up promising CSA practices (CIAT, 2015). Knowledge profiles have to be developed at country level or even at the local level if there is a wide variation in the agro-ecological and socio-economic conditions. A reliable knowledge base should be taking inputs from scientific and informal sources, good cases, champion or lead farmers, women farmers and civil society organizations. Each country needs to develop a dynamic knowledge base or platform that is accessible and open to all. Evidence-based CSA practices that constitute part of the knowledge base can attract investments from different sources.

The agriculture sector has not been the priority for investments for a majority of the governments in developing countries so far. This could change, as CSA is now seen as a business model that can potentially increase agricultural output while maintaining or lowering amounts of inputs, such as land, water or fertiliser (IDB, 2015). Investment banks and industries have started to invest in CSA-based projects in sub-Saharan Africa and other regions. CSA, however, could face a number of hurdles to access finance, due to lengthy payback periods, as well as significant barriers to information on sustainable practices. Scaling out CSA through smallholders would need risk tolerant or even risk-free capital investments. All CSA projects may not be viable from a business point of view, but governments need to balance the trade-offs. Smallholders constitute the majority of farmers in developing countries. They can make a difference in scaling out CSA if properly supported and their needs are met. Very often smallholders are cash constrained, and lack other assets, including education or training required for adoption. On-farm testing by farmers, combined with farmer-farmer learning and extension among communities is important for successful scaling out CSA practices.

Social inclusion and balancing trade-offs

The effects of global climate change will not be distributed evenly around the world (Nagel *et.al,* 2010). According to the latest IPCC report (IPCC, 2014), people who are socially, economically, culturally, politically, institutionally, or otherwise marginalised are especially vulnerable to climate change and also to some adaptation and mitigation responses (*medium evidence, high agreement*). Vulnerabilities to climate change will also differ with social class and age in both developed and developing countries. These inequalities will be further exacerbated by the unequal burdens inflicted by climate-related disasters. Hence, while introducing new systems of agriculture, the fundamental importance of social inclusion, looking beyond the technical and economic aspects is of utmost importance. Accordingly, including social involvement and empowerment of smallholders and women within indicators for climate-resilient agriculture is a necessity for ensuring successful implementation. The agriculture sector should prepare for and remedy specific challenges posed by social inequalities magnified by global climate change both within and among states.

The food sovereignty agenda that has emerged from the farmer movements emphasises the right of local people to control their own agricultural and food production systems including local markets and food cultures (Rosset, 2008; La Via Campesina, 2009). Industrial agriculture largely has ignored the social priorities, and we cannot repeat the same mistake once again. The food sovereignty movement advocates the rebuilding of food production systems that include farmers and farmer organisations as owners of the public services and goods they generate. The movement in recent years has endorsed the agro-ecological farming approach and the social process methodologies within the approach as most suitable for smallholders (Holt-Giménez and Altieri, 2012). It further argues that national policies and reforms to promote CSA should, in fact, respect and favour the local culture and social needs.

There will be synergies and trade-offs around sustainable improvements in productivity, improving resilience and reducing net GHG emissions. It is important to account for these trade-offs as CSA programs are being designed and implemented in the future. Governments must buffer farmers from the trade-offs. Identifying appropriate ways to incentivise the uptake of climate-smart alternatives should be a key priority, as the farmers might not see the immediate benefits of these practices that could involve trade-offs in resource allocation.

Gender and vulnerability

Women constitute over 40% of the agricultural labour force, although historically, women have not been fully recognised as farmers, a situation that still prevails (Jones, 2012). The 'invisibility' of women's contributions to agriculture has been an important hindrance to their participation in decision-making (IAASTD, 2008). Often women are sidelined while designing new agricultural adaptation and mitigation programs or strategies. This is despite the fact that they contribute equally or even more in certain contexts (Edmunds *et al.*, 2013). Also, women face a plethora of gender-specific hindrances in decision-making about basic resources for production and face greater challenges compared to men in accessing productive resources, markets and services (IAASTD, 2008). Gender, in fact, needs to be mainstreamed as it serves as a lens for understanding and responding to the needs of men and women with diverse identities in almost all societies. To achieve mainstreaming requires more targeted research, as relevant studies with a gender focus in the CSA context are still rare, and there is a large knowledge gap in this field (Farnworth *et al.*, 2015; Muriel *et al.*, 2014). Though gender is getting more attention in several global initiatives in recent years, it still does not get adequate importance in the national or local programs as observed in a majority of the developing countries.

Gurung and Biggs (2010) observed that even in global programs such as REDD or REDD+, there is an absence of gender analysis that can lead to distributional injustices for women, the poor, and other marginalised groups. According to Otzelberger (2011), the 'gender blindness' that characterises many agricultural and environmental ministries in developing countries will likely lead to mitigation projects that have particular, and largely negative, impacts on women.

A gender perspective is, therefore, important and critical for successful implementation of CSA technologies. CSA interventions will influence the differentiated role of men and women who share and take responsibility for different tasks in agriculture or livestock production in developing countries (CCAFS, 2014). With the out-migration of men from rural areas for various reasons, one of them being increasing climate risks, the burden on women and their role in agriculture will increase in the future. A first and useful step here would be sensitization of scientists, managers and policymakers about the role and importance of mainstreaming gender in any future program or policy related to climate adaptation and mitigation. Moving beyond the consultation or participation element of gender mainstreaming to ensuring an equal space and voice in CSA development will also be a key for better results. A toolbox developed by CCAFS in 2014 enables development practitioners to mainstream gender and social inequality approaches (Jost *et al.*, 2014). Special efforts are required to support women farmers to lead CSA innovations together with men as this helps towards a better understanding of their roles and contributions to food and nutritional security and mitigation (CCAFS, 2014).

Opportunities and barriers for adoption

Opportunities

Although scientific understanding is rapidly evolving, critics undercut agreement about the seriousness of the climate crisis. It is obvious that critics who do not believe in climate change will not advocate CSA or other similar approaches. It will be the responsibility of the scientific community and state policy leadership to counter the criticism and motivate local adoption of CSA (Homsy and Warner, 2015).

At a local level, CSA could help to protect agriculture production systems from the adverse effects of climate change, improve farm yields and household incomes, and build stronger and more resilient communities (CCAFS, 2014). At a national level, the approach could help achieving food security and development goals while reducing GHG emissions. Moreover, CSA could lead to changes in agriculture technologies and practices that could improve the livelihoods for those locked in current, and prone to future, food insecurity and poverty cycles. Since agriculture affects and is directly affected by climate change, there are many opportunities for capturing synergies between the three main elements of CSA (adaptation, mitigation and productivity), nevertheless, in many situations trade-offs are inevitable (FAO, 2011).

There is a scope for enhancing low emission development trajectories without compromising development and food security (Behnassi *et al.*, 2014a). Specific interventions, however, only make sense within a broader, holistic approach to agricultural management. There may not be any net mitigation effect if larger on-farm efficiency shifts emissions to other parts of the landscape or food chain. Similarly, one might want to avoid emissions reductions at the expense of, for example, biodiversity conservation (Campbell *et al.*, 2011). For mitigation, it is crucial to understand both the expected increase in emissions under a conventional agricultural growth strategy, and the degree to which such business-as-usual baseline emissions could be reduced under alternative agricultural growth strategies (Thornton and Lipper, 2014). However, more targeted research and experience is needed to better understand the benefits and synergies, trade-offs and limitations of major mitigation and adaptation options, beside their implications for sustainable and equitable development that could facilitate decision-making about climate resilient pathways (IPCC, 2014).

Most climate vulnerable people live in developing countries, especially in dry and tropical regions of South Asia and Sub-Saharan Africa where agriculture productivity is projected to decrease. It is estimated that 74% of total agricultural emissions comes from developing countries, and that these same countries have 70% of the agricultural mitigation potential (Gattinger *et al.*, 2011). Accordingly, most developing countries will need to both mitigate and adapt. This should involve smallholders to ensure a climate-friendly and equitable agriculture approach with a pro-poor focus. Many societies are not even adapted to existing climatic conditions and variability, suggesting the existence of an 'adaptation-deficit' that needs to be addressed even before any agenda addressing adaptation to future climate changes (Mearns and Norton, 2010).

Barriers

In order to move towards more sustainable development models, a significant paradigm shift will be needed, particularly from the current economic models that assumes that a society can only develop by expanding its use of resources and increasing per capita consumption patterns despite the related long-term negative effects (Behnassi *et al.*, 2014a). From a climate-justice perspective, the resource-poor farmers who are most affected by climate change and have contributed least to it, need particular attention so that developing countries can improve their food security and speed their economic growth (Campbell *et al.*, 2014). Smallholders' adoption of sustainable agriculture practices, including CSA, has however been slow (Farnworth *et al.*, 2015; Zurek *et al.*, 2014). Barriers to the diffusion of new technologies and improved practices can occur at all stages, from inception to uptake of agricultural innovations and marketing. While all farmers could face obstacles, resource-poor smallholders are particularly constrained (McCarthy *et al.*, 2011; Streck, 2012). Table 1.1 lists barriers to CSA adaptation that smallholders in developing countries often encounter.

Effective indigenous, as well as modern, CSA technologies already exist but have not yet been applied at a larger scale. Scaling out the CSA approach is challenging, partly due to a lack of adequate experience as well as unrealistic timeframes to achieve the scale set by donors or governments, especially given that the focus is mainly on the hardest to reach sector of agriculture. Up to now, there are no universally accepted guidelines, or even recommendations to the implementation of CSA, nor will there be, as CSA as such is not a prescriptive approach. In the short term, some CSA practices can even lead to negative effects on yields and increased variability, and risk-averse farmers and farmers in risk-prone regions are less likely to invest in new technologies and practices with uncertain and delayed returns (McCarthy *et al.*, 2011, Streck, 2012, Branca *et al.*, 2011). In many countries, agricultural policy is inseparably linked with economic

Table 1.1 Barriers for CSA adoption

Investment barriers	Social/institutional barriers	Technological barriers
• Lack of assets, savings, credit services, and insurance • Poor access to extension services • Lack of infrastructure and equipment • Lack of reliable CSA knowledge base	• No access or limited access to markets and market information • Weak land tenure security • Lack of safety nets, social safeguards and transparency • Lack of gender perspective	• Lack of technical expertise • Existing resource degradation (for example soil or water) • Lack of baseline data (for example on forest or soil carbon content)

Source: Adapted from Streck (2012)

support for rural communities, and studies show that up-front investment costs can be a major obstacle to adoption (Garido, 2005; McCarthy *et al.*, 2011).

An example of adaptation barriers can be found in an assessment of farmers' preferences in the Indo-Gangetic Plain (Taneja *et al.*, 2014). The study indicated that even though the farmers showed interest to adopt new CSA technologies and interventions, the existence of new technologies alone was not a sufficient condition to bring about the change, as effective institutions and sustained policy support to bring the technologies within the reach of farmers were equally important for adoption on a larger scale. Smallholders should be enabled to make informed choices according to their specific needs (Streck, 2012). There is an emerging body of literature related to barriers to adaptation to climate change. In a meta-study of 81 papers, Biesbroek *et al.* (2013) identify a wide range of barriers to uptake, most of which are related to institutional and social factors of adaptation. Institutions have the potential to raise awareness, strengthen technical capacities, create enabling policy environments for action and encourage cross-sectoral and landscape approaches to climate change. Therefore, better aligned policy approaches across agricultural, environmental and financial boundaries and innovative institutional arrangements to promote their implementation will be needed (Behnassi *et al.*, 2014b). Another constraint for CSA adoption by smallholders is the size of land holdings and lack of land tenure. As the benefits of many CSA interventions, such as convervation agriculture (CA) are not realised in the first few years smallholders with little land security have no motivation to invest. Similarly, there is a lack of coherent policies, as evident from the subsidies given to fertilisers and electricity in many developing countries. These often do not encourage farmers to switch to water saving technologies or CSA practices. Often the political interests in these can be conflicting with the environmental goals. Cross-site and cross-country transfer of best practices will be critical to effective adaptation. However, scaling up will be limited by several factors, including possible nonlinearity of climate impacts and the complexity of agricultural decision making within contrasting economic and cultural settings (Campbell *et al.*, 2011). Sustained integration of research, development, farmer-led innovation, capacity building and policy is essential for creating climate-smart outputs, since CSA is meant to have impacts far beyond the farm gate, and is influenced by policy and the enabling environment.

Final remarks

It is only in recent years that CSA, as a concept, has started to gain importance within the scientific community and policy makers at the global level. In this chapter, attempts were made to review the ongoing debate on CSA and other approaches, identify opportunities, barriers and variables needed to make the CSA framework more robust. The chapter has given an account of the role of the CSA and other approaches such as SI and agroecology in fighting climate change risks affecting food security and the sustainability of agricultural production systems, as the world's population continues to grow. Some groups see CSA as merely another intangible paradigm that various global, national and local actors are attempting to push to promote their own interests. However, for smallholders and rural people facing climate change in their daily life, the success of any climate change combating initiatives, could be the difference between life and death. If the debate is to move forward, a better understanding of the criteria of CSA needs to be developed. Better clarification of the CSA concept and approach could potentially help to establish a common agenda and a greater degree of commitment both globally and locally. This could attract investments in the agriculture sector that have been limited so far as compared to other sectors such as energy and transport.

A review of the literature suggests that CSA lacks clarity, but if implemented in its true sense, it has a significant role in promoting food security and contributing to a cross-cutting range of development goals. Barriers to adoption of CSA technologies must be met with an enabling environment where farmer-led decision-making together with an evidence-based and context-specific approach need to be applied. Evidence-based gendered policies that promote secure access to land, capacity building, capital and strengthen poor farmers' ability to tackle climate change and uncertainty, and also improve productivity, are urgently needed. Linkages must be made between local, regional and national policy needs and enabling actors. We certainly have to build on experiences of farmers and the knowledge gained from decades of work on various approaches to sustainable agriculture in our efforts towards developing new paradigms of agricultural and food security systems. This requires a completely new set of interventions, ranging from community-based scenario planning and other tools to build adaptive capacity to enhancing climate information services for farmers and livestock keepers. There is no need to be disheartened by the emerging polarised debate amongst scientists on the new paradigms of agricultural development. Criticisms are useful if they are constructive and contribute to making the approaches more robust and clearer. Relevant stakeholders including farmer's organisations, governments, the scientific community and the private sector all need to find ways to work together, engage and influence the outcomes.

Notes

1 Adaptation aims at reducing the vulnerability of human or natural systems to the impacts of climate change and climate-related risks, by maintaining or increasing adaptive capacity and resilience at multiple levels. While mitigation activities should help to stabilise GHG concentrations in the atmosphere at a level that would prevent dangerous anthropogenic interference with the climate system by promoting efforts to reduce or limit GHG emissions or to enhance carbon sequestration (OECD-DAC, 2011).

2 Climate change impacts refer to 'the effects on natural and human systems of extreme weather and climate events and of climate change' (IPCC, 2014).

3 UNEP (2015) define green economy as one that results in improved human well-being and social equity, while significantly reducing environmental risks and ecological scarcities. Basically, green economy can be understood as one which is low carbon, resource efficient and socially inclusive.

References

ActionAid, (2012) Climate resilient sustainable agriculture: a real alternative to false solutions. (A backgrounder June 2012). Available at: http://www.actionaid.org/sites/files/actionaid/crsa_backgrounder_june_2012_design.pdf (accessed 15 May 2015).

Adger, W.N., Brown, K., Nelson, D.R., Berkes, F., Eakin, H., Folke, C., Galvin, K., Gunderson, L., Goulden, M., O'Brien, K., Ruitenbeek, J. and Tompkins, E.L. (2011) Resilience implications of policy responses to climate change. *Wiley Interdisciplinary Reviews, Climate Change*, 2 (5): 757–766.

Behnassi, M., Boussaid, M. and Gopichandran, R. (2014a) Achieving food security in a changing climate: the potential of climate-smart agriculture. In *Environmental Cost and Face of Agriculture in the 27 Gulf Cooperation Council Countries: Fostering Agriculture in the Context of Climate Change* (pp. 27–43) Gulf Research Centre: Cambridge. Available at: http://link.springer.com/10.1007/978-3-319-05768-2_2 (accessed 4 November 2015).

Behnassi, M., Boussaid, M. and Gopichandran, R. (2014b) *Environmental Cost and Face of Agriculture in the Gulf Cooperation Council Countries* (pp. 27–43). Available at: http://link.springer.com/10.1007/978-3-319-05768-2 (accessed 4 November 2015).

Biesbroek, G. R., Klostermann, J. E., Termeer, C. J. and Kabat, P. (2013) On the nature of barriers to climate change adaptation, *Regional Environmental Change*, 13 (5): 1119–1129.

Blackwater, B. (2012) The contradictions of environmental Keynesianism, online, http://climateandcapitalism.com/2012/06/14/the-contradictions-of-environmental-keynesianism/ (accessed 27 September 2015).

Branca, G., McCarthy, N. Lipper, L. and Jolejole, M.C. (2011) Climate smart agriculture: A synthesis of empirical evidence of food security and mitigation benefits from improved cropland management, online, http://www.fao.org/climatechange/29764-aa5796a4fb093b6cfdf05558c6dd20bb.pdf (accessed 28 October 2015).

Campbell, B., Mann, W., Melendez-Ortiz, R., Streck, C. and Tennigkeit, T. (2011) *Addressing Agriculture in Climate Change: A Scoping Report*. Available at: https://cgspace.cgiar.org/handle/10568/10306 (accessed 23 November 2015).

Campbell, B., Thornton P., Zourgmore, R. van Asten, P. and Lipper, L. (2014) Sustainable intensification: What is its role in climate smart agriculture? *Current Opinion in Environmental Sustainability*, 8, 39–43. Available at: http://linkinghub.elsevier.com/retrieve/pii/S1877343514000359 (accessed 23 November 2015).

CIDSE (2015) Climate smart agriculture: the emperor's new clothes?, online, http://www.cidse.org/publications/just-food/food-and-climate/csa-the-emperor-s-new-clothes.html (accessed 23 September 2015).

Climate Change Agriculture and Food Security (CCAFS) (2014) *Climate-smart Agriculture. Acting Locally, Informing Globally*. The CCAFS 2014 annual report, Available at: https://cgspace.cgiar.org/bitstream/handle/10568/65717/CCAFS_2014_Annual_Report.pdf.

Edmunds, D., Sasser, J. and Wollenberg, E. (2013) A gender strategy for pro-poor climate change mitigation. CCAFS Working Paper no. 36. Available at: http://ccafs.cgiar.org/publications/gender-strategy-pro-poor-climate-change-mitigation (accessed 3 November 2015).

Farnworth, C.R., Baudron, F., Andersson, J.A., Misiko, M., Badstue, L. and Stirling C.M. (2015) Gender and conservation agriculture in East and Southern Africa: towards a research agenda. *International Journal of Agricultural Sustainability*, 14 (2): 142–165. http://dx.doi.org/10.1080/14735903.2015.1065602 (accessed 4 November 2015).

Food and Agricultural Organization of the United Nations (FAO) (2009a) *Food Security and Agricultural Mitigation in Developing Countries: Options for Capturing Synergies*. October, p.82. Available at: http://www.fao.org/docrep/012/i1318e/i1318e00.pdf (accessed 4 November 2015).

Food and Agricultural Organization of the United Nations (FAO) (2009b) *Harvesting Agriculture's Multiple Benefits : Mitigation, Adaptation, Development and Food Security*. Available at: http://www.fao.org/3/a-ak914e.pdf (accessed 3 September 2015).

Food and Agricultural Organization of the United Nations (FAO) (2011) *Climate-Smart Agriculture: Managing Ecosystems for Sustainable Livelihoods*. Available at: http://www.fao.org/docrep/015/an177e/an177e00.pdf (accessed 3 November 2015).

Food and Agricultural Organization of the United Nations (FAO) (2012a) Developing a climate-smart agriculture strategy at the country level: lessons from recent experience. Second Global Conference on Agriculture, Food Security and Climate Change. Available at: http://www.fao.org/docrep/016/ap401e/ap401e.pdf (accessed 1 November 2015).

Food and Agricultural Organization of the United Nations (FAO) (2012b) *Greening the Economy with Climate Smart Agriculture*. Available at: http://www.fao.org/docrep/016/ap403e/ap403e.pdf (accessed 4 November 2015).

Food and Agricultural Organization of the United Nations (FAO) (2013) *Climate-Smart Agriculture Sourcebook*, Rome, Italy. Available at: http://www.fao.org/docrep/018/i3325e/i3325e00.htm (accessed 3 November 2015).

Garido, A. (2005) Using good economic principles to make irrigators become true partners of water and environment policies. Organization for Economic Co-operation and Development Workshop on Agriculture and Water: Sustainability, Markets and Policies, OECD.

Garnett, T. (2015) Gut feelings and possible tomorrows: (where) does animal farming fit?, Food Climate Research Network, online, http://www.fcrn.org.uk/sites/default/files/fcrn_gut_feelings.pdf (accessed 22 September 2015).

Gattinger, A. Jawtusch, J. Muller, A. and Mader, P. (2011) No-till agriculture–a climate smart solution. Available at: http://orgprints.org/id/eprint/20302 (accessed 19 February 2015).

Gurung, B. and Biggs, S. (2010) Institutional change: the unanticipated consequences of action, *Development in Practice*, 20 (8): 1014–1026.

Harvey, C. A., Chacón, M., Donatti, C. I., Garen, E., Hannah, L., Andrade, A., and Wollenberg, E. (2014) Climate-smart landscapes: opportunities and challenges for integrating adaptation and mitigation in tropical agriculture, *Conservation Letters*, 7 (2): 77–90, http://doi.org/10.1111/conl.12066 (accessed 29 January 2016).

Holt-Giménez, E. and Altieri, M. (2012) Agroecology, food sovereignty and the new green revolution, *Agroecology and Sustainable Food Systems,* 37: 90–102.

Homsy, C.G. and Warner, M.E. (2015) Cities and sustainability: polycentric action and multilevel governance, *Urban Affairs Review*, 51 (1): 46–73.

Inter-American Development Bank (IDB) (2015) IDB approves climate-smart agriculture fund, online, http://www.iadb.org/en/news/news-releases/2015-07-16/climate-smart-agriculture-fund,11207.html (accessed 23 September 2015).

Intergovernmental Panel for Climate Change (IPCC) (2014) *Climate Change 2014: Impacts, Adaptation, Vulnerability. Part A: Global and Sectoral Aspects*. Available at: http://ipcc-wg2.gov/AR5/report/final-drafts/ (accessed 4 November 2015).

International Assessment of Agricultural Knowledge, Science and Technology for Development (IAASTD) (2008) Agriculture at a crossroads: international synthesis report. In McIntyre, B.D. *et al.* (eds), *A Synthesis of the Global and Sub-global IAASTD Reports*, Washington DC, Island Press.

International Centre for Tropical Agriculture (CIAT) (2015) World Bank: Smart investments in "triple win" agriculture, online, http://ciatblogs.cgiar.org/support/world-bank-smart-investments-in-triple-win-agriculture/ (accessed 23 September 2015).

Jones, M. (2012) First Global Conference on Women in Agriculture (GCWA): Empowering women in agriculture: rethinking the agricultural needs and actions through the eyes of women, *Food Security*, 4 (2): 305–306.

Jost, C., Ferdous, N. and Spicer, T.D. (2014) *Gender and Inclusion Toolbox: Participatory Research in Climate Change and Agriculture*, Copenhagen: Denmark.

La Via Campesina (2009) Small scale sustainable farmers are cooling down the earth, online, http://viacampesina.net/downloads/PAPER5/EN/paper5-EN.pdf (accessed 15 October 2015).

La Via Campesina (2015) We want system change, not climate change. Available at: http://viacampesina.org/en/index.php/actions-and-events-mainmenu-26/-climate-change-and-agrofuels-mainmenu-75/1751-we-want-system-change-not-climate-change (accessed 04 March 2015).

McCarthy, N., Lipper, L. and Branca, G. (2011) Climate-smart agriculture: smallholder adoption and implications for climate change adaptation and mitigation. Available at: http://www.indiaenvironmentportal.org.in/files/file/Climate Smart Agriculture Smallholder Adoption and Implications for Climate Change Adaptation and Mitigation.pdf (accessed 19 February 2015).

Mearns, R. and Norton, A. (eds) (2010) *Social Dimensions of Climate Change: Equity and Vulnerability in a Warming World*, World Bank: Washington, DC.

Meinzen-Dick, R., Bernier, Q., Haglund, E., Markelova, H. and Moore, K. (2012) *Identifying the Institutions for Climate-Smart Agriculture*. pp. 1–21. Available at: http://dlc.dlib.indiana.edu/dlc/bitstream/handle/10535/8906/HAGLUND_0975.pdf?sequence=1 (accessed 4 November 2015).

Miller, C., Richter, R., McNeillis, P. and Mhlanga, N. (2010) Agricultural investment funds for developing countries. Available at: http://190.11.224.74:8080/jspui/handle/123456789/517 (accessed 23 February 2015).

Mosley, P. (2003) *A Painful Ascent: The Green Revolution in Africa*. London: Routledge.

Muriel, J., Twyman, J. and Gonzalez, C. (2014) Women's roles in rice production in Northern Peru. Available at: http://www.slideshare.net/CIAT/womens-roles-in-rice-production-in-northern-peru (accessed 9 July 2015).

Naess, L.O. (2011) Climate smart agriculture: the new holy grail of agricultural development? Available at: http://www.future-agricultures.org/component/content/article?id=7643:climate-smart-agriculture-the-new-holy-grail-of-agricultural-development (accessed 30 April 2015).

Nagel, J. Dietz, T., and Broadbent, J. (2010) Sociological perspectives on global climate change, online, http://www.asanet.org/research/NSFClimateChangeWorkshop_120109.pdf (accessed 27 September 2015)

Neufeldt, H., Jahn, M., Campbell, B. M., Beddington, J. R., DeClerck, F., Pinto, A. D. and Zougmoré, R. (2013) Beyond climate-smart agriculture: toward safe operating spaces for global food systems. *Agriculture & Food Security*, 2 (12). Available at: http://www.biomedcentral.com/content/pdf/2048-7010-2-12.pdf (accessed 17 March 2015).

OECD-DAC (2011) Tracking aid in support of climate change mitigation and adaptation in developing countries. In *Identifying Opportunities for Climate-smart Agriculture Investments in Africa*. Available at: http://www.oecd.org/development/stats/rioconventions.htm (accessed 24 September 2015).

Otzelberger, A. (2011) Gender-responsive strategies on climate change: recent progress and ways forward for donors. *Institute of Development Studies*, (June), pp. 1–58. Available at http://www.eldis.org/vfile/upload/1/document/1107/Gender%20responsive%20strategies%20onclimate%20change%20progress%20and%20ways%20forward%20for%20donors.pdf (accessed 2 November 2015).

Patel, R. (2009) Food sovereignty, *The Journal of Peasant Studies*, 36 (1): 663–706.

Patel, R. (2012) The long green revolution, *Journal of Peasant Studies*, 40 (1): 1–63.

Pearce, F. (2011) Can 'climate-smart' agriculture help both africa and the planet. Available at: http://e360.yale.edu/feature/after_durban_can_climate_smart_farming_help_africa_and_the_planet/2477/ (accessed 30 April 2015).

Peters, I., Ackerman, F. and Bernow, S. (1999) Economic theory and climate change policy, *Energy Policy*, 27 (9): 501–504.

Pingali, P.L. (2012) Green revolution: impacts, limits, and the path ahead, online, http://www.ncbi.nlm.nih.gov/pmc/articles/PMC3411969/ (accessed 27 April 2014).

Powlson, D.S, Stirling, C.M., Jat, M.L., Gerard, B.G., Palm, C.A., Sanchez, P.A and Cassman, K.G. (2014) Limited potential of no-till agriculture for climate change mitigation, *Nature Climate Change,* 4: 678–683.

Rosset, P. (2008) Food sovereignty and the contemporary food crisis, *Development*, 51 (4): 460–463.

Scherr, S., Shames, S. and Friedman, R. (2012) From climate-smart agriculture to climate-smart landscapes, *Agriculture & Food Security* Available at: http://www.biomedcentral.com/content/pdf/2048-7010-1-12.pdf (accessed 19 February 2015).

Selvaraju, R. (2010) Climate risk assessment and management in agriculture. Available at: http://www.fao.org/3/a-i3084e/i3084e06.pdf (accessed 28 October 2015).

Sharma, S. and Suppan, S. (2011) *Elusive Promises of the Kenya Agricultural Carbon Project*. (September 2011), p.5. Available at: http://www.iatp.org/documents/elusive-promises-of-the-kenya-agricultural-carbon-project (accessed 3 November 2015).

Streck, C. (2012) Towards policies for climate change mitigation: Incentives and benefits for smallholder farmers. *Security*, (7), 36. Available at: http://re.indiaenvironmentportal.org.in/files/file/smallholder_farmer_finance.pdf (accessed 15 March 2015).

Swaminathan, M.S. (2012) *Green economy and inclusive growth*, online, http://www.uncsd2012.org/content/documents/Plenary3%20Day1%20M%20S%20Swaminathan%20Whole.pdf (accessed 30 April 2014).

Taneja, G. Pal, B.D. Joshi, P. Aggarwal, P.K. and Tyagi, N.K. (2014) Farmers' preferences for climate-smart agriculture: an assessment in the Indo-Gangetic plain. Available at: http://cdm15738.contentdm.oclc.org/utils/getfile/collection/p15738coll2/id/128116/filename/128327.pdf (accessed 19 February 2015).

Thornton, P.K. and Lipper, L. (2014) How does climate change alter agricultural strategies to support food security?, Available at: http://ebrary.ifpri.org/utils/getfile/collection/p15738coll2/id/128124/filename/128335.pdf (accessed 02 November 2015).

United Nations Environmental Program (UNEP) (2015) About GEI. Available at http://www.unep.org/greeneconomy/AboutGEI/WhatisGEI/tabid/29784/Default.aspx (accessed 04 March 2015).

Zurek, M., Streck, C., Roe, S. and Haupt, F. (2014) Climate readiness in smallholder agricultural systems: Lessons learned from REDD+. Working Paper. Available at: http://www.climatefocus.com/sites/default/files/climate_readiness_in_smallholder_agricultural_systems_lessons_learned_from_redd.pdf (accessed 04 November 2015).

2 Climate extremes, climate variability and climate smart agriculture

Molly E. Brown

Introduction

One of the most immediate and obvious impacts of climate change is on the weather-sensitive agriculture sector. Both local and global impacts of climate on production of food will have a negative effect on the ability of humanity to meet its growing food demands. Many factors are affecting agriculture – trends in rainfall and temperature, changes in ecological resilience and the increasing need for ecosystem services such as fresh water provision and cycling of nutrient waste, changes in consumer demand, commodity prices and the role of the retail sector. Extreme events such as multi-year droughts, flooding, late frosts, and severe storms such as hurricanes, tornados, and heat waves affect agricultural productivity across the world. As the context in which agriculture is being conducted changes, the ability of farmers from both developed and developing countries to sustain and grow the amount of food they produce is challenged (Shepherd *et al.*, 2013).

A growing population, with expanding middle-class incomes, has resulted in significant increase in demand for agricultural food and fibre products that will challenge our ability to meet it, even though yields continue to rise and are projected to continue to grow as a result of technological innovation (IAASTD, 2008; Nelson *et al.*, 2013). The ability to adopt technology in the agriculture sector depends on having supportive economic, and financial policies, and infrastructure like roads, electricity, railways and ports (Briceño-Garmendia *et al.*, 2004; IAC, 2004). Many developing countries that are likely to have the largest population growth have the weakest infrastructure, reducing their ability to adopt agricultural technology (Ezeh *et al.*, 2012; Harding and Wantchekon, 2012). Having adequate weather during the growing season, affects how well crops grow, and ultimately the income of farmers, traders, wage labourers and others in the agriculture sector, affecting the ability of the entire sector to grow. Regions with high rates of poverty and high reliance on the agriculture sector, also have a high incidence of natural hazards such as earthquakes, droughts, floods, high temperatures, and tropical cyclones (Shepherd *et al.*, 2013).

Climate, climate variability, and global environmental change can affect food production and the wellbeing of agricultural communities in multiple

Table 2.1 Stressors to small-scale agriculture

Stressors to small-scale agriculture	Source
Population increase driving fragmentation of landholding	(Bilsborrow and Okoth-Ogendo 1992; Walker and Jodha 1986)
Impact of climatic extremes, including drought, floods and storms on productive capacity	(Ahmed *et al.* 2007; Shepherd *et al.* 2013)
Environmental degradation stemming from population, poverty, ill-defined property rights	(Grimble *et al.* 2002; Hamukwala *et al.* 2010)
Regionalizing and globalized markets, and regulatory regimes, increasingly concerned with issues of food quality and food safety	(Lee *et al.* 2012; Reardon *et al.* 2003)
Market failures interrupt input supply following withdrawal of government	(Haile and Kalkhul 2013; Kherallah *et al.* 2002)
Continued protectionist agriculture policies in developed countries and continued declines and unpredictability in the world prices of many agricultural commodities of developing countries	(Lipton 2004; Mano *et al.* 2003)
Health impacts of malaria, human immunodeficiency virus (HIV) and/or acquired immunodeficiency syndrome (AIDS) and other infectious diseases affect agriculture by diverting labour away from farming, eroding household assets, disrupting knowledge transfer and reducing capacity of agricultural service providers	(Barnett and Whiteside 2002; Chima *et al.* 2003)
For pastoralists, encroachment on grazing lands and failure to maintain traditional natural resource management.	(McPeak *et al.* 2011)
State fragility, lack of governance, and armed conflict in some regions	(Brinkman and Hendrix 2011; Turner 1999)

Source: Adapted from IPCC (2007).

ways (Crane *et al.,* 2011; Vermeulen *et al.,* 2012). Rising temperatures, altered precipitation patterns, and extreme weather events affect agricultural yields, the geographical distribution of food and water-borne diseases, and trade patterns (Schmidhuber and Tubiello, 2007). Smallholder farmers and others who live in rural, isolated places are particularly vulnerable to food insecurity due to inter-annual variations in income. This chapter will focus on rainfed agricultural areas and describe current climate extremes using satellite remote sensing data records over the past three decades. A review of the literature on projected changes in climate extremes in the coming decades will be presented, along with potential shifts in crop suitability in these regions. The chapter will finish with a review of the adaptive capacity of agriculture to climate extremes through climate-smart agriculture and other programs. This includes improved seeds and cultivars for smallholder agriculture, policy interventions such as agricultural insurance that will improve access to these tools, value chain interventions to ensure new markets, and improved efficiencies that ensure that smallholder farmers grow their income even in the face of a changing climate.

Smallholder farmers

Smallholders, defined as those farms with fewer than 2 ha, dominate the agricultural systems of food insecure nations (Nagayets, 2005). Statistics from the Food and Agricultural Organization (FAO) show that 85 percent of all farms have fewer than 2 hectares of land and 87 percent of these small farms are located in Asia (FAO, 2012). Small farmers are an important focus for responding to climate change and climate variability, given their lack of resources, marginalization, and prevalence in less developed countries around the world (Table 2.1). Analyses have shown that in order to meet our food needs in the coming decades, we must increase the productivity and resilience to extreme events of small farmers. There is a positive relationship between average farm size and the level of economic development, as represented by gross domestic product (GDP) per capita, where the higher the per capita GDP, the larger the average farm size in a nation (Eastwood, 2014). Although small farms can be just as efficient as larger sized farms, the larger the farm the more likely the farmer will have access to appropriate financial, technological and transportation infrastructure that will raise the farm's overall productivity (Lund and Hill, 1979).

Societies have many methods of increasing agricultural productivity, including conversion of land from forests and other ecosystems to agriculture, as well as various agricultural intensification practices, including tilling and fertilizing the soil, using improved seed varieties, and crop rotations (Matson *et al.,* 1997). Technology is a critical part of these approaches; therefore adoption by farmers of new technological approaches has a big impact on the yields farmers attain. Besides technology, economic and demographic changes influence the performance of the agriculture sector.

Figure 2.1 shows the relationship between the GDP from the agriculture sector and the national income (NI) for the 36 least developed countries. As income rises, the proportion of the GDP from agriculture declines due to an expanding economy that includes much-expanded industrial, retail and financial sectors that are fed and supported by agriculture. At the same time, employment in the agriculture sector declines as people are drawn into other occupations less related to agriculture. Thus, funding and supporting agriculture is often seen as the first step towards reducing its role in the country's economy.

Poor infrastructure, lack of education and skills, and weak institutions to support the use of technology are important factors in explaining the lack of technology adaptation in low income developing countries such as Sierra Leone, Liberia, Niger, Mali, and Burkina Faso, among others (AGRA, 2013). These limitations are likely to continue to affect smallholders across the developing world, affecting the ability of the global food system to meet the demand for

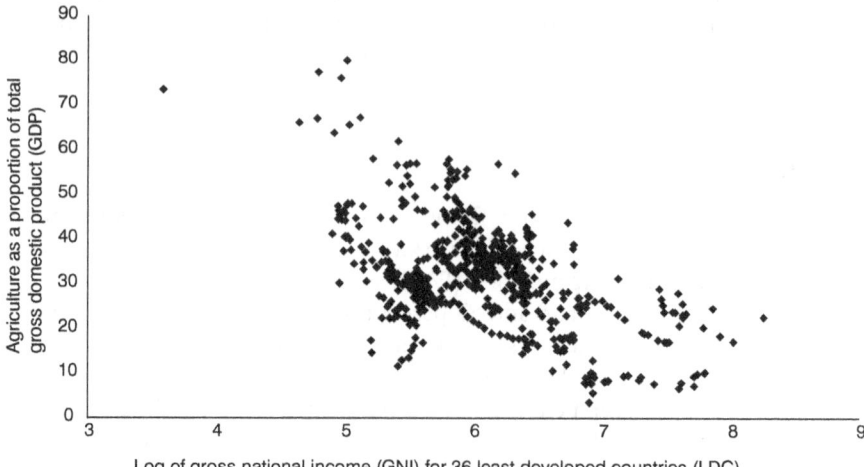

Figure 2.1 Proportion of GDP from the agriculture sector (circles) and proportion of total employment in agriculture (x) vs the log of gross national income (GNI) for the 36 least developed countries 1960–2013

Source: Based on data from the World Bank (2015)

food. Since it is far less expensive to increase yields in low productivity systems with yields at one quarter or less of potential than to increase the already high yields in developed countries, smallholders in developing countries are a critical focus of attention (IAASTD, 2008; Lee *et al.*, 2012).

Climate variability, extreme events, and agriculture

Climate variability refers to weather variations from mean climate conditions, including variations in rainfall, temperature, winds, and other weather events over space and time beyond a few days. Variability in the climate may result from natural internal processes within the climate system (internal variability), from variations in natural or anthropogenic external forces (external variability), and from the interactions between these (IPCC, 2007). Although people and societies have adjusted to and coped with climate variability they experienced in the past, the impact of climate variability continues to challenge weather-sensitive economic sectors such as agriculture in recent years (NAS, 2004).

According to Field *et al.* (2012) climate extremes are defined as 'the occurrence of a weather or climate variable above (or below) a threshold value near the upper (or lower) ends of the range of observed values of the variable'. For the poor and vulnerable, prolonged and severe droughts, extreme, intense rainfall and flooding, and severe heat waves threaten lives and livelihoods as well as hard-won improvements in the standards of living of the poorest populations. Up to 118 million extremely poor people in sub-Saharan Africa will be exposed to drought, flood and extreme heat hazards by 2030 (Field *et al.*, 2012).

Although all hazards threaten economic development and infrastructure, here we are concerned about agriculturally relevant hazards that affect the productive capacity of millions of farmers (Beddington *et al.,* 2012).

Research on heat stress on agricultural production is particularly urgent, given the consensus of the scientific community on the likelihood of higher temperatures in the coming decades, regardless of mitigation (IPCC, 2014). By 2080, most cropping areas in the world are likely to be exposed to record high air temperatures during the growing season. Higher 'average' seasonal temperatures increase the risk of drought because they limit photosynthesis rates and reduce light interception by accelerating the rate at which the plant moves through each phase of its development (Tubiello *et al.,* 2007). Most agricultural yield losses, however, come from extreme events such as heat waves (Teixeira *et al.* 2013). In 2010, unprecedented extreme high temperatures, reducing wheat yields, affected Russian agricultural producing areas (Asseng *et al.,* 2010). This resulted in a significant crisis in the international wheat market, with the price of wheat increasing by up to 50 percent (AMIS, 2013).

Teixeira *et al.* (2013) found in their analysis that most heat stress is in the sub-tropical and temperate latitudes, where the optimum crop calendar puts the crop period of highest vulnerability to heat stress (the reproductive stage) at the hottest part of the summer. The probability that these areas, which include the continental areas of Central Asia and Central North America, will experience yield impacts due to high temperatures increase dramatically after 2050. The impact of heat stress from extreme high temperatures also depends on the crop, with wetland rice the crop most intensely affected (Teixeira *et al.,* 2013). The paper reports that for each extra one degree-day accumulation above a base temperature of 30 degrees Celsius there was one percent decline in maize yield.

Climate variability can also bring weather that changes agricultural productivity through abnormally dry conditions that affect the growth of the crop. High temperatures increase evaporative stress and increase the demand for soil moisture, which if additional moisture over the average precipitation is not forthcoming, reduces the ability of plants to grow and thrive. Analyses show that even in developed agricultural systems with a high level of adaptation, climate variability and the resulting weather fluctuations will affect agricultural profitability (Deschenes and Greenstone, 2007). In tropical regions that house many of the world's poorest countries, climate impacts on agricultural productivity are expected to be particularly harmful, particularly in light of technological, resource, and institutional constraints that exist in these regions (Kurukulasuriya and Rosenthal, 2003). Households that rely upon agriculture for their livelihoods will be especially negatively affected, including reduced food access, poor nutrition outcomes and overall low productivity of the population (Grace *et al.,* 2013; Johnson and Brown, 2014).

Climate smart agriculture (CSA) is a response to the growing need of the agriculture sector in developing countries to respond to multiple stresses, including transforming domestic consumer demand, changing growing season weather, and a degrading resource base. The CSA approach, as defined by the

FAO, is focused on providing technical, policy, and investment conditions necessary to achieve agricultural development required to increase incomes, improve agricultural productivity, adapt to climate change, and reduce greenhouse gas emissions simultaneously, where possible (FAO, 2013a). The focus of CSA is to provide site-specific assessments to identify agricultural production technologies that will allow for increasing of crop yields, farmer incomes' and resilience and rapid adaptation to a changing environment. Although the approach incorporates agricultural development strategies from the previous five decades, it adds a strong climate and trend analysis component that will integrate adaptation and mitigation strategies and funding streams into a rapidly changing agricultural system.

Although agriculture is very weather sensitive, the major differences between commercial agriculture systems and those in developing countries is not the weather or environmental characteristics – it is the ability to use technology and provide appropriate economic, financial, and physical infrastructure necessary to use it (Briceño-Garmendia *et al.*, 2004). Thus to be effective, the response to climate variability and extreme events needs to be connected to the broader food system that links agricultural production to consumption (Vermeulen *et al.*, 2012).

Evidence from the climate data record

Climate data records are often used to detect the impact of climate variability and extremes on ecosystems, and are derived from satellite and ground observations of sufficient length, consistency, and continuity (NAS 2004). Measuring how severe a heat wave was involves comparison of the event to a long-term mean derived from climate data records. The relatively short record of satellite rainfall estimates presents a limitation to understanding the range of drought events required for a risk reduction model, but many records now cover 30 years, and can be used to estimate changes in growing conditions through time.

The Advanced Very High Resolution Radiometer (AVHRR) satellite data product has been used for decades to estimate changes in growing season length and variability (Cracknell, 2001). The AVHRR sensor has been placed on weather satellites since the late 1970s and has been used to measure directly the photosynthetic absorption of light by plants (Tucker, 1979). These observations have been incorporated into a wide variety of models and weather models to improve our understanding of land-atmosphere interactions (Collatz *et al.*, 1991; Neigh *et al.*, 2008; Pettorelli *et al.*, 2005; Sellers *et al.*, 1996).

We can use satellite remote sensing information to characterize the variability and trends in land surface global vegetation dynamics. Research using the Advanced Very High Resolution Radiometer (AVHRR) has shown that more than half of all vegetated land areas exhibit a significant trend in seasonality over the 30 year time period measured (Eastman *et al.*, 2013). Over the 30–year AVHRR normalized difference vegetation index (NDVI) record, forested areas have been shown to be experiencing a consistent increase in productivity throughout the year due to a lengthening of the growing season, both starting early and ending late (Gunderson *et al.,* 2012). Grassland and savannah ecosystems are also experiencing increased productivity, but do not show significant lengthening of the growing season. Grassland ecosystems that have changed over the past 30 years have been shown to have increased access to water during the wet or warm season (Eastman *et al.,* 2013).

Given that the impacts of drought on regional food production are a function of the severity and spatial extent of dryness, there is a need to improve our understanding of the probability of dry events in a way that captures not only the likelihood of drought at a single location but also the likelihood of regions experiencing dry conditions. Husak *et al.* (2013) present a technique to create seasonal rainfall scenarios from existing satellite rainfall estimates that further clarifies the likelihood of extreme events. This technique is effective at a 10–day timescale, where the autocorrelation of individual periods is insignificant, and the data maintains reasonable spatial patterns based on existing gridded estimates. The result is a suite of rainfall simulations that better capture the range and likelihood of conditions in areas where the short satellite records prove inadequate on their own (Husak *et al.,* 2013).

These simulated datasets allow for a more refined understanding of drought events than can be achieved from only 10 or 30 years of observations for a given point. Applying the vulnerability model to the different simulations reveals the likelihood of specific losses at a given point, but also identifies the likelihood of large regions experiencing drought. Similarly, it may be that a mild, but extremely widespread drought results in larger losses at a national level than an extreme, but localized drought event. Loss exceedance curves developed from these simulations are the likelihood of specific crop loss event in a specific location. They can be used by decision makers to respond in an effective and timely way.

Alexander *et al.* (2006) set out a series of analyses of both temperature and precipitation observations from 1951 that demonstrate the impact of a changing climate on agriculturally-relevant weather, such as the number of wet days, warm nights, and hot spells. Table 2.2 below shows the trends reported from 1951 to 2003 for these data, as reported by Alexander *et al.* (2006). The paper uses 2223 temperature stations and 5948 precipitation station data records and grids them to 2291 land-only grids globally. Trends are considered significant if they are significant at the 5 percent level.

Looking at satellite-derived rainfall data, Figure 2.2 shows the coefficient of variation plotted against the mean rainfall for 1200 locations across Africa. The results show that the higher the mean, the less variable the rainfall is in tropical ecosystems. Thus, the arid and semi-arid areas may be more susceptible to inter-annual variation that affects agricultural production. This is not to say that the trend in these places is overwhelmingly negative. Fensholt *et al.* (2012) found that when the satellite data record for vegetation was examined over the past thirty years, a greening trend was found for most arid and semi-arid zones.

Table 2.2 Percentage of land area sampled showing significant annual trends for each indicator from daily data 1951–2003; indicates field is significant at 5% level

Indicator	Total number of grid points with significant trend, out of 2,291 land points	Significant positive trend, percent	Significant negative trend, percent
Maximum daily maximum temp. °C	1,028	11.6	2.7
Maximum daily minimum temp. °C	997	24.5	1.1
Minimum daily maximum temp. °C	1,325	29.3	2.6
Minimum daily maximum temp. °C	1,379	45.0	1.9
Cold spell duration	1,010	10.2	26.0
Warm spell duration	871	28.8	0.6
Maximum 1–day precipitation	824	7.0	2.7
Maximum 5–day precipitation	514	6.0	2.1
Growing season length	761	16.8	0.3
Consecutive dry days	816	4.0	6.9
Consecutive wet days	323	3.1	4.6
Very wet days (> 95th percentile)	653	11.2	2.5
Extremely wet days (> 99th percentile)	426	6.1	2.8
Annual precipitation total	1115	10.1	7.2

Source: Adapted from Alexander *et al.*, (2006)

Impact of climate extremes on agriculture

The 2014 US Climate Assessment identified the following elements as key climate impacts on agriculture:

- Temperature increase impacts on crop production through increased evaporative stress and the lack of cold temperatures on perennial crop production, among many other diverse impacts (Hatfield *et al.*, 2014);
- Temperature extremes impact on the animal system, through feed grain production, pastures and forage crop production, animal health, growth and reproduction and through enhancing disease and pest virulence and distribution (Rötter and Geijn, 1999);
- Weeds, diseases, and pests will continue to be a major threat to agricultural production, with uncertain and growing losses as the climate changes (Oerke, 2006);

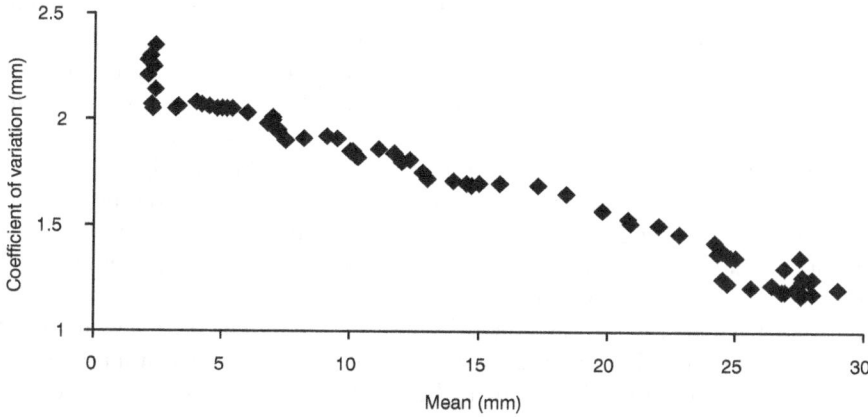

Figure 2.2 Mean verses the coefficient of variation (CoV) for 10-day rainfall data for 1200 locations across Africa

Source: Author's own calculations, adapted from Funk *et.al.*, (2014)

- Increasing intensity of rainfall and frequency of extreme rainfall events is degrading soil and increasing erosion as well as affecting crop growth when they occur (Kunkel *et al.*, 2013);
- Increasing average temperatures (Alexander *et al.*, 2006) will amplify natural variability in rainfall, amplifying future drought events and will expose livestock to extreme heat events exceeding the maximum thresholds for which those production systems were developed.

These changes will affect all agricultural systems, increasing the importance of adaptation to climate (Ezeh *et al.*, 2012). Many international reports have stressed the importance of meeting not only the needs of more people by 2100, but also the growing demand for meat, dairy, and oils from an increasingly wealthy population (IAASTD, 2008; Reid *et al.*, 2005). It is likely that increasing incomes of the middle class in large population centres as Asia will become more important than overall population numbers in coming decades (Tilman *et al.*, 2011). Thus, responding to the threat of climate variability on agriculture is critical in order to ensure food security.

Often, crop failures tend to be the result of dry spells (a period of 10 or more days without rain during a rainy season). And occasionally an abnormal planting date, usually caused by early or late rains that hinder the farmer's ability to plant on time, put the demands of the crop out of phase with the available moisture (Araya and Stroosnijder, 2011). Crop varieties that are drought resistant, heat resistant and are able to cope with pests with less pesticide will be in great demand in the future. Huge efforts to continue to increase the productivity of rice and wheat have been underway for the past decade. For example, Fischer (2011) describes the research that seeks to link the genetic expression of traits such as time to flowering, aluminium, and salt tolerance to a small number of genes. Improvement of our understanding of the functional basis of these links is likely to lead to their use in improved breeding programs (Fischer, 2011). We still have a long way to go in order to significantly improve the potential yield in situations of optimal management. Huge productivity gains are possible, however, by improving the application of known technologies to existing agricultural lands that currently have poor yields (IAASTD, 2008).

With better information about rainfall and temperature in the coming growing seasons, investment in higher yielding seeds, high value vegetables and fruit stock, and an augmented use of inorganic fertilizer can increase yields and farm incomes (Crane *et al.*, 2011; IPCC, 2012; Scheel, 2012). Improved local governance, reduced developed-world agricultural subsidies, and more nuanced food aid policies that protect local markets could together produce rapid improvements in food access and availability, reducing hunger while providing for more people.

Changes in climate extremes

A review of the literature on extreme events was conducted for the 2014 Intergovernmental Panel on Climate Change (IPCC) report. Seneviratne *et al.* (2012) summarizes that there is very likely to be increases in warm days and nights, and a decrease in cold days and nights on a global scale. It is likely that more regions have experienced increases in heavy precipitation events, with more rain falling over shorter periods. A 1-in-20 year annual maximum daily precipitation event is likely to become a 1-in-5 to 1-in-15 year event by the end of the twenty-first century in many regions. There also was a medium confidence that some regions have experienced more intense and longer droughts, but in other regions, droughts have become less frequent, less intense or shorter. There are consistent projections of dry soil moisture anomalies and consecutive dry days at the end of the century for regions that are already dry, particularly the Mediterranean region, central Europe, central North America, Central America and Mexico, Northeast Brazil and southern Africa (Seneviratne *et al.*, 2012).

These changes will have significant agricultural impacts for farmers, particularly changes in temperature and precipitation. Even with the same precipitation, research shows that increasing temperatures increase the evaporative stress on plants, increasing the respiration and moisture needs (Hatfield *et al.*, 2011). Interactions between water-stressed plants and increasing carbon dioxide (Hatfield, 2013), changes in pests and plant diseases (Oerke, 2006), and changes in soil biome (Graaff *et al.*, 2006) have a great deal of uncertainty and variability across different agro-ecosystems (Hatfield *et al.*, 2014). As we learn more about how climate changes will affect agriculture, we will be able to mitigate and adapt to it more effectively.

Adjusting to future climate in smallholder systems

Climate variability and climate change pose a range of hazards to agricultural systems, in particular smallholders that constitute a majority. While region-dependent, these hazards are likely to include more frequent and extreme heat events, more intense precipitation, sea level rise and enhanced coastal flooding, as well as changes in seasonal and long-term water availability. These changing climate hazards may challenge key elements of the food system by threatening operations and damaging critical infrastructure. Specific climate change impacts include: shifting reliability and increasing costs of water and energy, and changes in safety and operations related to extreme events such as fire and floods. By developing climate change adaptation strategies tailored to the specific impacts that are anticipated, decision makers will be able to minimize negative effects of climate and climate change, while leveraging positive outcomes.

Yield gap and climate futures

The 'yield gap' is the difference between actual yields obtained by a farmer and the potential achievable given the use of all available known technologies, seeds, inputs, and techniques. The suitability of a region to grow a particular crop can be estimated using models, such as the FAO's Global Agro-ecological Zone (GAEZ) mapping system (FAO, 2013b). The model uses climate characteristics, including radiation, temperature, and average soil moisture in a simplified land productivity model to calculate potential biomass production and yield by crop. The potential agro-climatic yields are combined with a number of other factors related to climate, including pests, diseases, soil, and terrain conditions such as slope. The model produces suitability assessments related to production that is achievable on a long-term basis, and assumes specific fallow periods and removes areas with terrain slopes that are too steep for optimal cultivation.

Studies demonstrate that adaptation to climate change and variability holds considerable promise for minimizing yield decreases, and even increasing yields in some circumstances. For example, Valdivia *et al.* (2012) and Claessens *et al.* (2012) demonstrate in two regions in Kenya that the use of new crop varieties and intensive agricultural systems could raise overall productivity and ameliorate climate change, even in a high emissions scenario. There is considerable potential for improving crop yields using existing agronomic technologies, but application depends upon individual financial capacity and surrounding infrastructure in order to access and employ technologies.

Climate smart agriculture interventions in smallholder systems

The increasing likelihood of climate extremes that will affect agriculture means that adaptation is important for both retaining livelihoods as well as responding appropriately to shocks. Extreme climate events can be very harmful, as they can disrupt livelihoods across multiple regions and undermine development progress that took years to achieve. Thomas *et al.* (2007) describes how climate is affecting rainfall regimes in South Africa and the ways local communities were coping with the change. Three regions in South Africa had changing onset and timing of rains, changing rainfall totals and intensities, and number of dry days. Farmers would either cope with the changes in water availability through short-term decisions, or make larger adaptive changes that may reduce the farmers' livelihood exposure to future changes. For example, breaks in the rainy season that affected yields and pasture productivity could cause livestock to die, seeds to be lost, and require the farmer take on debt to replant or replace cereals in the family diet (Thomas *et al.*, 2007). To adapt to increasing climate variability, farmers may switch to new crop varieties, take small stock such as goats and sheep to new areas with better pasture, take up wage labour, migrate during the dry season to make more money, or start new businesses, among other responses. Table 2.3 describers other adaptations that may be part of CSA.

In order to increase productivity in tropical agriculture while still reducing greenhouse gases and ensuring long term sustainability, the two goals need to

Table 2.3 Adaptation to climate variability and change by category

Category of action	Type of adaptation	Examples
Addressing the adaptation deficit	Resilience building	• Diversify livelihoods to reduce poverty in the face of weather-related income declines • Crop insurance, seasonal forecasting and irrigation to reduce risk to farmers • Early warning systems for country-level adaptation
Adapting to incremental changes	Climate proofing	• Upgrading infrastructure to be more resilient to extreme high water and temperature events • Adapting cropping systems to shorter growing seasons, switching crops to those productive at higher water and temperature stresses • Improving disaster planning to speed response during more frequent and severe extremes
Adapting to qualitative changes	Transformational change	• Phased relocation of at-risk communities in regions with severely reduced rainfall or pasture productivity • Shifts in emphasis in large-scale economic activity away from areas and resources threatened by climate change (i.e. water intensive crops, climate-sensitive tourism, etc.) • Transformation of agriculture to be less climate sensitive, such as the use of irrigation or switching crops to those that are less-water intensive

Source: after Brooks *et al.* (2011)

be integrated (Harvey *et al.* 2014). Smallholders and poor farmers who have limited capacity to adapt to climate change can benefit from funding supporting mitigation. Adaptation of existing agricultural systems can have significant benefits to maintaining agricultural production in the face of increasing temperatures (Howden *et al.* 2007). In smallholder systems where yields are typically less than a fifth of what is obtainable in developed agriculture, doubling production is attainable in the short term with multidisciplinary approaches that include policymakers, scientists, farmers and business (IFAD, 2012; McIntosh *et al.,* 2013).

Policy interventions to support adoption of new technologies

New technologies and interventions will be needed to increase overall food production to meet increasing global demand for food in the coming decades, while at the same time improving farmers' resilience to extreme events. Harvey *et al.* (2014) describe the policy interventions that are likely to be necessary in tropical agricultural systems most affected by rising temperatures:

- Engineering solutions to increase water availability and reduce flood risk;
- Crop breeding programs to increase high temperature, drought, and salt tolerance, among other environmental stresses;
- Develop financial and environmental risk management tools at the household, community, government, and regional scales to ensure appropriate and least expensive response to extreme events; and
- Adapting agricultural management practices to be more drought and heat resistant, such as no-till agriculture, water conservation practices, and crop diversification (Howden *et al.,* 2007).

Understanding the impact of climate includes the identification of who and what is vulnerable to climate change; the assessment of the capacity to adapt to observed and perceived threats; and the equity and justice of the distribution of these impacts. To do this effectively, strong collaborations between the data users, data providers and the stakeholders who will be affected by the decisions taken is essential. Adaptation and mitigation in the agriculture sector must be done with a broad conception of both the ultimate cost to society and the economy as well as the cost of poor decisions for policy makers, emergency responders, businesses, and individuals. Drought and widespread storms, heat waves and other climate disasters can have lasting impact on a broad section of society.

Multiple economic and government agencies and sectors will be affected by climate change. As climate change is experienced at the local level, the information required by decision makers will need to be specific and flexible enough to quantify climate impacts specifically enough to be integrated into multiple policy and decision making tools (Dunford *et al.,* 2014; Krishnamurthy *et al.,* 2014). The spatial, temporal and process-based relationships among different types and resolutions of data require that significant investment is needed in transforming the raw science data products currently available into usable products, which can be used directly by decision makers. Decision-support tools are an element of the broader decision-making context or Decision-Support System (DSS) (Rice *et al.*, 2012; Workman, 2004; Zhang and Wilhelm, 2011). These systems include not just computer tools but the institutional, managerial, financial, and other systems involved in the decision-making process.

A key aspect of the transformation of data into information includes increasing the geographical scale resolution of the data products. Because all climate impacts are local, understanding the local geography, climate, and context is critical to understanding the impact of a larger climate issue. Only through investment in science can downscaled, high-resolution climate models be developed for every region that needs them. It will be more difficult to provide this information for some regions than others, due to government and institutional barriers to change and existing capabilities. Because of the current limitations to climate change information, both in terms of resolution as well as uncertainty, very few decision support tools in operation in the United States explicitly incorporate climate change scenarios (Brown *et al.*, 2008). As models improve and the impact of a changing climate becomes evident in the weather, the demand for improved information will increase. This plan will enable demand for information to be met with highly reliable and high quality datasets, in a format that can be integrated by data users.

Adaptation to extreme events will be needed across all sectors of the economy. In developed countries such as the United Kingdom, by far the vast majority of documented climate adaptations are coming from the public sector, confirming that adaptation is both within the competence of government and within its responsibility. Governments can work to improve their adaptation to both observed and anticipated climate impacts (Brown *et al.*, 2010), while increasing opportunities for the private sector to both be transparent about their industry's risk to climate extremes and to adapt to climate events as they become known (CDP, 2015).

Conclusions

Climate extremes due to a warming climate are an important factor in understanding how climate change will affect agriculture in the coming decades. Agricultural communities will experience climate change as a series of extreme events: heat waves, droughts, cold winters, large storms, intense rainfall, and of course, fewer cold days and more warm ones. Severe weather that affects agricultural yields can be recovered from if they do not occur too frequently, however, with a warming climate, it may require significant adaptation in order to continue to have profitable agricultural systems in temperate and sub-tropical regions.

CSA is one approach to dealing simultaneously with multiple challenges, by focusing on integrating institutional responses with adaption and mitigation to climate change. To overcome the multiple challenges of growing extreme events, growing population and expanding demand for food, we need to engage with the government, international development, private companies, and farmers to work together to continue to increase agricultural productivity. This will require a real focus on infrastructure, both financial and physical, in order to provide farmers with the inputs needed to increase productivity in good years and reduce their risk in bad years.

In this chapter, the impact of rainfall variability and warming temperatures on agricultural production were described, along with the probability that climate change will exacerbate these affects. Smallholders, who make up 75 percent of all farms around the world, have the fewest resources to respond to climate change (Lee *et al.*, 2012) and will need the most help in overcoming it to raise their standards of living.

References

Agricultural Market Information System (AMIS) (2013) Agricultural market information system market monitor. http://www.amis-outlook.org/amis-monitoring

Ahmed, A.U., Hill, R.V., Smith, L.C., Wiesmann, D.M. and Frankenberger, T. (2007) The world's most deprived: characteristics and causes of extreme poverty and hunger, *International Food Policy Research Institute* (pp. 148), Washington DC.

Alexander, L.V., Zhang, X., Peterson, T.C., Caesar, J., Gleason, B., Tank, A.M.G.K., Haylock, M., Collins, D., Trewin, B., Rahimzadeh, F., Tagipour, A., Kumar, K.R., Revadekar, J., Griffiths, G., Vincent, L., Stephenson, D.B., Burn, J., Aguilar, E., Brunet, M., Taylor, M., New, M., Zhai, P., Rusticucci, M. and Vazquez-Aguirre, J.L. (2006) Global observed changes in daily climate extremes of temperature and precipitation, *Journal of Geophysical Research-Atmospheres*, 111: D05109.

Alliance for a Green Revolution in Africa (AGRA) (2013) *Africa Agriculture Status Report: Focus on Staple Crops*. Nairobi, Kenya. Available at: http://www.agra.org/download/533977a50dbc7/ [Accessed 2 April 2015].

Araya, A. and Stroosnijder, L. (2011) Assessing drought risk and irrigation need in northern Ethiopia, *Agricultural and Forest Meteorology*, 151: 425–436.

Asseng, S., Foster, I. and Turner, N.C. (2010) The impact of temperature variability on wheat yields, *Global Change Biology*, 17: 997–1012.

Barnett, A. and Whiteside, A. (2002) *AIDS in the Twenty-First Century; Disease and Globalization*, Palgrave Macmillan, Basingstoke and New York.

Beddington, J., Asaduzzaman, M., Clark, M., Fernández, A., Guillou, M., Jahn, M., Erda, L., Mamo, T., Van Bo, N., Nobre, C., Scholes, R., Sharma, R. and Wakhungu, J. (2012) *Achieving Food Security in the Face of Climate Change*. Final report from the Commission on Sustainable Agriculture and Climate Change. CGIAR Research Program on Climate Change, Agriculture and Food Security (CCAFS), Copenhagen. Available at: https://cgspace.cgiar.org/bitstream/handle/10568/35589/climate_food_commission-final-mar2012.pdf?sequence=1 [Accessed 15 April 2015].

Bilsborrow, R.E. and Okoth-Ogendo, H.W.O. (1992) Population-driven changes in land use in developing countries, *Ambio*, 21: 37–45.

Briceño-Garmendia, C., Estache, A. and Shafik, N. (2004) *Infrastructure Services In Developing Countries: Access, Quality, Costs and Policy Reform*, World Bank Reports WPS 3468, Washington DC.

Brinkman, H. and Hendrix, C.S. (2011) *Food Insecurity and Violent Conflict: Causes, Consequences and Addressing the Challenges*, United Nations World Food Programme, Rome, Italy.

Brooks, N., Anderson, S., Ayers, J., Burton, I. and Tellam, I. (2011) *Tracking Adaptation and Measuring Development,* International Institute for Environment and Development, London.

Brown, M.E., de Beurs, K. and Vrieling, A. (2010) The response of African land surface phenology to large scale climate oscillations, *Remote Sensing of Environment*, 114: 2286–2296.

Brown, M.E., Lary, D., Vrieling, A., Stathakis, D. and Mussa, H. (2008) Neural networks as a tool for constructing continuous NDVI time series from AVHRR and MODIS, *International Journal of Remote Sensing*, 29: 7141–7158.

Carbon Disclosure Project (CDP) (2015) *Supply Chain Sustainability Revealed: A Country Comparison*, online, https://www.accenture.com/t20150523T015757__w__/kr-en/_acnmedia/Accenture/Conversion-Assets/DotCom/Documents/About-Accenture/PDF/2/Accenture-CDP-Supply-Chain-Report-2015.pdf [Accessed 15 September 2015].

Chima, R.I., Goodman, C.A. and Mills, A. (2003) The economic impact of malaria in Africa: a critical review of the evidence, *Health Policy*, 63: 17–36.

Claessens, L., Antle, J.M., Stoorvogel, J.J., Valdivia, R.O., Thornton, P.K. and Herrero, M. (2012) A method for evaluating climate change adaptation strategies for small-scale farmers using survey, experimental and modelled data, *Agricultural Systems,* 111: 85–95.

Collatz, G.J., Ball, J.T., Grivet, C. and Berry, J.A. (1991) Physiological and environmental regulation of stomatal conductance, photosynthesis and transpiration: a model that includes a laminar boundary layer, *Agricultural and Forest Meteorology*, 54: 107–136.

Cracknell, A.P. (2001) The exciting and totally unanticipated success of the AVHRR in applications for which it was never intended, *Advanced Space Research*, 28: 233–240.

Crane, T.A., Roncoli, C. and Hoogenboom, G. (2011) Adaptation to climate change and climate variability: the importance of understanding agriculture as performance, *Wageningen Journal of Life Sciences*, 57: 179–185.

Deschenes, O. and Greenstone, M. (2007) The economic impacts of climate change: Evidence from agricultural output and random fluctuations in weather, *American Economic Review*, 97: 354–385.

Dunford, R., Harrison, P.A., Jäger, J., Rounsevell, M.D.A. and Tinch, R. (2014) Exploring climate change vulnerability across sectors and scenarios using indicators of impacts and coping capacity, *Climatic Change*, 128: 1–16.

Eastman, J.R., Sangermano, F., Machado, E.A., Rogan, J. and Anyamba, A. (2013) Global trends in seasonality of normalized difference vegetation index (NDVI), *Remote Sensing Journal*, 5: 4799–4818.

Eastwood, R., Lipton, M. and Newell, A. (2014) Farm size. In: Pingali, P. and Evenson, R. eds. *Handbook of Agricultural Economics*. Elsevier, North Holland.

Ezeh, A.C., Bongaarts, J. and Mberu, B. (2012) Global population trends and policy options, *The Lancet,* 380: 142–148.

Fensholt, R. Langanke, T., Rasmussen, K., Reenberg, A., Prince, S.D., Tucker, C., Scholes, R.J., Le, Q.B., Bondeau, A., Eastman, R., Epstein, H., Gaughan, A.E., Hellden, U., Mbow, C., Olsson, L., Paruelo, J., Schweitzer, C., Seaquist, J. and Wessels, K. (2012). Greenness in semi-arid areas across the globe 1981–2007: an Earth observing satellite based analysis of trends and drivers, *Remote Sensing of Environment*, 121: 144–158.

Field, C.B. Barros, V., Stocker, T.F., Qin, D., Dokken, D.J., Ebi, K.L., Mastrandrea, M.D., Mach, K.J., Plattner, G.-K., S.K. Allen, Tignor, M. and Midgley, P.M. (2012) *Managing the Risks of Extreme Events and Disasters to Advance Climate Change Adaptation.*

A Special Report of Working Groups I and II of the Intergovernmental Panel on Climate Change. Cambridge University Press, Cambridge.

Fischer, R.A. (2011) Wheat physiology: a review of recent developments, *Crop & Pasture Science*, 62, 95–114.

Food and Agriculture Organization (FAO) (2012) *The State of Food Insecurity in the World: Multiple Dimensions of Food Security*, United Nations FAO, Rome, Italy.

Food and Agriculture Organization (FAO) (2013a) *Climate-Smart Agriculture Sourcebook*. United Nations FAO, Rome, Italy.

Food and Agriculture Organization (FAO) (2013b) *Global Agro-ecological Zones (GAEZ v3.0)*. United Nations FAO and IIASA, Rome, Italy.

Funk, C.C., Peterson, P.J., Landsfeld, M.F., Pedreros, D.H., Verdin, J.P., Rowland, J.D., Romero, B.E., Husak, G.J., Michaelsen, J.C. and Verdin, A.P. (2014) *A Quasi-Global Precipitation Time Series for Drought Monitoring* U.S. Geological Survey Data Series 832. http://dx.doi.org/10.3133/ds832

Graaff, M.A.D., Groenigen, K.J.V., Six, J., Hungate, B. and Kessel, C.V. (2006) Interactions between plant growth and soil nutrient cycling under elevated CO_2: a meta-analysis, *Global Change Biology*, 12: 2077–2091.

Grace, K., Davenport, F., Funk, C. and Lerner, A. (2013) Child malnutrition and climate conditions in Kenya, *Applied Geography*, 11: 164–177.

Grimble, R., Cardoso, C. and Omar-Chowdhury, S. (2002) *Poor People and the Environment: Issues and Linkages*, Natural Resources Institute, Chatham.

Gunderson, C.A., Edwards, N.T., Walker, A.V., O'Hara, K.H., Campion, C.M. and Hanson, P.J. (2012) Forest phenology and a warmer climate – growing season extension in relation to climatic provenance, *Global Change Biology*, 18: 2008–2025.

Haile, M.G. and Kalkhul, M. (2013) Volatility in the international food markets: implications for global agricultural supply and for market and price policy. 53rd Annual Conference of the German Society of Economic and Social Sciences in Agriculture (GEWISOLA), Berlin, Germany.

Hamukwala, P., Tembo, G., Larson, D. and Erbaugh, M. (2010) *Sorghum and Pearl Millet Improved Seed Value Chains in Zambia: Challenges and Opportunities for Smallholder Farmers*. USAID and International Sorghum and Millet Collaborative Research Support Program (INTSORMIL/CRSP), Washington DC.

Harding, R. and Wantchekon, L. (2012) *Food Security and Public Investment in Rural Infrastructure: Some Political Economy Considerations*, United Nations Development Programme, Regional Bureau for Africa, Working Paper 2012–017, New York.

Harvey, C.A. Chacón, M., Donatti, C.I., Garen, E., Hannah, L., Andrade, A., Bede, L., Brown, D., Calle, A., Chará, J., Clement, C., Gray, E., Hoang, M.H., Minang, P., Rodríguez, A.M., Seeberg-Elverfeldt, C., Semroc, B., Shames, S., Smukler, S., Somarriba, E., Torquebiau, E., Etten, J.V. and Wollenberg, E. (2014) Climate-smart landscapes: opportunities and challenges for integrating adaptation and mitigation in tropical agriculture, *Conservation Letters*, 7: 77–90.

Hatfield, J.L. (2013) Climate change: challenges for future crop adjustments. In: Tuteja, N., and Gill, S.S. eds. *Climate Change and Plant Abiotic Stress Tolerance*. Wiley Verlag, Weinheim, Germany.

Hatfield, J.L., and Prueger., J.H. (2015) Temperature extremes: Effect on plant growth and development, *Weather and Climate Extremes*, 10 (2015): 4–10.

Hatfield, J.L., Booteb, K.J., Kimball, B.A., Ziska, L.H., Izaurralde, R.C., Ort, D., Thomson, A.M. and Wolfe, D. (2011) Climate impacts on agriculture: implications for crop production, *Agronomy Journal*, 103 (2): 351–370.

Hatfield, J., Takle, G., Grotjahn, R., Holden, P., Izaurralde, R.C., Mader, T., Marshall, E. and Liverman, D. (2014) Agriculture. In: Melillo J.M. eds. *Climate Change Impacts in the United States: The Third National Climate Assessment*. Global Change Research Program, Washington DC.

Howden, S.M., Soussana, J.F., Tubiello, F.N., Chhetri, N., Dunlop, M. and Meinke, H. (2007) Adapting agriculture to climate change, *Proceedings of the National Academy of Sciences*, 104: 19691–19696.

Husak, G.J., Funk, C.C., Michaelsen, J., Magadzire, T. and Goldsberry, K.P. (2013) Developing seasonal rainfall scenarios for food security early warning, *Theoretical and Applied Climatology*, 111: 1–12.

International Assessment of Agricultural Knowledge (IAASTD) (2008) *International Assessment of Agricultural Knowledge, Science and Technology for Development*, Island Press, London.

InterAcademy Council (IAC) (2004) *Realizing the Promise and Potential of African Agriculture – Science and Technology Strategies for Improving Agricultural Productivity and Food Security in Africa*, IAC Secretariat the Netherlands, Amsterdam.

Intergovernmental Panel on Climate Change (IPCC) (2007) *The Effects of Climate Change on Agriculture, Land Resources, Water Resources and Biodiversity*, Washington DC.

Intergovernmental Panel on Climate Change (IPCC) (2012) *Managing the Risks of Extreme Events and Disasters to Advance Climate Change Adaptation: Special Report of the Intergovernmental Panel on Climate Change*, Cambridge University Press, Cambridge.

International Fund for Agricultural Development (IFAD) (2012) *Sustainable Smallholder Agriculture: Feeding the World, Protecting the Planet*, IFAD, Rome, Italy.

Intergovernmental Panel on Climate Change (IPCC). (2014) Summary for policymakers. *Climate change 2014: Impacts, Adaptation, and vulnerability*. Part A: Global and Sectoral aspects. Cambridge University Press, Cambridge.

Johnson, K.B. and Brown, M.E. (2014) Environmental risk factors and child nutritional status and survival in a context of climate variability and change, *Applied Geography*, 54: 209–221.

Kherallah, M., Delgado, C., Gabre-Medhin, E., Minot, N. and Johnson, M. (2002) *Reforming Agricultural Markets in Africa*, Johns Hopkins University Press, Baltimore, MD.

Krishnamurthy, P.K., Lewis, K. and Choularton, R.J. (2014) A methodological framework for rapidly assessing the impacts of climate risk on national-level food security through a vulnerability index, *Global Environment Change*, 25: 121–132.

Kunkel, K.E., Karl, T.R., Brooks, H., Kossin, J., Lawrimore, J.H., Arndt, D., Bosart, L., Changnon, D., Cutter, S.L., Doesken, N., Emanuel, K., Groisman, P.Y., Katz, R.W., Knutson, T., O'Brien, J., Paciorek, C.J., Peterson, T.C., Redmond, K., Robinson, D., Trapp, J., Vose, R., Weaver, S., Wehner, M., Wolter, K. and Wuebbles, D. (2013) Monitoring and understanding trends in extreme storms: state of knowledge, *Bulletin of the American Meteorological Society*, 94: 499–514.

Kurukulasuriya, P. and Rosenthal, S. (2003) *Climate Change and Agriculture: A Review of Impacts and Adaptations*, The World Bank, Washington DC.

Lee, J., Gereffi, G. and Beauvais, J. (2012) Global value chains and agrifood standards: Challenges and possibilities for smallholders in developing countries, *Proceedings of the National Academy of Sciences*, 109: 12326–12331.

Lipton, M. (2004) Crop science, poverty and the family farming in a globalising world, 4th International Crop Science Congress: Brisbane, Australia, online, https://michaellipton.files.wordpress.com/2012/02/crop_science_poverty2004.pdf [Accessed 15 September 2015].

Lund, P.J. and Hill, P.G. (1979) Farm size, efficiency and economies of size, *Journal of Agricultural Economics*, 30: 145–158.

Mano, R., Isaacson, B. and Dardel, P. (2003) *Identifying Policy Determinants of Food Security Response and Recovery in the SADC Region: The Case of the 2002 Food Emergency*, FANRPAN (Food, Agriculture and Natural Resources Policy Network), Gaborone, Botswana.

Matson, P.A., Parton, W.J., Power, A.G. and Swift, M.J. (1997) Agricultural intensification and ecosystem properties, *Science*, 277: 504–509.

McIntosh, C., Sarris, A. and Papadopoulos, F. (2013) Productivity, credit, risk, and the demand for weather index insurance in smallholder agriculture in Ethiopia, *Agricultural Economics*, 44: 399–417.

McPeak, J., Little, P.D. and Doss, C.R. (2011) *Risk and Social Change in an African Rural Economy: Livelihoods in Pastoralist Communities*. Routledge Press, Abingdon and New York.

Nagayets, O. (2005) Small farms: current status and key trends, Research Workshop on the Future of Small Farms, IFPRI: Wye, UK. Available at: http://citeseerx.ist.psu.edu/viewdoc/download?doi=10.1.1.146.4632&rep=rep1&type=pdf [Accessed 22 April 2015].

National Academy of Sciences (NAS) (2004) *Climate Data Records from Environmental Satellites: Interim Report*, Board on Atmospheric Sciences and Climate. Available at: http://www.nap.edu/catalog/10944/climate-data-records-from-environmental-satellites-interim-report [Accessed April 29 2015].

Neigh, C., Tucker, C. and Townshend, J. (2008) North American vegetation dynamics observed with multi-resolution satellite data, *Remote Sensing of Environment*, 112: 1749–1772.

Nelson, G.C., Valin, H., Sands, R.D., Havlík, P., Ahammad, H., Deryng, D., Elliott, J., Fujimori, S., Hasegawa, T., Heyhoe, E., Kyle, P., Lampe, M.V., Lotze-Campen, H., d'Croz, D.M., Meijl, H.V., Mensbrugghe, D.V.D., Müller, C., Popp, A., Robertson, R., Robinson, S., Schmid, E., Schmitz, C., Tabeau, A. and Willenbockel, D. (2013) Climate change effects on agriculture: Economic responses to biophysical shocks, *Proceedings of the National Academy of Sciences*, 111: 3274–3279.

Oerke, E.C. (2006) Crop losses to pests, *The Journal of Agricultural Science*, 144: 31–43.

Pettorelli, N., Vik, J.O., Mysterud, A., Gaillard, J.-M., Tucker, C.J. and Stenseth, N.C. (2005) Using the satellite-derived NDVI to assess ecological responses to environmental change, *Trends in Ecology & Evolution*, 20: 503–510.

Reardon, T., Timmer, C.P., Barrett, C.B. and Berdegué, j.a. (2003) The rise of supermarkets in Africa, Asia, and Latin America. *American Journal of Agricultural Economics*. 85: 1140–1146.

Reid, W.V., Mooney, H.A., Cropper, A., Capistrano, D., Carptenter, S.R., Chopra, K., Dasgupta, P., Dietz, T., Kuraiappah, A.K., Hassan, R., Kasperson, R.E., Leemans, R., May, R.M., McMichael, A.J., Pingali, P., Samper, C., Scholes, R.J., Watson, R.T., Zakri, A.H., Shidong, Z., Ash, N.J., Bennett, E., Kumar, P., Lee, M.J., Raudsepp-Hearne, C., Simons, H., Thonell, J. and Zurek, M.B. (2005) *Millennium Ecosystem Assessment Synthesis Report*. Island Press: London.

Rice, J., Moss, R., Runci, P., Anderson, K. and Malone, E. (2012) Incorporating stakeholder decision support needs into an integrated regional Earth system model, *Mitigation and Adaptation Strategies for Global Change*, 17: 805–819.

Rötter, R. and Geijn, S.C.v.d. (1999) Climate change effects on plant growth, crop yield and livestock, *Climatic Change*, 43: 651–681.

Scheel, M.L.M. (2012) Early warning systems: the "last mile" of adaptation, *EOS Transactions of the American Geophysical Union*, 93: 209–210.

Schmidhuber, J. and Tubiello, F.N. (2007). Global food security under climate change, *Proceedings of the National Academy of Sciences*, 104: 19703–19708.

Sellers, P., Los, S.O., Tucker, C.J., Justice, C.O., Dazlich, D.A., Collatz, G.J. and Randall, D.A. (1996) A revised land surface parameterization (SiB2) for Atmospheric GCMs: Part 2, *Journal of Climate*, 9: 706.

Seneviratne, S., Nicholls, N., Easterling, D., Goodess, C., Kanae, S., Kossin, J., Luo, Y., Marengo, J., McInnes, K., Rahimi, M., Reichstein, M., Sorteberg, A., Vera, C. and Zhang, X. (2012) Changes in climate extremes and their impacts on the natural physical environment. In: Field, C.B., V. Barros, T.F. Stocker, D. Qin, D.J. Dokken, K.L. Ebi, M.D. Mastrandrea, K.J. Mach, G.-K. Plattner, S.K. Allen, M. Tignor, & P.M. Midgley (Eds.) *Managing the Risks of Extreme Events and Disasters to Advance Climate Change Adaptation*, Cambridge University Press, Cambridge.

Shepherd, A., Mitchell, T., Lewis, K., Lenhardt, A., Jones, L., Scott, L. and Muir-Wood, R. (2013) *The Geography of Poverty, Disasters and Climate Extremes in 2030*, Overseas Development Institute: London.

Teixeira, E.I., Fischer, G., Velthuizen, H.V., Walter, C. and Ewert, F. (2013) Global hotspots of heat stress on agricultural crops due to climate change, *Agricultural and Forest Meteorology*, 170: 206–215.

Thomas, D.S., Twyman, C., Osbahr, H. and Hewitson, B. (2007) Adaptation to climate change and variability: farmer responses to intra-seasonal precipitation trends in South Africa, *Climatic Change*, 83: 301–322.

Tilman, D., Balze, C., Hill, J., and Befort, B.L. (2011) Global food demand and the sustainable intensification of agriculture, *Proceedings of the National Academy of Sciences*, 108: 20260–20264.

Tubiello, F.N., Soussana, J.-F.O. and Howden, S.M. (2007) Crop and pasture response to climate change, *Proceedings of the National Academy of Sciences*, 104: 19686–19690.

Tucker, C.J. (1979) Red and photographic infrared linear combinations for monitoring vegetation, *Remote Sensing of Environment*, 8: 127–150.

Turner, M.D. (1999) Conflict, environmental change and social institutions in dryland africa: limitations of the community resource management approach, *Society and Natural Resources*, 12: 643–657.

Valdivia, R.O., Antle, J.M. and Stoorvogel, J.J. (2012) Coupling the trade-off analysis model with a market equilibrium model to analyse economic and environmental outcomes of agricultural production systems, *Agricultural Systems*, 110:17–29.

Vermeulen, S.J., Aggarwal, P.K., Ainslie, A., Angelone, C., Campbell, B.M., Challinor, A.J., Hansen, J.W., Ingram, J.S.I., Jarvis, A., Kristjanson, P., Lau, C., Nelson, G.C., Thornton, P.K. and Wollenberg, E. (2012) Options for support to agriculture and food security under climate change, *Environmental Science and Policy*, 15: 136–144.

Walker, T.S. and Jodha, N.S. (1986) How small farm households adapt to risk. In: Hazell, P. *et al.* eds. *Crop Insurance for Agricultural Development*, Johns Hopkins University Press, Baltimore, MD.

Workman, M. (2004) Expert decision support system use, disuse and misuse: a study using the theory of planned behaviour, *Computers in Human Behaviour*, 21: 211–231.

Zhang, W. and Wilhelm, W.E. (2011) OR/MS decision support models for the specialty crops industry: a literature review, *Annals of Operations Research*, 190: 131–148.

3 Building climate resilience through smart water and irrigation management systems

Experiences from southeast Asia

Matthew P. McCartney, Robyn Johnston and Guillaume Lacombe

Introduction

Over the next few decades, agriculture in southeast Asia (SEA) must undergo a significant transformation in order to meet the dual challenges of achieving food security and responding to climate change. Climate change further complicates an already difficult situation in which agriculture in the region must keep pace with growing food demand driven by a rising and increasingly prosperous population.

In SEA, climate change threatens agricultural production stability and productivity. In the long-term, the likely consequences of climate change – higher temperatures, changes in rainfall patterns and more frequent climate extremes – are expected to shift production seasons and pest and disease patterns, affecting production, prices, incomes and ultimately, livelihoods and people's wellbeing (IPCC, 2014).

Against a backdrop of enhanced climate risks, and associated agro-ecological and socio-economic threats, preserving and enhancing food security requires higher agricultural productivity and, importantly, lower yield variability. In order to increase and stabilize output, agricultural production systems in SEA must become more resilient. That means they must have a greater ability to perform well in the face of, and recover quickly from, disruptive events. More productive and resilient agriculture, so called 'climate smart agriculture' requires changes in the management of natural resources and greater efficiency in the use of those resources, including water. Thus, 'water smart agriculture' is best viewed as a subset of activities that contribute to overall climate smart agriculture.

In this chapter, we highlight how different approaches to improve water management can be used to ameliorate some of the major impacts of climate change on agricultural production in SEA and, in some cases, generate significant mitigation benefits by reducing greenhouse gas (GHG) emissions and increasing carbon sinks. However, it is important to note that the most appropriate interventions must be selected carefully to match the specific conditions in any given context, and the extent to which a range of objectives can be met with single interventions is limited. As a result, in most situations, suites of interventions are generally required. Furthermore, increasing yields, necessary for food security and socio-economic development in the short-term, are not always congruent with mitigating climate change impacts necessary for resilience and sustainability in the long-term. The interrelationships between adaptation and mitigation need to be carefully considered and an appropriate balance struck.

Illustrated with case studies from the region, this chapter provides an overview of how agricultural water management can contribute to climate smart agriculture in SEA. Currently, the food security of the poor in SEA depends critically on both rainfed and irrigated rice, which is the region's staple crop. For this reason, the chapter focuses primarily, though not exclusively, on rice. Section 2 describes the background and provides the context of agriculture in SEA, including the need for agricultural water management and the contribution of rice to GHG emissions. Section 3 briefly describes a range of interventions that can contribute to climate smart water interventions through mitigation and adaptation, and provides examples from SEA. Finally, section 4 presents some conclusions and recommendations.

Features of southeast Asia

Agriculture, livelihoods and economy

SEA is very diverse physically, socially, culturally, politically and economically, and is rapidly changing. In recent decades, significant economic growth – in part driven by growth in agricultural output[1] – has contributed positively to the alleviation of poverty and improvement in living conditions. Despite relatively rapid population growth (i.e. 1–2 percent in Thailand, Vietnam, Cambodia and Myanmar and 2–3 percent in Lao PDR), increases in per capita GDP have translated into increases in the human development index (HDI), and declines in infant mortality and chronic poverty. However, despite these positive indicators, significant segments of the total population, especially those living in rural areas, continue to suffer from poverty and food insecurity. More than 40 percent of the population of Vietnam and Indonesia and more than 50 percent of the population of Cambodia, Lao PDR and Myanmar continue to live on less than US $2 a day. The number of malnourished infants continues to be unacceptably high (Table 3.1).

Though industry and services are gradually increasing in importance, the economy of SEA remains largely dominated by agriculture. About 60 percent of the population of SEA is rural (Table 3.1) and, to a large extent, dependent on agriculture. Since the 1960s, agriculture has gradually been shifting from traditional subsistence to modern commercial farming, although the pace of change has varied considerably between the different countries. Rice is by far the most important food crop, grown for local consumption throughout the region and, increasingly, for export. Thailand and Vietnam are major global exporters.

Table 3.1 Indicators of development

	World Bank– income group	Population (million)	Rural population % of total	GDP/capita (US $)	Human development index	percent of population living < $2/day	% under 5 malnourished*
Thailand	Upper Middle	69.52	52	4,972	0.682	4.6	7
Malaysia	Upper Middle	29.72	27	10,538	0.773	2.3	13
Philippines	Lower Middle	101.65	55	2,737	0.660	21.5	26
Indonesia	Lower Middle	242.33	48	3,495	0.617	46.1	19
Viet Nam	Lower Middle	87.84	68	1,407	0.593	43.4	20
Lao PDR	Lower Middle	6.29	64	1,320	0.524	66.0	32
Cambodia	Low	14.31	80	897	0.523	53.3	29
Myanmar	Low	48.34	67	890	0.483	NA	23

* Based on weight

Source: Adapted from a variety of sources including World Bank (WB), UNDP and Asia Development Bank).

The need for agricultural water management

The climate of SEA is governed predominantly by alternating monsoons. On mainland SEA (i.e. Myanmar, Cambodia, Lao PDR, Thailand and Vietnam), the wet season is characterized by the southwest monsoon (May to October) with humid winds coming from the southwest, induced by the development of circulation features and convective activity in the tropical East Indian Ocean and the Bay of Bengal (Yihui and Chan, 2005). The dry season (November to April) is due to the East Asian Winter Monsoon bringing dry and cold northeasterly winds, along the coast of East Asia (Zhou, 2011). Overall 80–90 percent of the total rainfall occurs during the wet season.

In maritime SEA (i.e. Indonesia, Malaysia and Philippines) the reverse occurs. Rainfall distribution within the wet season is bimodal with a first peak in late May and a second (the highest) in late August-early September. The onset of the monsoon is related to sea surface temperature of the Pacific and Indian Oceans (Lau and Yang, 1997; Singhrattna *et al.*, 2005) and to snow cover in Eurasia and the Tibetan Plateau (Yihui and Chan, 2005). The multi-decadal variability of rainfall is partly due to the El Nino–Southern Oscillation (Kripalani and Kulkarni, 1997; Xu *et al.*, 2004) and the North Pacific Oscillation (Wang *et al.*, 2007).

Rice is critically dependent on water (Box 3.1) and rainfall variability is a central driver, not just of the livelihoods and wellbeing of many millions of the region's people but also national economies. Dry years negatively impact grain production, food security and GDP (Wassman *et al.*, 2010). Against this background, agricultural water management – particularly irrigation – has been widely promoted as an important contributor to development, food security and poverty reduction in all SEA countries (Table 3.2).

Water for agriculture in SEA is becoming increasingly scarce. The causes are diverse and location-specific but include decreasing physical availability as a consequence of falling groundwater tables and silting of reservoirs, decreasing quality as a consequence of chemical pollution and salinization, malfunctioning of irrigation systems and increased competition from other sectors like urban and industrial users. Climate change will exacerbate these problems and further complicate rice growing and other agriculture. Against this background, it is essential to develop and promote strategies and technologies, which farmers can adapt to help them improve their water management and productivity.

Impacts of climate change

Analyses of historic rainfall time series across the central Mekong region has found wetting trends in both the dry and wet seasons (Lacombe *et al.* 2013). These trends are consistent with rainfall alteration in the neighboring southeastern part of China over the last half century (Zhou, 2011). While trends in seasonal rainfall averages remain moderate in magnitude, a much stronger trend in inter-annual and inter-seasonal rainfall variability is anticipated in the future.

Box 3.1 Water and rice

Rice *(Oryza sativa)* is the staple crop in SEA. Rice depends on water for its growth, development, and to produce good yields. Most rice is grown in lowland, where flooded rice paddies are either rainfed or irrigated. Upland rice, which is less common, is grown on non-flooded rainfed fields, at altitudes up to 2,000 m and with rainfall ranging from 1,000 to 4,500 mm annually.

There is an estimated 48 million ha of rice land in SEA, of which overall approximately 45 percent is irrigated, usually with continuous flooding for most of the crop season. In many irrigated areas, rice is grown as a monoculture with two (sometimes three) rice crops per year. Irrigated rice yields are high, typically 6–8 tha-1y-1 compared to 2.5 tha-1y-1 for rainfed rice.

Source: Redfern *et al.,* (2012)

Table 3.2 Area irrigated and volumes of water withdrawal

	Area equipped for full control irrigation (1000 ha)	*percent of cultivated area*	*Water withdrawal by agriculture (Km3)*
Thailand	6,415	33.8	51.8
Malaysia	380	5.1	2.51
Philippines	1,879	18.3	67.1
Indonesia	6,722	16.0	92.8
Viet Nam	4,585	48.7	77.8
Lao PDR	310	23.2	3.2
Cambodia	354	8.9	2.1
Myanmar	2,083	18.1	29.57

Source: Based on FAO (2015)

Several studies have downscaled global climate scenarios for the SEA region. Many of these predict longer and warmer summers, shorter and warmer winters, wetter rainy seasons and increasing inter-annual variability (IPCC, 2014). There remains considerable uncertainty on the impacts of climate change on both high and low river discharge (Thompson *et al.,* 2014) with consequent ambiguity on the implications for irrigation supplied by rivers.

Greater variability in the onset and end of the monsoon, as well as increased frequency of droughts and floods, all increase uncertainty for farmers and intensify the difficulty of agriculture and water management. Throughout the region, overall water security is deemed low and generally worsening (Figure 3.1). Countries in regions with low levels of water security are expected to be disproportionately affected by the impacts of climate change (ADB, 2013).

Rice production systems of the region are at risk from the effects of climate change. Large rice growing areas are located in especially vulnerable regions (e.g. lowlands, deltas and coastal regions) (Masutomi *et al.*, 2009). Changes in temperature regimes greatly influence not only the growth duration, but also the growth pattern and the productivity of rice crops. A decrease of 10 percent in rice yield was found to be associated with every 1 °C rise in temperature (ADB, 2009). However, high temperatures alone are not likely to become a major constraint to rice production within the next 20 years. Rather the immediate impacts of climate change on rice production is observed primarily in the form of extreme weather events (Haefele *et al.*, 2010).

Drought has long been recognized as the primary constraint to rainfed rice production in SEA. With some exceptions (e.g. the Dry Zone of Myanmar), the impact of drought on rice productivity generally depends more on rainfall distribution than on total seasonal rainfall (Wassman *et al.*, 2010). Drought reduces rice yields directly by reducing transpiration or causing spikelet sterility, and indirectly by impeding management operations such as crop establishment or weeding. Even limited temperature stress can be aggravated by drought because plants lose their ability to cool through transpiration (Haefele *et al.*, 2010). Many areas of rainfed lowland rice are located in drought prone areas where yields are regularly reduced as a consequence of low rainfall: Thailand (8.2 million ha), Indonesia (4.0 million ha), Vietnam (2.9 million ha), Myanmar

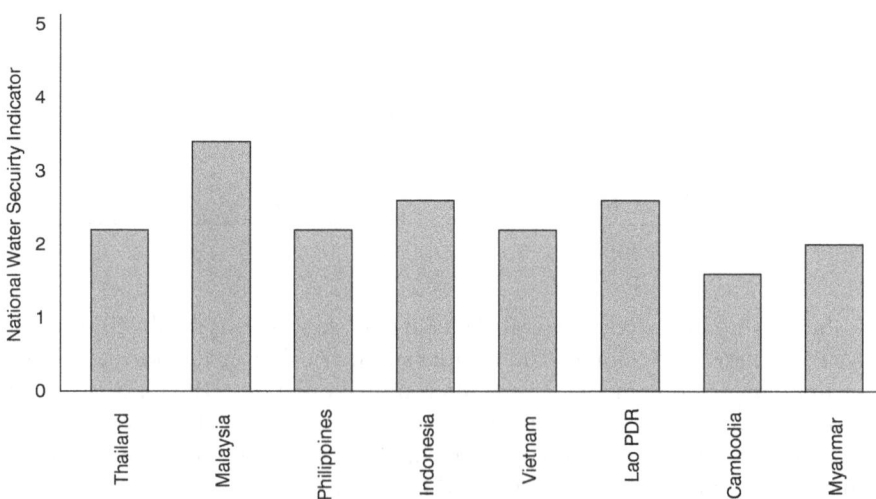

Figure 3.1 National water Security, estimated on a scale 0 to 5 (i.e. no to high security)

Source: Adapted from ADB (2014)

(2.4 million ha), Cambodia (1.6 million ha) and the Philippines (1.3 million ha) (Haefele and Hijmans, 2007).

Flooding is another major threat to rice production that is likely to worsen as a consequence of climate change. More frequent high-intensity rains will cause more floods in the lower areas of landscapes that tend to be dominated by rice production. Furthermore, coastal regions will increasingly be affected by sea level rise (anticipated to be as much as 70cm globally by the end of the century), exacerbated by sinking land surfaces in most of the large Asian deltas (Syvitski *et al.*, 2009). Increased inundation, erosion and salinization around the coasts of SEA will have particularly severe implications for the many millions of people and the major rice growing areas in the deltaic and coastal lowlands (ADB, 2009).

Contribution to greenhouse gas emissions

Across SEA, agriculture is a major contributor of GHG emissions. Agricultural emissions have been steadily increasing in recent years and in 2012 the total carbon dioxide equivalent (CO_2-e) emissions from agriculture were approximately 453 Mt CO_2-e (Figure 3.2). Of this, 45 percent (204 Mt CO_2-e) was from lowland irrigated and rainfed rice cultivation, and 5 percent (22 Mt CO_2-e) was from energy use in machinery (i.e. burning of fossil fuels) (Figure 3.2). A significant proportion of the latter was pumping of surface and groundwater for irrigation.

Like all wetlands, flooded rice fields both sequester carbon and emit greenhouse gases (GHGs). Rice fields, and in particular intensive rice production methods, contribute to climate change in several ways: i) they produce CO_2 and methane (CH_4) emissions as a consequence of the anaerobic decomposition

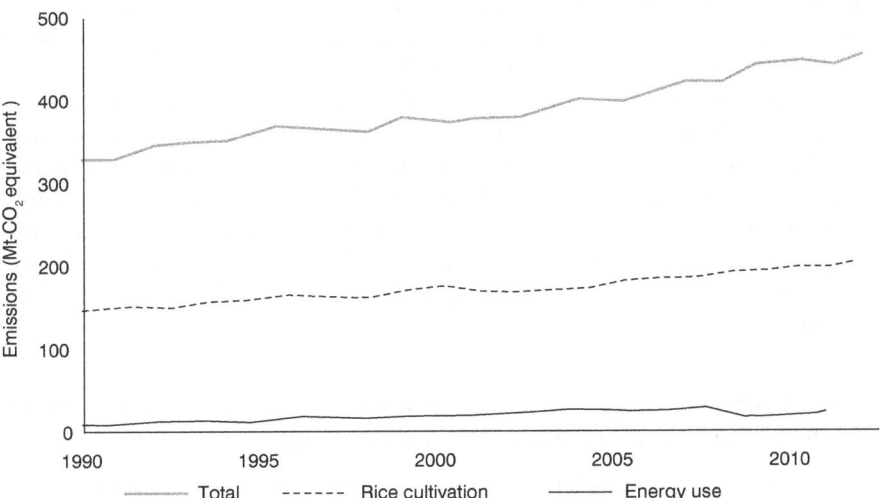

Figure 3.2 Agricultural GHG emissions across SEA, 1990–2012

Source: Adapted from FAOSTAT, 2015

of organic rice residues; ii) intensive rice cropping typically uses considerable amounts of nitrogen fertilizer (urea), which is partly converted to nitrous oxide (N_2O), another potent GHG; iii) the widely used practice of burning field residues contributes to the black carbon (soot) in the atmosphere, another major contributor to global warming, especially in tropical Asia; and iv) the burning of fossil fuels to pump and convey water for irrigation results in CO_2 emissions.

In traditional rainfed rice systems, still typical for many places across SEA, only one rice crop is grown in a year, and only a little nitrogen fertilizer is used. Because of the low intensity and limited use of external inputs, rice yields are relatively low in these systems but the fields are only flooded for short periods of the year, so most organic matter decomposes aerobically, mainly to CO_2 rather than the more potent CH_4. Emissions of GHGs from low intensity rainfed systems are therefore limited, but they generally increase as a result of irrigation and interventions to intensify production.

Agricultural water management in the region

Today, much agricultural policy in the region emphasizes agricultural intensification and irrigation development to meet rising food demand and spur economic growth. For example, Lao PDR plans to ensure two seasons of rice production on 500,000 ha of land by investing in new irrigation (200,000 ha) and significantly increasing the use of chemical fertilizers (LPDR, 2014). Major irrigation expansion is also proposed in Cambodia, linked to investments in flood control in the undeveloped Cambodian delta and elsewhere linked to hydropower development (Young, 2009). In Thailand, major water transfers from the Mekong have long been considered to complement national approaches to alleviate drought in the northeast of the country (Sneddon, 2003). Overall, the irrigated area in the lower Mekong basin is anticipated to triple from 66,000 to 186,000 km² by the 2060s (MRC, 2011).

Whilst irrigation expansion is certainly important in some contexts, most farming, including most rice farming in SEA remains rainfed, and will do so for the foreseeable future. From field to basin scales, there are many interventions used to manage rainfall efficiently and productively in smallholder farming systems, which can achieve significant long-term impacts. In the near future, improvements in water management in rainfed systems are likely to be the most cost-effective strategies for increasing food production and building resilience (Johnston *et al.*, 2012).

'Water smart' interventions: options to mitigate, adapt to and cope with the adversities of climate change

'Climate smart' agriculture requires practices that sustainably increase productivity and resilience, remove/reduce GHG emissions and simultaneously enhance achievement of national food security and development goals (FAO, 2010). Thus, ideally, climate smart agriculture contributes to climate change adaptation and mitigation as well as food security and development. Within this context, there are numerous possible interventions that improve agricultural water management and can be classified as 'water smart' interventions.

Mitigation

Improved water management is usually associated with adaptation to climate change, not mitigation. However, as discussed earlier in the chapter, irrigated rice fields are a major source of GHG emissions, particularly CH_4. GHG emissions from rice are influenced by a range of factors including temperature, root material, soil and water organic residues (including fertilizers and other chemical additives), plant physiology and the physical, chemical and biological properties of soil, as well as, very importantly, in-field water regime and management.

Recently, scientists have used computer models to estimate GHG emissions from alternative cropping systems. Studies have shown that intermittent irrigation, or alternate wetting and drying of lowland rice fields, can significantly reduce CH_4 emissions. Care is needed because shifting the water management from continuous flooding to mid-season drainage whilst reducing CH_4 can simultaneously increase N_2O fluxes. However, a study conducted in China indicated that water management without the use of inorganic fertilizer is an effective option for not only saving water but also mitigating the combined climatic impacts from CH_4 and N_2O in paddy rice production (Zou *et al.,* 2005). In the Philippines, some rice schemes have been modified to enable alternate wetting and drying (AWD) (Box 3.2).

GHG emissions can be further reduced through techniques introduced in tandem with water management. For example, deep placement of urea (i.e. burying urea granules at soil depth of 7–10 cm, rather than broadcasting it in fields) shortly after rice is transplanted, not only increases yields, but also reduces the nitrogen lost to the air and surface runoff (FAO, 2011). Finally, charring (i.e. partly burning) rice residues and adding the obtained black carbon or 'biochar' to rice fields as an alternative to fully burning or incorporating untreated harvest residue in the soil has been shown to reduce field CH_4 emissions by up to 90 percent (IRRI, 2015a).

The other way water management can contribute to mitigation is by decreasing the use of fossil fuels in the pumping of water. Large-scale pumped irrigation schemes are common in many places in SEA. For example, in Myanmar, irrigation development since 1995 has focused primarily on pumped schemes and it is estimated that approximately 201,000 ha is currently under lift irrigation (i.e. 10 percent of all formal irrigated area in the country), with most water pumped (up to 100 m) from the country's major rivers (i.e. the Irrawaddy, Chindwin and Salween). Similarly, in Lao PDR pumped irrigation is practiced across 28,824 ha of the Vientiane plains, one of the major areas of rice cultivation in the country, with water pumped from the Nam Ngum River, a tributary of the Mekong (Lacombe *et al.,* 2014; Box 3.5).

In many places the performance of lift irrigation schemes is undermined by a range of deficiencies, including: i) lack of electricity to operate the pumps

Box 3.2 Mitigating methane emissions through new irrigation schemes (Bohol, Philippines)

Bohol Island is one of the biggest rice growing areas in the Philippines. Before the completion of the Bohol Integrated Irrigation System (BIIS) in 2007, two old reservoirs were unable to ensure sufficient water for the year's second crop (November to April) for farmers who live the farthest downstream from the dam. This problem was aggravated by unequal water distribution and a preference by farmers for continuously flooded rice fields.

In the face of declining rice production, the National Irrigation Administration created an action plan that included the construction of a new dam and the implementation of AWD. The visible success of AWD in pilot farms, and specific training programs for farmers, was able to dispel the widely held perception of yield losses from non-flooded rice fields. Widespread adoption of AWD improved the use of irrigation water, so that the cropping intensity increased from about 119 percent to about 160 percent (relative to the maximum of 200 percent in these double-cropping systems). Moreover, it is estimated that CH_4 emissions have been reduced by 48 percent compared to continuous flooding of rice fields. There is no report of changes in N_2O emissions, but the AWD is reported as a great success with multiple benefits: CH_4 emission reduction (mitigation), reduced water use (adaptation), increased productivity and greater food security.

Source: Bouman, *et al.,* (2007)

either because supply is insufficient or because the cost of electricity is too high; ii) considerable losses due to canal seepage; iii) poor canal and embankment construction and maintenance; iv) sediment deposition in the canals; v) insufficient drainage that causes waterlogging and, in places build-up of soil salinity; and vi) poor timing in relation to the delivery of water. These factors, and others, all adversely affect crop yields, with significant implications not only for the economic viability of the schemes but also GHG emissions. In many cases neither water nor energy are being used efficiently: GHG emissions are effectively increased without resultant benefits.

Studies of pumped irrigation schemes in Central Asia, China and elsewhere have shown that improvements in scheme water use efficiency translate into energy savings and so can be an effective way of reducing emissions (Zou *et al.,* 2012). Options to improve scheme performance include: i) changes in cropping patterns, ii) changes in soil tillage practices; iii) conjunctive use of groundwater and surface water; iv) optimizing the timing of pump use; v) within scheme water storage; and vi) recycling of water.

Box 3.3 Solar powered pump irrigation

Solar powered pumps are increasingly popular globally for domestic use, livestock watering and irrigation. However, despite recent innovations and the increasing mass production of photovoltaic panels, which has meant their cost has declined significantly, they remain generally too expensive for smallholder farmers in SEA. Consequently, they are not widely used in the region. In contrast, in India where the government subsidizes their cost (typically by 80–90 percent) smallholder farmers increasingly use them. Care is needed to avoid over abstraction and ensure sustainability, but policies are currently being developed to try and ensure this.

Source: Weerasekera (2012)

In SEA, smallholder farmers are increasingly using low-lift diesel pumps to tap shallow groundwater aquifers. They also use pumps for surface irrigation and for draining excess water. Data specifically on GHG emissions from these pumps is scarce, but with increasing numbers being used it is undoubtedly increasing. In this case, solar pumps offer a potential approach to reduce emissions from informal irrigation (Box 3.3).

Adaptation

Depending on local contexts, needs and interests, there are opportunities for improving water management that can significantly contribute to people's livelihoods and make them more resilient to the adverse impacts of climate change. Many of the challenges that SEA farmers face in adapting to climate change relate to capturing and storing uncertain rainfall, managing water resources, enhancing soil moisture retention and improving water use efficiency.

Surface water irrigation and drainage

The most obvious, and most common, adaptation to variable rainfall in SEA is the establishment of irrigation. As noted above this is the intervention commonly favored by governments and over the past few decades considerable capital investments have been made in formal irrigation schemes (Table 3.2). Irrigation practiced in many previously purely rainfed lowland systems has undoubtedly contributed to large productivity increases (Box 3.4). The widespread construction of water storage reservoirs for hydropower also increases opportunities for dry season irrigation in some places (Box 3.5).

Throughout the region, most irrigation supports the monsoon crop (i.e. supplements rainfall) and typically only 15–35 percent of irrigated areas are used for double cropping. Dry-season crops are rarely fully irrigated, but are planted to take advantage of the beginning or end of the wet-season rains or utilize

Box 3.4 Irrigation in Vietnam
In Vietnam, rice is cultivated on 82 percent of the arable land, largely in the deltas of the Red and Mekong rivers, in the north and south of the country respectively. In the 1980s, Vietnam was a net importer of rice but since the 1990s it has been one of the biggest exporters in the world. Total annual paddy rice production grew from 19.2 million tons in 1990 to 35.8 million tons in 2005. Currently productivity in Vietnam exceeds 5.5 t/ha and it exports around 7 million t/year. This turn in fortunes was achieved through a shift from self-supporting production to highly intensified cropland systems, supported by significant investment in irrigation infrastructure: at least US$ 725 million between 1995 and 2000. Half of Vietnam's rice production area is irrigated, though competition for water resources is growing rapidly.

Source: IRRI (2015b)

Box 3.5 Synergies between irrigation and hydropower
Since hydropower dams generally increase dry season river flows in SEA, the construction of dams for hydropower presents possible opportunities for increased downstream irrigation in some places. Lacombe *et al.* (2014) investigated how the building of dams will likely modify the balance between water demand and water supply along the Nam Ngum River, a Mekong tributary in Lao PDR. The basin already contains both irrigation schemes and hydropower dams, and has the potential for significantly enlarged river-fed irrigation, as well as additional upstream hydropower. By the 2030s, if, as planned, eight hydropower dams are completed in the Nam Ngum Basin, dry season river flow could increase by more than 200 percent, allowing current irrigation water demand to triple, sufficient for full irrigation development, whilst maintaining environmental flows in all years. However, beyond the effects on water resources, there are a number of other impacts on fisheries, sediment, biodiversity, ecosystems, and population resettlement that must also be considered.

Source: Lacombe *et.al.*, (2014)

residual soil moisture. In the Mekong Delta, growing two irrigated dry-season crops a year has progressively replaced traditional wet-season rice, but both receive a significant amount of rainfall. It is estimated that in Cambodian and Vietnamese sub-basins of the southern Mekong Basin, more than 60 percent of total evapotranspiration from irrigated crops is derived from rainfall (Kirby *et al.*, 2010). Provision of drainage to extend the period of cultivation around the wet season, as in the polders of the Red River valley in Vietnam and the colmatage systems of southern Cambodia, can be as important as provision of water.

In the past, irrigation investment has focused mainly on large formal canal command systems, drawing water from reservoirs or pumping directly from rivers. However, there are opportunities for water-management interventions at a range of scales. Interventions can vary from large-scale, formal irrigation and drainage schemes that are planned and managed by governments or corporations, to small-scale, informal water systems managed at farm or community level. The two are not necessarily mutually exclusive, as small-scale water management efforts may complement large-scale systems, providing a greater degree of flexibility. Collective small-scale interventions may be able to achieve similar outcomes to large-scale projects, but with lower costs and greater ownership and involvement by farmers.

Experience from Vietnam and elsewhere suggests that irrigation alone is unlikely to make a large difference to the incomes and livelihoods (and hence resilience) of farmers. Investments are needed to structure input and output market chains, so that farmers have access to high-quality seeds, fertilizers and pesticides, and are able to procure a fair price for their crops. Furthermore, if farmers are to make the best use of irrigation through good crop choices and the most appropriate in-field soil and water management techniques, they require extension services providing sound agronomic advice.

Groundwater

The importance of groundwater is growing as surface water resources come under increasing pressure, and improved and more cost effective pumping and drilling technologies facilitate access to sub-surface supplies. Globally groundwater use is a key component of adaptation strategies in many areas, and dependence on groundwater as a buffer to climate variability is likely to increase under climate change. However, widespread uptake of groundwater irrigation has not yet been documented in SEA.

It is estimated that the area irrigated by groundwater is only 5 percent of the total area equipped for irrigation in mainland SEA and 4 percent in maritime SEA (Siebert *et al.,* 2010). Anecdotal evidence suggests that this may be changing. For example, recent studies report increasing evidence of small-scale shallow groundwater pumping in the Cambodian Mekong Delta (de Silva *et al.,* 2014), the Central Plain of Thailand (Pavelic *et al.,* 2012) and Myanmar's Dry Zone (IWMI, 2015). Groundwater has underpinned Vietnam's rise to one of the largest exporters of coffee (Box 3.6).

Where reliable groundwater supplies are available, the advantages over surface water for irrigation can be very significant. Aquifers provide both storage and transmission of water, reducing the need for large-scale infrastructure, and irrigation can be developed quickly and incrementally, with low capital cost. In contrast to most large canal systems, water is available directly on-demand, allowing farmers' flexibility and control over timing and quantity of supply.

Box 3.6 Groundwater and coffee in Vietnam

Vietnam is the world's leading Robusta coffee producer and coffee is the second largest export-earning crop, generating revenues of US$ 3.74 billion (3 percent of national GDP) in 2012. Coffee supports livelihoods of around 2 million rural people. Dac Lac province in the Central Highlands accounts for around half of Vietnam's coffee, produced mainly by smallholders. Irrigation is crucial for growth of coffee during the dry season (i.e. January to April) and groundwater is the major source of water. However, excessive groundwater pumping has caused declining groundwater levels, threatening sustainability of production in the region. Climate change may further exacerbate the water scarcity, since projections indicate that the dry season in the Central Highlands may start earlier and last longer.

Studies indicate that smallholders, with limited access to information, irrigate more than twice the recommended amount believing that it will increase yields. In fact, field experiments have shown that water stress can improve yields, and reducing irrigation to 70 percent of the locally recommended level, synchronized with management of other inputs, could increase average yield up to 4,000 kgha^{-1}, from the present level of 2,400 kgha^{-1}. Since irrigation accounts for 15–20 percent of total production costs, improving irrigation water management not only addresses the groundwater issues but also reduces the costs of production and increases profits.

Source: D'haeze (2004); Haggar and Schepp (2011); Amarasinghe *et al.*, (2015)

Conjunctive use of surface and groundwater offers major opportunities for irrigated agriculture, and is a realistic adaptation strategy for climate change (GWP, 2012). Groundwater development of large alluvial aquifers, closely linked to rivers and replenished annually by the monsoon, can potentially replace surface storage (with reduced costs and evaporative losses), if managed sustainably. Appropriate physical conditions exist in the large alluvial systems of SEA, such as the Irrawaddy, Chao Phraya and Mekong. In other places, managed aquifer recharge (MAR), a technology applied to capture and store wet season flows, could be used to both mitigate the impacts of flooding and enhance water security, offsetting potential groundwater overuse by enhancing aquifer storage (Box 3.7). 'Informal' conjunctive use also provides opportunities for increasing the overall performance of formal irrigation schemes. For example, in the Dry Zone of Myanmar, farmers with plots located in the tail end of some irrigation schemes have invested in shallow tubewells and low-lift diesel pumps to supplement water provided by the scheme. They have resorted to this largely because, as a consequence of limited power supply and pumping duration, the scheme fails to deliver sufficient water to their fields. The informal irrigation

Box 3.7 Underground taming of floods for irrigation (UTFI)

A novel technological intervention called 'Underground Taming of Floods for Irrigation' (UTFI) is being proposed to simultaneously reduce flooding issues in low lying areas and enhance groundwater resources for dry season irrigation. Using principles of conjunctive management of surface and groundwater at the river basin scale, UTFI involves selected harvesting and storage of surplus wet season flows with groundwater recharge structures in upstream areas, to protect urban and other high valued infrastructure downstream, and later recovery of stored groundwater for irrigation.

Analysis of the Chao Phraya River Basin in Thailand indicates that 28 percent of the wet season discharges into the Gulf of Thailand from the basin could be harvested without significant impact on water use from existing surface storage or the riverine and marine ecosystems. Capturing peak flows in wet years requires dedicating around 200 km^2 of land within the basin to groundwater recharge. This would not only reduce the magnitude of flooding, but generate US$ 140 million per year in additional agricultural production through irrigation.

Source: Pavelic *et al.,* (2012); Brindha and Pavelic (2016)

taps shallow aquifers, which are at least partly recharged through canal leakage from the formal scheme (IWMI, 2015).

Groundwater irrigation thus offers opportunities for adaptation to climate change through stabilization of water supply; intensification of production by enabling a dry season crop; and diversification to high value production (Shah, 2014). However, two sets of risks must be taken into account before promoting expansion of groundwater use in SEA as an adaptation strategy.

First, the largely unregulated nature of groundwater use has already resulted in over-exploitation in many places, and there are concerns over long-term sustainability (Taylor *et al.,* 2012). A range of potential responses have been suggested, including: i) demand reduction through crop choice and water saving technologies; ii) supply augmentation through MAR; iii) tradable property rights; iv) community aquifer management; and v) indirect approaches such as energy pricing and rationing (to control pumping) or land use regulation. There are no simple solutions: governance and management practices must be customized for each case, depending on both hydrogeological and socio-economic contexts (Shah, 2014).

Second, the potential impacts of climate change on groundwater availability and demand must be considered. For example, changes in precipitation patterns will impact groundwater recharge; and sea-level rise may result in salinization of aquifers in delta areas. The likely impacts are highly uncertain and will be specific to each region and aquifer system (Holman *et al.,* 2012). Although the buffering capacity of groundwater to climate change is typically higher than that of surface waters drawn from rivers and ponds, groundwater is not exempt from drought and water table declines can undermine irrigation performance. Good understanding of aquifer characteristics and functions is essential for effective management of groundwater resources. Irrigation is the dominant cause of groundwater depletion, and the impacts of future climate variability may be felt mainly as the indirect effects of increasing irrigation demand (Taylor *et al.*, 2012).

Rainwater harvesting

In some areas, access to fresh water resources is limited. Aquifers are too deep or contaminated by salinity or other pollution (e.g. arsenic). Streams may be depleted during the dry season, or located too far from where the water is needed. In this situation, small on-farm water storage can be beneficial to deal with the risks of droughts at the onset or during the wet season, or to allow limited irrigation at the beginning of the dry season.

Collecting water from a small catchment area and storing it in village and farm ponds is a common solution throughout SEA. Since they store relatively small volumes of water, most ponds empty every year. They tend to be shallow, with relatively large surface areas, so that, in most cases, a significant proportion of the water is lost through evaporation. Nevertheless, they are useful for livestock watering, domestic purposes and sometimes small-scale irrigation. One major advantage is that they represent a decentralized system that enables individuals and communities to manage their own water for their own purposes (Barron, 2009). In some instances, small ponds may also be used for aquaculture, though integration of fish and irrigation requirements may not be easy. In some regions of the Philippines, Indonesia, northeast Thailand and the Myanmar Dry Zone, considerable numbers of rainwater harvesting ponds have been established.

Another form of water harvesting that is increasingly popular in SEA is roof-top rainwater harvesting. Such structures are primarily used for domestic needs, but can also be used to irrigate small vegetable gardens, typically for household consumption. Ideally, storage capacity should be large enough to cover irrigation demand for two to three weeks (Box 3.8).

Agronomic approaches

The frontline of adaptation by farmers to climate change will be through improved farming practices. A wide range of agronomic measures at field-to-farm level are available to increase water productivity and reduce risks in cropping systems. Agronomic measures are related to soil management (erosion prevention and control; improving fertility, structure, organic matter), soil cover, crop varieties, crop mixtures and rotations.

Many agronomic measures are encapsulated within the principles of conservation agriculture (CA): minimum soil disturbance through no-till or minimal tillage systems; permanent soil cover using crop residues and cover

Box 3.8 Rooftop rainwater harvesting and storage in jars in northeast Thailand

Harvesting of water from rooftops for domestic use is probably more widespread in Thailand than in any other country in the world. With an average annual rainfall of 1000–2000 mm, harvesting rainwater for domestic use is economically viable provided the right technologies are adopted. A construction boom for rainwater jars followed the announcement of a nationwide rainwater jar construction programme by the Ministry of Interior in 1985. The government, working jointly with local NGOs, supported the manufacture of several million jars. Although it was envisioned that householders would construct their own jars, the small-scale private sector became very active in construction. At the peak of the programme, small village-based manufacturing companies were turning out around 30 jars per day. The immense success of the jar programme arose from the fact that the technology met a real need, was affordable, and invited community participation. The programme involved a broad range of stakeholders, including households, communities, NGOs, universities and the private sector.

Source: University of Warwick (2002); Wang (2013)

crops; and crop rotations. CA also promotes precision placement of inputs to reduce use of agricultural chemicals (FAO, 2011). In SEA, a basket of CA technologies has been developed in different countries, and demonstrated to regenerate fertility of degraded soils, provide livestock with high-quality forage and increase soil carbon sequestration (Legoupil *et al.*, 2014).

There are many approaches to managing soil that directly affect agricultural water use. Rice transplanting is the most common method of rice establishment in lowland areas, usually on well tilled and puddled soil. Puddling, done in standing water of 50–100 mm depth, is associated with churning of the soil to change its physical properties (i.e. bulk density, percolation rate and porosity) for weed control, to reduce percolation, to make the soil soft for transplanting, and to increase nutrient (e.g. iron, zinc, phosphorous) availability (Farooq *et al.*, 2011). In the traditional transplanting system, puddling creates a hard pan below the plough zone and reduces soil permeability. However, puddling not only consumes a lot of time and energy but also a large quantity (up to 20 percent) of the total water requirement in rice (Adhikari *et al.*, 2007).

Non-puddled direct seeded rice is an approach where dry seed is drilled into the soil, thereby avoiding repeated puddling. In recent years there has been a shift – principally brought about by labour shortage – from traditional transplanting to direct seeding in several countries of SEA (Pandey and Velasco, 2002). However, probably as a consequence of better water control, it is more often adopted in the dry season than in the wet season, and currently dry season rice accounts for less than 25 percent of rice production throughout the region (Farooq *et al.,* 2007).

Since they are not supressed by flooding, weeds that compete with the rice for nutrients, light, space and moisture are the major problem for direct seeded rice. Herbicides are widely used in Malaysia, Vietnam and Thailand to control weeds. However, incidence of weeds becoming herbicide resistant is on the rise (Watanabe *et al.,* 1997). Integrated weed management involving cultural practices, crop rotation, stale seedbed practices and use of herbicide mixes is recommended for direct seeded rice (Maity and Mukherjee, 2008).

The system of rice intensification (SRI) is a widely promoted, though quite controversial, agro-ecological methodology aimed at increasing the yield of rice produced in irrigated farming by changing a suite of agronomic practices. SRI is presented not as a technology but rather as an approach based on a set of ideas and principles that are to be translated into specific practices. The methodology is presented to farmers as a set of techniques that reduce inputs, including water and increase yields (see chapter 11). With the increased impacts of climate change, increasing variability of rainfall, and the growing competition for water and land, SRI offers an opportunity for increasing the water productivity of rice (World Bank Institute, 2008). As rice cultivated under SRI grows with stronger stalks and longer roots, it is also more resistant to episodes of drought, waterlogging, storm and typhoons (Africare, Oxfam America, WWF-ICRISAT Project, 2010).

Improved crop varieties played a critical role in Asia's 'green revolution' and it is likely that they will be similarly important in securing food production under changing climates (Mackill *et al.,* 2010). In addition to high-yielding varieties and drought-resistant varieties, short duration and non-photoperiod-sensitive varieties can reduce risk by allowing shorter, more flexible cropping seasons. Development of submergence-tolerant rice varieties (Box 3.9) is particularly

---

Box 3.9 Submergence tolerant rice for Indonesia

Although rice in SEA is mostly grown in flooded paddies, complete submergence can kill the crop within a few days. To reduce this risk, researchers have developed rice varieties that can withstand submergence for a limited period. When the plants are submerged, they effectively become dormant. The development of new varieties, known as 'scuba rice', is underway in Cambodia, Indonesia, Lao PDR, Myanmar, Thailand, Vietnam and the Philippines. In Indonesia, farmer field tests have been carried out that have shown that average yields of submergence tolerant rice are comparable to that of normal rice under non stress conditions and almost double after 15 days of submergence. Since seed germination may be adversely affected by flooding, a new variant that will tolerate anaerobic germination is currently under development, to allow direct seeding.

Source: Adapted from Septiningsih *et al.* (2014) and Toledo *et al.* (2015)

important for SEA, where much of the productive land is in low-lying flood-prone deltas and coastal zones, at risk of increased incidence of flooding.

Ecosystem-based adaptation

Building resilience is not just about increasing yields and productivity. For many poor farmers the ecosystem services of rice systems provide vital livelihood and welfare benefits. Hence, the ability of people to adapt to climate change is inextricably linked to the condition of ecosystems. Healthy, well-functioning ecosystems enhance natural resilience to the adverse impacts of climate change and reduce the vulnerability of people. 'Ecosystem-based Adaptation' (EbA), is promoted as an approach that uses biodiversity and ecosystem services as part of an overall adaptation strategy to help people and communities adapt to the negative effects of climate change.

The biodiversity that rice agro-ecosystems support is central to the livelihoods of the households that farm them. Traditionally aquatic biodiversity has provided a food safety net during periods of crop and other food shortages (Box 3.10). In some places, traditional governance systems have enabled rice fields to be cultivated, and a range of ecosystem benefits to be derived, sustainably for many hundreds of years. It is clear that in these systems people and nature are intrinsically linked. However, these close relationships are under threat by the push to intensify rice production. The reasons to intensify rice production may be rational at the national level but clearly at the scale of a community or household the costs and benefits must be evaluated very carefully before deciding to follow the path of intensification. It is important that the development opportunities are realized without undermining the living aquatic resources on which so

Box 3.10 Importance of biodiversity in rice fields

Rice ecosystems throughout SEA often harbour a highly diverse set of organisms that provide multiple benefits, including pest control and maintenance of soil fertility, as well as being an important food source in their own right. From an ecosystems services perspective, some rice-based ecosystems contain more than 100 useful species: fish, crustaceans, molluscs, reptiles, amphibians, insects and plants.

In relation to food, the individual 'catch' from rice fields is usually modest. Much of what is caught is consumed and it is rarely sold. Consequently, it often goes unreported in official statistics. Yet this 'invisible' fishery can be vitally important for livelihoods and people's wellbeing. In Laos, they account for a large share of many people's intake of protein, micronutrients and essential fatty acids.

Source: Halwart (2008); Garaway *et al.*, (2013)

many people currently depend, and which make a significant contribution to the resilience of rice systems.

Conclusion and recommendations

In SEA, climate change is occurring in, and adding to, a highly dynamic and uncertain agricultural context. Improved management of water is key to the significant transformation that agriculture in the region must undergo in order to meet the inter-linked challenges of increasing food demand and climate change.

Managing current climate variability is the best indicator of the ability to manage future variability and, as this chapter has highlighted, effective technologies and practices – some of which are highly innovative – already exist. The case studies presented demonstrate how water smart interventions can reduce the risks associated with climate variability, improve yields and increase incomes, in both irrigated and rainfed systems, across SEA. However, there are no simple generic solutions.

Interventions must be formulated in the context of the whole range of impacts and drivers. What is needed are robust, no-regret interventions that address current problems in a manner that ensures sustainability and builds resilience regardless of the direction of hydroclimatic trends. To be effective, multi-faceted interventions must be tailored to local conditions and all costs and benefits fully evaluated. Effective implementation of such multi-faceted approaches can improve carbon sinks and significantly reduce energy requirements, with knock-on implications for GHG emissions.

One of the major factors determining resilience is economic status. Therefore, poverty reduction is critical to underpinning adaptation. Experience from many developing agrarian economies shows that improving water management is an important first step to increasing agricultural production and reducing poverty. Better water management reduces the risk of crop failure, enables the cultivation of more than one crop a year, and facilitates farmers' investment in improved varieties and fertilizers. In this sense, there are no boundaries between climate-specific interventions and those that simply target socio-economic development. However, care is needed to ensure that the latter are placed in the context of future climate change to ensure that they do not increase long-term vulnerability.

Boosting resilience and productivity across SEA is about transforming landscapes and rural livelihoods. As elsewhere, it is likely that throughout SEA future water management by smallholders will overtake the public irrigation sector, in terms of the number of farmers involved, the area covered and the value of production. In this context, the capacity of farmers to make both short and long-term planning decisions and technological choices is critical. Agricultural extension services are the main conduit for disseminating information required to make such choices but throughout SEA they are under resourced and in decline. The imperative of climate change requires that governments significantly bolster investment in extension and ensure that adequate resources are available to provide services to farmers. Governments must also increase support to research to better understand the practical implications of changes in water management at local, national, regional and global levels.

Throughout SEA, the majority of the impacts of climate change on agriculture and rural livelihoods will result from changes in the water cycle. For this reason, water must be central to strategies for climate change adaptation in agriculture. Systematically applied, water smart agricultural interventions can increase the resilience of communities in the region and contribute to sustainability.

Note

1 Since the 1970s average growth has been about 3.5 percent per annum which is significantly above the global average

References

Adhikari, U., Justice, S., Tripathi, J., Bhatta, M.R. and Khan, S. (2007) Evaluation of non-puddled and zero till rice transplanting methods in Monsoon rice, online, http://www.naef-nepal.org/ZeroTillTranspRice_IAEC_AAAE2007.pdf (accessed 15/08/15).

Africare, Oxfam America, WWF-ICRISAT Project (2010) *More Rice for People, More Water for the Planet*. Hyderabad, India: WWF-ICRISAT Project.

Amarasinghe, U.A., Hoanh, C.T., D'haeze, D. and Hung, T.Q. (2015) Towards sustainable coffee production in Vietnam: more coffee with less water, *Agricultural Systems*, 136: 96–105. doi: http://dx.doi.org/10.1016/j.agsy.2015.02.008

Asian Development Bank (ADB) (2009) *The economics of climate change in southeast Asia: a regional review*. Asian Development Bank, Manila, Philippines.

Asian Development Bank (ADB) (2013) *Asian water development outlook 2013: Measuring water scarcity in Asia and the Pacific*. Asian Development Bank, Manila, Philippines.

Asian Development Bank (ADB) (2014) *Climate change and rural communities in the Greater Mekong Subregion: a framework for assessing vulnerability and adaptation options*. Asian Development Bank, Manila, Philippines.

Barron, J. (ed.) (2009) *Rainwater harvesting: a lifeline for human wellbeing*. United Nations Environment Programme. Nairobi/Stockholm Environment Institute: Stockholm pp. 69.

Bouman, B.A.M., Lampayan, R.M. and Toung, T.P., (2007) *Water management in irrigated rice: coping with water scarcity*. Los Banos, Philippines: International Rice Research Institute, pp. 54.

Brindha, K. and Pavelic, P. (2016) Identifying priority watersheds to mitigate flood and drought impacts by novel conjunctive water use management strategies, *Environmental Earth Sciences*, 75: 399. doi: 10.1007/s12665-015-4989-z.

D'haeze, D. (2004) Water management and land use planning in the Central Highlands of Vietnam, the case of Coffea canephora in Dak Lak province, unpublished PhD thesis. University of Leuven, Belgium.

de Silva, S., Johnston, R. and Senaratna Sellamuttu, S. (2014) *Agriculture, irrigation and poverty reduction in Cambodia: Policy narratives and ground realities compared*. Working Paper: AAS-2014-13. Penang, Malaysia: CGIAR Research Program on Aquatic Agricultural Systems.

Farooq, M., Siddique, K.H.M., Rehman, H., Aziz, T., Lee, D-J. and Wahid, A. (2011) Rice direct seeding: experiences, challenges and opportunities, *Soil and Tillage Research*, 111: 87–98.

Farooq, M., Basra, S.M.A. and Ahmad, N. (2007) Improving the performance of transplanted rice by seed priming, *Plant Growth Regulation*, 51: 129–137.

Food and Agriculture Organization of the United Nations (FAO) (2010) *'Climate Smart' agriculture: policies, practices and financing for food security, adaptation and mitigation*, Rome: FAO, pp. 41.

Food and Agriculture Organization of the United Nations (FAO) (2011) *Save and grow – a policymaker's guide to the sustainable intensification of smallholder crop production*, Rome: FAO, pp. 112.

Food and Agriculture Organization of the United Nations (FAO) (2015) FA)stat data, http://faostat3.fao.org/browse/G1/*/E (accessed 14 July 2015).

Garaway, C.J., Photitay, C., Roger, K., Khamsivilay, L. and Halwart, M. (2013) Biodiversity and nutrition in rice-based ecosystems: a case of Lao PDR, *Human Ecology*, 41: 457–562.

Global Water Partnership (GWP) (2012) Groundwater resources and irrigated agriculture. Global Water Partnership Perspectives Paper, Stockholm: Global Water Partnership, 19 pp.

Haefele S.M. and Hijmans R.J. (2007) Soil quality in rice-based rainfed lowlands of Asia: characterization and distribution. In: Aggarwal *et al.* (eds.) *Science, technology, and trade for peace and prosperity. Proceedings of the 26th International Rice Research Conference*, New Delhi, 9–12 Oct 2006, Indian Council of Agricultural Research, pp. 297–308.

Haefele, S.M., Ismail, A.M., Jonson, D.E., Vera Cruz, C. and Samson, B. (2010) Crop and natural resource management for climate-ready rice in unfavorable environments: coping with adverse conditions and creating opportunities. In: Wassmann R, (ed.) *Advanced technologies of rice production for coping with climate change: 'no regret' options for adaptation and mitigation and their potential uptake*, online, http://www.agsri.com/images/documents/Producingpercent20morepercent20withpercent20lesspercent20–percent20IRRI_FAO.pdf (accessed 29/09/15)

Haggar, J. and Schepp, K. (2011) *Coffee and climate change desk study: impacts of climate change in four pilot countries*, Hamburg, Germany: The Coffee & Climate, c/o E.D.E. Consulting.

Halwart, M. (2008) Biodiversity, nutrition and livelihoods in aquatic rice-based ecosystems, *Biodiversity*, 9 (1 & 2): 36–40.

Holman, I.P., Allen, D.M., Cuthbert, M.O. and Goderniaux, P. (2012) Towards best practice for assessing the impacts of climate change on groundwater, *Hydrogeology Journal*, 20: 1–4.

Intergovernmental Panel on Climate Change (IPCC) (2014) *Climate Change 2014: Synthesis report. Contribution of Working Groups I, II and III to the Fifth Assessment Report of the Intergovernmental Panel on Climate Change* [Core Writing Team, R.K. Pachauri and L.A. Meyer (eds.)], Geneva: IPCC.

International Rice Research Institute (IRRI) (2015a) *Rice and climate change: reducing emissions from rice*, online, http://irri.org/news/hot-topics/rice-and-climate-change (accessed 29/09/15).

International Rice Research Institute (IRRI) (2015b) *Vietnam and IRRI* http://books.irri.org/Vietnam_IRRI_brochure.pdf (accessed 29/09/15).

International Water Management Institute (IWMI) (2015) *Improving water management in Myanmar's dry zone for food security, livelihoods and health*. Colombo, Sri Lanka: IWMI, doi: 10.5337/2015.213.

Johnston, R. M., Hoanh, C. T., Lacombe, G., Lefroy, R., Pavelic, P. and Fry, C. (2012) *Improving water use in rainfed agriculture in the Greater Mekong Subregion:*

Summary report. Swedish International Development Cooperation Agency (SIDA). doi:10.5337/2012.200.

Kirby, M., Mainuddin, M. and Eastham, J. (2010) Water-use accounts in CPWF basins: Simple water-use accounting of the Mekong Basin. CPWF Working Paper: Basin Focal Project series, Colombo, Sri Lanka: The CGIAR Challenge Program on Water and Food.

Kripalani R.H. and Kulkarni, A. (1997) Rainfall variability over South-East Asia – connections with Indian monsoon and ENSO extremes: new perspectives, *International Journal of Climatology,* 17: 1155–1168.

Lacombe G., Smakhtin V., and Hoanh C.T. (2013) Wetting tendency in the Central Mekong Basin consistent with climate-change-induced atmospheric disturbances already observed in East Asia, *Theoretical and Applied Climatology,* 111 (1–2): 251–263.

Lacombe G., Douangsavanh S., Baker J., Hoanh C.T., Bartlett R., Jeuland M. and Phongpachith, C. (2014) Are hydropower and irrigation development complements or substitutes? The example of the Nam Ngum River in the Mekong Basin, *Water International*, 39 (5): 649–670.

Lao People's Democratic Republic (LPDR) (2014) National Rice Policy for food security (draft), Vientiane, Lao: Ministry of Agriculture and Forestry, PDR 43 pp.

Lau, K.M. and Yang, S. (1997) Climatology and inter-annual variability of the southeast Asian Summer Monsoon, *Advances in Atmospheric Science,* 14 (2): 141–162.

Legoupil, J.-C., Lienhard, P. and Khamhong, A. (2014) Conservation Agriculture in southeast Asia. In: Farooq, M. and Siddique, K.H.M. (eds.) *Conservation Agriculture,* Heidelberg New York Dordrecht London: Springer Cham, pp. 285–310.

MacKill, D.J., Ismail, A.M., Pamplona, A.M., Sanchez, D.L., Carandang, J.J. and Septninsih, E.M. (2010) *Stress tolerant rice varieties for adaptation to a changing climate*, International Rice Research Institute, online, http://web.tari.gov.tw/csam/CEB/ member/publication/7(4)/004.pdf (accessed 21/08/15).

Maity, S.K. and Mukherjee, P.K. (2008) Integrated weed management in dry direct-seeded rainy season rice (*Oryza sativa*), *Indian Journal Agronomy,* 53: 116–120.

Masutomi, Y., Takahashi, K., Harasawa, H. and Matsuoka, Y. (2009) Impact assessment of climate change on rice production in Asia in comprehensive consideration of process/parameter uncertainty in general circulation models, *Agriculture, Ecosystems, and Environment,* 131 (3–4): 281–291.

Mekong River Commission (MRC) (2011) *Basin Development Plan Programme, Phase 2: Assessment of basin-wide development scenarios, Main Report,* Phnom Penh, Vientiane.

Pandey, S. and Velasco, L. (2002) Economics of direct seeding in Asia: patterns of adoption and research priorities. In: Pandey, S. *et al.* (eds), *Direct Seeding: Research strategies and opportunities*. Los Banos, Philippines: International Rice Research Institute, pp. 139–160.

Pavelic, P., Srisuk, K., Saraphirom, P., Nadee, S., Pholkern, K., Chusanathas, S., Munyou, S., Tangsutthinon, T., Intarasut, T. and Smakhtin, V. (2012) Balancing-out floods and droughts: Opportunities to utilize floodwater harvesting and groundwater storage for agricultural development in Thailand, *Journal of Hydrology,* 55–64: 470–471.

Redfern, S.K., Azzu, N. and Binamira, J.S. (2012) Rice in southeast Asia: facing risks and vulnerabilities to respond to climate change. In: Meybeck, A. *et al.* (eds.) *Building resilience for adaptation to climate change in the agricultural sector,* Proceedings of a joint FAO/ OECD Workshop 23–24 April 2012. Rome, pp. 295–314.

Septiningsih, E.M., Hidayatun, N., Sanchez, D.L., Nugraha, Y., Carandang, J., Pamplona, A.M., Collard, B.C.Y., Ismail, A.M. and Mackill, D.J. (2014) Accelerating the development of new submergence tolerant rice varieties: the case of Ciherang-Sub1 and PSB Rc18–Sub1, *Euphytica*, doi: 10.1007/s10681-014-1287-x.

Shah, T. (2014) Groundwater governance and irrigated agriculture. TEC Background Paper 19. Stockholm: Global Water Partnership.

Siebert, S., Burke, J., Faures, J.M., Frenken, K., Hoogeveen, J., Doll, P. and Portmann, F.T. (2010) Groundwater use for irrigation – a global inventory, *Hydrology and Earth System Sciences*, 14: 1863–1880.

Singhrattna N., Rajagopalan B., Kumar, K.K. and Clark, M. (2005) Inter-annual and inter-decadal variability of Thailand summer monsoon season, *Journal of Climate*, 18: 1697–1708.

Sneddon, C. (2003) Reconfiguring scale and power: the Khong-Chi Mun project in northeast Thailand, *Environment and Planning A*, 35: 2229–2250.

Syvitski, J.P.M., Kettner, A.J., Overeem, I., Hutton, E.W.H., Hannon, M.T., Brakenridge, G.R., Day, J., Vörösmarty, C., Saito, Y., Giosan, L. and Nicholls, R.J. (2009) Sinking deltas due to human activities, *Nature Geoscience*, 2 :681–686.

Taylor, R. G. *et al.* (2012) Ground water and climate change, *Nature Climate Change*, 3: 322–329. doi:10.1038/nclimate1744.

Thompson, J.R., Laizé, C.L.R., Green, A.J., Acreman, M.C. and Kingston, D.G. (2014) Climate change uncertainty in environmental flows for the Mekong River, *Hydrological Sciences Journal*, 59: (3–4): 935–954, doi: 10.1080/02626667.2013.842074.

Toledo, A.M.U., Ignacio, J.C.I., Casal, C., Gonzaga, Z.J., Mendioro, M.S. and Septiningsih, E.M. (2015) Development of improved Ciherang-Sub1 having tolerance to anaerobic germination conditions, *Plant Breeding and Biotechnology*, 3: 77–87.

University of Warwick (2002) *Very-low-cost domestic roof-water harvesting in the humid tropics: existing practice*. Warwick: School of Engineering, University of Warwick.

Wang, M.J. (2013) *Report 2: water harvesting and water management options for agricultural communities in Laos* (IRAS project): improving the resilience of the agriculture sector in Lao PDR to Climate Change Impacts, Vientiane.

Wang, L., Chen, W. and Huang, R. (2007) Changes in the variability of North Pacific Oscillation around 1975/1976 and its relationship with East Asian winter climate, *Journal of Geophysical Research*, 112: D11110. doi:10.1029/2006JD008054 (accessed 29/09/15).

Wassman, R., Jagadish, S.B., Peng, K., Hosen, Y. and Sander, B.O. (2010) Rice production and global climate change: scope for adaptation and mitigation activities. In: Wassmann R. (ed.) (2010) *Advanced technologies of rice production for coping with climate change: 'no regret' options for adaptation and mitigation and their potential uptake*, online, http://www.agsri.com/images/documents/Producingpercent20morepercent20withpercent20lesspercent20-percent20IRRI_FAO.pdf (accessed 29/09/15).

Watanabe, H., Azimi, M. and Zuik, I.M. (1997) Emergence of major weeds and their population change in the Muda area, Peninsular, Malaysia. In: *Proceedings of 16th Asian Pacific Weed Science Society*, pp. 246–250.

Weerasekera, D. (2012) Solar pumps lift more than just groundwater in parts of India. http://wle.cgiar.org/blogs/2012/11/28/solar-pumps-lift-more-than-just-groundwater-in-parts-of-india/ (accessed 13/08/15).

World Bank Institute (2008) System of rice intensification, online, http://info.worldbank.org/etools/docs/library/245848/index.html (accessed 29/09/15).

Xu, Z.X., Takeuchi K. and Ishidaira, H. (2004) Correlation between El Nino-Southern Oscillation (ENSO) and precipitation in South-east Asia and the Pacific region, *Hydrological Processes*, 18: 107–123.

Yihui, D. and Chan, J.C.L. (2005) The East Asian summer monsoon: an overview, *Meteorology and Atmospheric Physics*, 89: 117–142.

Young, A. (2009) *Regional irrigation sector review for joint basin planning processes.* Vientiane, Lao PDR: Mekong River Commission.

Zou, J., Huang, T., Jiang, J., Zheng, X. and Sass, R.L. (2005) A 3-year field measurement of methane and nitrous oxide emissions from rice paddies in China: effects of water regime, crop residue, and fertilizer application, *Global Biogeochemical Cycles*, 19: GB2021 doi:10.1029/2004GB002401, (accessed: 29/09/2015).

Zou, X., Li, Y., Gao, Q. and Wan, Y. (2012) How water saving irrigation contributes to climate change resilience – a case study of practices in China, *Mitigation and Adaptation Strategies for Global Change*, 17(2): 111–132.

Zhou, L.T. (2011) Impact of East Asian winter monsoon on rainfall over southeastern China and its dynamical process, *International Journal of Climatology*, 31 (5): 677–686.

4 Combating climate change

Developing sustainable intensive
farming systems in China

*Mei Xurong, Dong Hongmin, Li Yue,
Nicholas Clarke, Gong Daozhi, Hao Weiping,
Li Yingchun, Liu Buchun, Ma Xin and
Xiong Wei*

Introduction: climate and agriculture in China

Agriculture in China has been fundamental for socioeconomic development. In its 7000 years' history, China's traditional farming civilization was characterized by intensiveness, resource recycling at farm level, and harmony between nature and people. The kernel of the ideas and practices can be summarized in today's language as conforming to and adapting to nature, and effective use of resources. Since the 1950s, the Chinese government has launched various agricultural initiatives to develop food production and eradicate hunger. In particular, after the reform strategy launched in 1978, land tenure innovation, improved investments, and technology dissemination and application have tremendously improved agricultural productivity in the country. China's gross food production has risen from 113 million tons in 1949 to 607 million tons in 2014. This meant that per capita grain production has risen from 240 kg to 460 kg, a tremendous contribution to the food security of the country (Ministry of Agriculture, PRC, 1980–2013).

China has a typical monsoon climate and diversified topography. Due to the barrier created by the high Tibetan plateau, the arid and semiarid climate zone in China is located about 10 degrees further north compared with other parts of the globe, inducing arid, semiarid and sub-humid temperate regions in northern China, with a humid subtropical and tropical climate in southern China (Agroclimatic Teaching Group of Beijing Agricultural University, 1987). Climate, together with soil and water, consequently leads to agricultural diversification in species, cropping systems, and products, as well as agro-ecosystems such as the famous 'three plains' (the Northeast Plain, North China or Huang-Huai-Hai Plain, and the Yangtze River Plain, Figure 4.1).

However, the monsoon climate leads not only to high productivity due to a warm and wet growing season but also to high temporal and spatial variability. More frequent and intensified droughts, flooding and cold damage, and

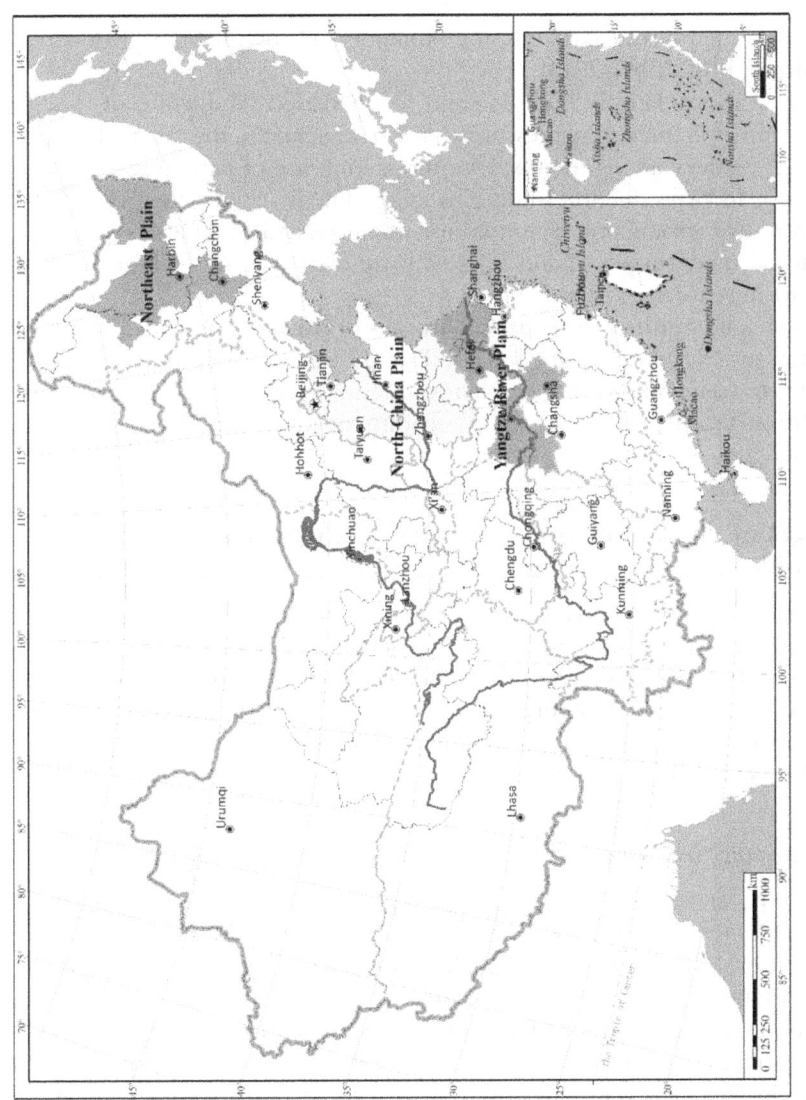

Figure 4.1 The location of Northeast Plain, North China Plain, and the Yangtze River Plain in China

Source: Adapted from the Group of Chinese Academy of Agricultural Sciences (1984)

increased agricultural monocultures have led to increased production losses as a percentage of gross production during the last decade compared to the 1980s. In addition, national mean crop yield has been estimated to be just 50–60 percent of the potential crop yield that could be obtained under favourable climatic conditions and good management. In modern times, land and water shortage and over-exploitation leading to agro-ecological degradation have been the major limitations to agricultural production and sustainability (Ministry of Agriculture, PRC, 1980–2013).

Thus, with a relatively high demand for increased production of food, in conjunction with increasing pressure on natural resources, the major challenges facing the country are: i) the need for integrating use of natural resources for sustainable food production under changing climatic conditions; ii) the need for innovative production patterns and their related technical systems for agricultural modernization; and iii) to ensure food security while conserving and restoring agro-ecosystems.

This chapter outlines the expected impacts of climate change and variability on agricultural production in China, followed by descriptions of interventions to promote resilience to climate change, including sustainable climate-smart agricultural practices. Finally, some recommendations are provided towards the end.

Impacts of climate change and variability on agricultural production

In order to develop sustainable agriculture systems that are also climate-smart, it is first necessary to understand the impacts that climate change is likely to have on agricultural production in China. This section provides a brief review of climate change and variability in China and its expected effects on soil, water, pests and diseases, cropping and livestock systems, and the implications for overall agricultural production.

Climate change in China

Since 1901, annual mean surface air temperature in mainland China has increased by 0.98°C with a warming rate of nearly 0.10°C per decade (Figure 4.2), slightly above the average global increase. The temperature in two periods (1920s–1940s and from the mid-1980s onwards) was apparently warmer. With the exception of the Sichuan basin and the northern Yunnan-Guizhou plateau where the temperature has declined slightly, the rest of China shows an increase in mean surface temperature, especially in north-western China and the northern Qinghai-Tibetan plateau, most of Inner Mongolia, most of north-eastern China, the northernmost part of China, and coastal regions in southern China (Lin *et al.*, 2005). Their study also concluded that depending on the level of future emissions, the average annual temperature increase in China by the end of the twenty-first century could be between 3 and 4°C, and there are likely

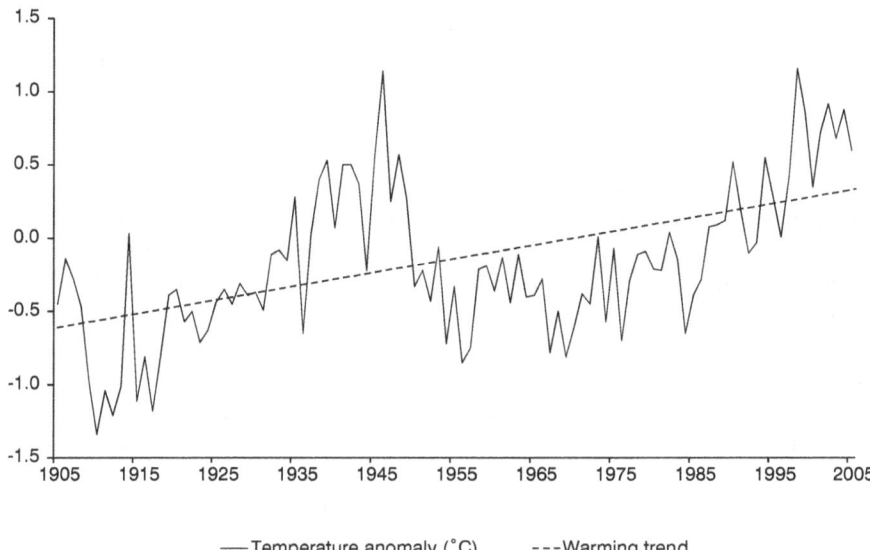

Figure 4.2 Changes in surface air temperature in China 1905–2005

Source: Tang and Ren (2005)

to be more extremely hot events during the summer and fewer extremely cold events during the winter.

Impacts of climate change and extreme climate events on crop production

Agricultural water resources

Mean annual rainfall in China may increase by about 10–13 percent, depending on emission scenario, for the 2070s compared with 1961–1990 (Lin *et al.*, 2005). The number of days with heavy rainfall is projected to increase and could result in more floods. The patterns indicate that the precipitation intensity is significantly increasing while precipitation frequency is significantly decreasing (Zhang and Cong, 2014).

Precipitation in general has greatly declined in the autumn and slightly increased in the winter periods. The frequency of both extreme rainfall events and droughts has increased. Studies show that the uneven spatial-temporal distribution of precipitation is leading to more water shortages as well as floods (Zhang *et al.*, 2014). Northeast China, north China, and the Yellow River basin show a decreasing trend of extreme rainfall events from 1961 to 2009, while the Yangtze River basin, the southeast coast, south China, Inner Mongolia, northwest China, and the Tibetan plateau show an increasing trend of extreme rainfall events (Fu *et al.*, 2013).

Changes in precipitation, solar radiation, air temperature, and wind speed account for 56 percent, 14 percent, 13 percent, and 5 percent of an overall decrease in annual runoff respectively (Lei *et al.*, 2014). Increased surface runoff was observed in the mountainous areas of south-western and north-eastern China, and in some areas along the south coast (Tao *et al.*, 2003a). Soil moisture levels are determined by precipitation, evapotranspiration, runoff, and soil drainage, so changes in agricultural water demand will have significant implications for China's water supply, the potential for crop drought, and consequently food production. The North China Plain and Northeast Plain show a trend towards increasing agricultural water use, soil drying, and significant changes in soil moisture variability (Tao *et al.*, 2003b). However, southwest China shows a significant *decrease* in agricultural water use (a result of a transition from double cropping to single cropping due to labour shortage) and a significant increase in soil moisture levels, while south-eastern China indicates a generally insignificant change in agricultural water use and soil moisture levels (Tao *et al.*, 2003b). These changes have had corresponding impacts on soil water balance and are thus likely to have impacted agricultural production (Tao *et al.*, 2003b). The annual crop yield of China was reduced in total by 4–8 percent due to soil water deficit and atmospheric drought (again, despite the increase in extreme rainfall events) resulting from climate change during the period 1995–2005 (Zhai, 2011).

Soil fertility

Climate change will directly affect crop growth through effects on soil carbon (C), nitrogen (N), and microbial activities, and thus on soil fertility. Effects of climate change on soil C, N and microbial activity will vary across regions and cropping systems. Changes in availabilities of C and N suggest that adjustments in C and N inputs will be required in future grain cropping systems to maintain soil C and N availability and sustain grain yields.

Elevated CO_2 will significantly increase the total organic C and total N contents in rice paddy land (Guo *et al.*, 2015), mostly due to enhanced microbial growth and activity (Li *et al.*, 2004). However, elevated CO_2 will decrease the soil total N content in the semi-arid agroecosystems in northern China (Li *et al.*, 2013), because the increase in N demand for plant growth is satisfied mostly by indigenous mineral N (Lam *et al.*, 2012a, b). The water extractable organic C concentration of both rhizosphere and bulk soils also decreases due to changes in microbial activity elicited by elevated CO_2, which could be adverse to the accumulation and stability of soil carbon in northern China's semi-arid agroecosystems (Fang *et al.*, 2015). Warming and increased precipitation have differential effects on soil extracellular enzyme activities in a temperate grassland. Studies showed that warming significantly increased nitrogen-acquiring enzymes at 10–20 cm soil depth, but dramatically decreased carbon-acquiring enzymes in the subsurface (Zhou *et al.*, 2013). In contrast, increased precipitation significantly increased nitrogen-acquisition enzymes, but did not affect carbon-acquisition enzymes (Zhou *et al.*, 2013). Soil C storage is expected to decrease significantly by 129.3 g C/m² under warming conditions but increase significantly by 145.9 g C/m² under increased precipitation conditions in the semiarid Inner Mongolian grasslands (He *et al.*, 2012).

Cropping systems

Most agricultural areas in China accommodate various cropping systems including intercropping and multiple cropping (double or triple crop rotation). Climate warming has made it possible to increase national food production by replacing single or low rotational cropping systems with high rotational systems (Liu *et al.*, 2013). This is particularly true in middle/high latitude regions such as northern or north-eastern China. For example, the northern boundary of the double-cropping system has moved on average 44 km across the region during 1981–2007, compared to 1950–1980 (Yang *et al.*, 2010). In many central provinces such as Hubei, Jiangsu, and Zhejiang, triple cropping has replaced double cropping and become the dominant system. From 1986 to 1995, the multiple cropping index (i.e. the sum of areas planted to different crops harvested during the year divided by the total cultivated area) of China's arable land increased by 9.5 percent, and the index in north-eastern China has reached 10.2 percent (Du and Guan, 2007). Future warming and changes in precipitation will affect the length of the growing season: For a 1°C increase in annual average temperature, the number of successive days with ≥0°C accumulated thermal heat increases on average by 15 days in northern China (Committee of Second National Assessment Report on Climate Change, 2011; Wang, 1996). If only temperature is considered, without taking precipitation and soil condition into consideration, the regional system in the lower reach of the Yangtze River would shift from a double-cropping to a triple-cropping system and the triple-cropping system boundary would move northward by about 200 to 300 km by 2050, increasing grain yield per unit area (Yang *et al.*, 2011; Liu *et al.*, 2010a).

In fact, climate warming could play a positive role for some regions in China enabling a shift from single-cropping to a double-cropping system, and with lower impacts from cold spells countering increased survival of pests and diseases. The benefit from temperature rise must be evaluated together with other agro-environmental factors such as irrigation water availability and local cropland conditions. Climate change will increase crop water consumption, with severe challenges to implementation of multiple cropping systems. Structural adjustment to crop production systems will require detailed analysis, including the suitability of crop varieties for new locations and matching water availability for crops during the growing period (Yuan *et al.*, 2011; Li *et al.*, 2010).

Crop yield and quality

Due to China's large range of agro-ecological conditions, sensitivity, and exposure to climate change vary considerably, leading to complex spatial patterns of response. There is some tendency for wheat, rice, and maize yields to decrease in southern China and to increase in central, northern, and north-eastern China, which benefit from a longer growing season. Warming has a major impact on cropping systems, depending on whether current temperature is lower (e.g. parts of northern China) or higher (e.g. Guizhou, southern China) than the optimum temperature for yield (Li et al., 2011). Changes in growing-season radiation and diurnal temperature range, rather than growing-season temperatures *per se*, contributed most to simulated yield reduction (Yang et al., 2014). Climate change without CO_2 fertilization could reduce rice, maize, and wheat yields by up to 18–37 percent in the next 20–80 years (Lin et al., 2005). Combining output changes in irrigated areas and rain-fed regions, production of total major crops may be reduced by an average of 5–10 percent (Wang, 2003). Yields of winter wheat, rice, and summer corn in irrigated areas may change by 1.6 to 2.5 percent, 3.7 to 10.5 percent, and 11.6 to 0.7 percent respectively by the 2050s. If no measures are taken to adapt to changing climate, yields of irrigated wheat, rice and corn are projected to decrease by 2.2–6.7 percent, 4.3–12.4 percent and 0.4–11.9 percent, respectively by the 2050s, compared with the potential yields in 1961–1990 (Lin et al., 2005). The reason for this is that higher temperatures tend to shorten the growth period (Lin et al., 2005). In the case of rice, grain yield will decline by 10 percent for each 1°C increase in growing season minimum temperature in the dry season, although the effect of maximum temperature on crop yield is insignificant (Peng et al., 2004).

Elevated CO_2 can cause more or less deleterious effects on grain quality (Blumenthal et al., 1996; Rogers et al., 1996; Hakala, 1998; Monje and Bugbee, 1998). Several studies revealed that small increases in temperature had a larger effect than elevated CO_2 on grain quality, improving crop quality through enhanced grain protein content (Tester et al., 1995; Williams et al., 1995; Campbell et al., 1981; Benzian and Lane, 1986; Randall and Moss, 1990; Wrigley et al., 1994). However, it is unlikely that any high temperature effects will totally compensate for CO_2 enrichment (Lin et al., 2005). An even larger yield growth could be obtained with widespread adoption of climate-smart rice cultivars and management (Xiong et al., 2014).

Pests and diseases

One of the major problems due to climate change would be the increase in the incidence of pests and diseases. China's agricultural losses from pests and diseases currently account for about 20–25 percent of total agricultural output value (Committee of Second National Assessment Report on Climate Change, 2011). Ecological and environmental conditions for pests in farmlands will be altered by increasing temperature, precipitation variation, more extreme climate events, and changes in the system of cultivation, resulting in changes in the area and time suitable for disease (Zhang et al., 2012). Across China, there has been an increase in the area, intensity, and frequency of occurrence of pests and diseases over the years (Liu et al., 2010b). Some diseases and insects, never or seldom seen before, have emerged extensively, such as armyworm and corn borer in 2012 in north-eastern China, where there has been an increase in species quantity and distribution (Ju *et al.*, 2013). Milder and shorter winters mean that warm weather pests will start breeding sooner (Bale *et al.*, 2002). Other changes include expanded pest ranges, disruption of synchrony between pests and natural enemies, and increased frequency of pest outbreaks and upheavals (Parmesan, 2007; van Asch and Visser, 2007). Changing crops and increasing multiple cropping indexes also provide a more conducive environment for the spread of pests and pathogens (Ju *et al.*, 2013).

In the past 50 years, variations in meteorological conditions due to climate change have assisted expansion of the area of occurrence and intensity of diseases and pests. In the past 30 years, changes in meteorological factors have induced occurrence or increase of wheat powdery mildew and rice plant hopper to varying degrees in different provinces (Zhang *et al.*, 2012). Under CO_2 doubling, armyworm populations will double in a year according to a study by Wang *et al.* (2002). With emergence of new pests and diseases it is important to develop integrated pest management strategies including monitoring and forecasting services that can address multiple challenges that farmers will face.

Impacts of climate change and extreme climate events on livestock

Livestock health, diseases and productivity

Warming in temperate areas might enable the spread of more pathogenic species from tropical regions. Thus, it can be assumed that animals will tend to suffer increasingly high levels of infections (Cooper *et al.*, 2015). Behaviour and spatial distribution of vectors, especially arthropod vectors whose internal body temperature depends on the environmental temperature, are affected. Climate change can also cause heat stress in livestock (e.g. Nardone *et al.*, 2010), which will induce behavioural and metabolic changes, including altered heat exchange between animals and their environment. Feed intake, mortality, growth, reproduction, maintenance, and production are all potentially affected (Harle *et al.*, 2007; Nardone *et al.*, 2010; Silanikove and Koluman, 2015).

Grassland degradation and productivity

In the past 50 years, the annual mean temperature of China's grassland regions has generally risen, while precipitation changes have been more uncertain. Grassland degradation is a biotic disturbance in which grass struggles to grow or can no longer exist on a piece of land, so that it is extremely difficult to restore and reverse the direction of succession change (GB19377–2003[1]). Grassland degradation is caused by the joint action of natural, biological, and human factors, including climate change. Climate change may be the direct cause or exacerbate the grassland degradation process, or at least have a contributory role. In Qinghai province human activities are the main factors leading to grassland degradation rather than climate. Whereas in Inner Mongolia, high temperature and drought have exacerbated the area of grassland degradation. There is likely to be rising temperature and an unstable precipitation trend, a substantial increase in surface evaporation and increasing water demand in grassland ecosystems in the future. Water resources may become scarcer and grassland degradation more pronounced in the future.

There is a lot of uncertainty concerning the influence of climate change on plant phenology and regional differences in the net primary productivity (NPP) of grassland vegetation; warming may accelerate decomposition of soil C and reduce plant species diversity. Meanwhile, increasing precipitation could bring phenological development forward, prolonging the growing season and improving grassland NPP and species richness. However, vegetation growth has a certain hysteresis effect with respect to the variation in precipitation. Overall, climate change could promote growth of vegetation on China's grassland areas, while reducing vegetation growth in some other parts of China. Climate change induced changes in productivity and grassland degradation will lead to changes in feed resources linked to the carrying capacity of grasslands. This could result in decreased livestock production, and other indirect effects on livestock will be linked to the expected shortage of feed arising from the increasingly competitive demands of food, feed production, and land use systems (Thornton *et al.*, 2011).

Interventions to promote resilience to climate change and extreme climate events

Cropland management

Agricultural water management

In China, interventions such as rainwater harvesting, water-saving irrigation, soil evaporation control, and use of drought-tolerant varieties are among the means to improve agricultural water productivity and mitigate climate change impacts. Currently, contour planting, ridge planting, straw and plastic mulching, as well as their integration have been popularized in dryland agriculture to enhance water availability and minimize non-productive evaporation. Deficit irrigation, such as partial root zone irrigation for dryland crops, alternate wetting and drying irrigation for paddy rice, and drip irrigation under plastic mulch for cotton and cereals, has been applied widely in irrigated fields to improve water use efficiency and reduce greenhouse gas (GHG) emissions from agriculture. In addition, fertigation has been introduced to vegetable and fruit production in both greenhouses and open land (Chen *et al.*, 2013).

In order to disseminate such technologies, the Chinese government has launched a pilot project on water-saving agriculture, using subsidies so that: i) farmers can benefit from water-saving; ii) the effects of limited water supplies can be mitigated; and iii) adaptation to climate change is promoted.

Soil and nutrient management

Since agricultural soil is not only seriously affected by climate change but agriculture is also a major source of N_2O emissions, it is essential to introduce

Table 4.1 Mitigation potential of N fertilizer management practices in northern China

Treatment	Total N$_2$O emissions (kg N$_2$Ó hm^{-2})	Change relative to CK(%)	Change relative to normal N(%)
Control (CK)	0.565±0.069 d	–	–
Normal N	1.374±0.218 a	143	–
Optimized N (ON)	1.057±0.095 b	87	–23
ON80%+DCD	0.750±0.112 cd	33	–45
ON80%+NC	0.911±0.143 bc	61	–34

Note: Total N$_2$O emissions are expressed as mean ± standard deviation; different letters following numbers show significant difference at $p<0.05$ level

Source: Liu *et al.*, (2014)

adaptation and mitigation options to increase resilience. Globally, N fertilization of agricultural soils contributes about 36 percent of direct N$_2$O emissions (Mosier *et al.*, 1998). Several N fertilizer management practices are effective in decreasing N$_2$O emissions (Table 4.1). An optimized N (ON) practice which reduced N application by 21 percent decreased N$_2$O emissions by 23 percent compared with normal N practice (Liu *et al.*, 2014). Dicyandiamide (DCD) applied with 80 percent ON could reduce N$_2$O emissions by 45 percent. Nano-carbon (NC) is a kind of modified carbon with low ignition point and has non-conductive properties. Nano-carbon can filter toxic gases and harmful organisms, and is widely used in the research field of new fertilizers as it may improve crop yields and fertilizer use efficiency; its use together with ON reduced N$_2$O emissions by 34 percent. Slow-release fertilizers such as polythene-coated urea, sulphur-coated urea and urea formaldehyde can decrease N$_2$O emissions by 53, 79 and 65 percent, respectively (Wang, 2012).

Fertilization practices based on soil tests, with the aim of precise fertilization which can exactly meet crop demand, has been extensively investigated in China. This practice can reduce fertilizer requirements without decreasing yields. There are several ongoing projects funded by the Chinese government to study and demonstrate fertilizer management practices that can reduce the current rate of fertilizers used. However, it is necessary to scale up such management practices through increasing farmer awareness as well as sustained government support.

Need for improved crop varieties

Climate change enhanced biotic and abiotic crop stress, increased light and heat resources, and the extended growth period of crops, have all resulted in the breeding of new crop varieties (Zhou *et al.*, 2011). In order to reduce the negative impacts of climate change, China developed biological technologies such as somatic cell cloning (the transfer of a nucleus from a somatic cell into another cell from which the nucleus is removed), somatic cell embryo formation

(formation of a plant from a single somatic cell or group of somatic cells), plasmogamy (fusion of cell protoplasts), and DNA rearrangement (moving of DNA sequences from one position to another), and bred and selected improved crop varieties with resistance to drought, flooding, high temperature, and low temperature. In addition, significant efforts have been made in promoting capacity building in agricultural production (Committee of Second National Assessment Report on Climate Change, 2011).

According to the National Plan for the Development of Modern Agriculture, it is necessary to focus on major grain crops and economically important crops. To achieve this, it is important to collect, store, evaluate, and use relevant seed resources, and develop functional genes for high yield, high quality, disease and pest resistance, and high nutritional value. As of 2015, major grain crop seed production bases of varieties are to be established at national level, initially covering up to 96 percent of improved varieties of crops. The Plan also emphasizes the importance of genetic resources with outstanding yield and quality characteristics to be developed. By 2020, the Government's ambition is to: i) develop paddy rice parents with high adaptability, high resistance, high quality and high combined ability, a new 'three-line' sterile line and a 'two-line' sterile line with low sensitivity to low temperature; ii) establish a technical system of high-efficiency haploid breeding of corn and a technical platform for molecular marker-assisted breeding to form the backbone breeding group; and iii) produce water-saving wheat species with high yield, disease resistance and high stress resistance (General Office of the State Council, 2012).

Livestock, manure and grassland management

Improved management of livestock, manure, and grassland can contribute to mitigation of or adaptation to climate change. Diet can have a profound effect on methane (CH_4) emissions, as it drives the volume and composition of manure. For example, Na (2010) reported that corn straw, after treatment using silage technology, can help improve feed digestibility and reduce CH_4 production. With an identical ratio of fine feed to coarse feed, CH_4 emission decreased by 30 percent by feeding silage rather than dry corn.

GHG emissions during manure storage (especially slurry storage), in the form of CH_4 (in anaerobic conditions), but also NH_3 and N_2O, can be significant. One simple way to avoid cumulative GHG emissions is to reduce the time manure is stored (Costa *et al.*, 2012). Solid-liquid separation before storage and covering manure stores are other common options to reduce emissions. Anaerobic digester technology is considered as one of the solutions to environmental problems caused by manure management. Biogas is renewable, cheap and clean, and can be used for heating, cooking, and generating electricity, which saves trees, thus reducing deforestation (Thu *et al.*, 2012). This source of energy is also known to produce a biogas slurry with a high fertilizer value for crop production.

Sustainable grassland management can improve the utilization of grassland resources and promote soil carbon sequestration in grassland ecosystems, which can enhance the grassland ecosystem's capacity for climate change adaptation and mitigation. Overgrazed grassland may become a net carbon source, while moderate grazing and sustainable management can reduce carbon emissions and increase soil carbon storage in the grassland ecosystem (McSherry *et al.*, 2013). Overgrazing will on one hand decrease the carbon sequestration capacity, reduce soil porosity, increase soil bulk density, and decrease soil moisture permeability (Rong *et al.*, 2001; Zhang *et al.*, 2002). On the other side, overgrazing increases grassland soil respiration and carbon emissions (Klumpp *et al.*, 2009). It therefore becomes very important to seek a reasonable balance in different grazing intensities on grassland ecosystems (Li *et al.*, 1998; Han *et al.*, 2008). Grazing management is thus important for the stability of grassland ecosystems and the healthy development of livestock which can adapt and mitigate climate change. As the grazing effect is different in different regions, grazing management must be based on the regional characteristics of the grassland ecosystem (Bütof *et al.*, 2012).

Early warning systems and insurance

Early warning systems play a very important role in agricultural disaster emergency preparedness and mitigation. The China Meteorological Administration has built a long-term network to monitor weather events. It has also made significant progress in satellite remote sensing to monitor typhoons, regional floods, and drought in recent years (Ju *et al.*, 1996; Kuang *et al.*, 2007). Based on the national meteorological service platform, the China Climate Observing System and Arid Climate Observation System, early warning systems, assessment systems and decision management systems for weather and climate disasters have been established. With the current systems in place, it is possible to provide early warnings on extreme weather events, as well as daily comprehensive agricultural drought monitoring, real-time reports and one-week agricultural drought forecasts on the China Meteorological Society website.

The National Water Sector has developed a hydrological monitoring network for monitoring ground water levels, river and lake water levels, water flows, rainfall patterns, evaporation, and soil moisture content (He, 2010). The farmers can access this information through the website and use this information to assist with agriculture related planning and decision-making. The China Flood Control and Drought Relief Headquarters works together with the national meteorology, water resources, agriculture, and civil administration departments to implement the drought and flood monitoring and early warning systems.

Agricultural insurance is an efficient disaster risk transfer tool as well as a strategy for agriculture to adapt to climate change, but coverage was very low in China before 2004. It was first proposed in the No. 1 Document of the Central Government of China in 2004, facilitating the establishment of a policy-oriented agricultural insurance (PAI) system and the initial trial of selected products in some regions. Since then, PAI has been fully implemented. This means that PAI products cover all farmers who are voluntarily insured, for wheat, rice, corn, soybeans, cotton, and other important crops. PAI adopts traditional agricultural contracts, with wide coverage and low security, and the government is the provider of premium subsidies. The total of agricultural insurance premiums increased from 377 million RMB[2] to 32,570 million RMB and the average insurance loss ratio was 67.6 percent for the period 2004–2014 (Figure 4.3).

The government has encouraged innovative agricultural insurance products, such as weather index agriculture insurance (WIAI). The International Fund for Agricultural Development and the United Nations World Food Programme, in partnership with the Chinese Ministry of Agriculture, jointly launched the pilot project on WIAI in vulnerable rural areas in 2008. The Institute of Environment and Sustainable Development in Agriculture, Chinese Academy of Agricultural Sciences, was appointed to implement this project, to test if WIAI could offer an appropriate approach for sharing and mitigating weather risk in rural China (Liu *et al.*, 2010c).

WIAI has made tremendous progress, for example developing: i) the weather-based index product for citrus in Zhejiang province based on the annual extreme minimum temperature (Mao *et al.*, 2007); ii) the weather-based index insurance product for apple florescence freezing injury (Liu *et al.*, 2010d); iii) rainstorm index insurance for single-cropping rice (Lou *et al.*, 2010); and iv) the weather index insurance for wheat multiple disaster (Yang *et al.*, 2013). The insured farmers receive timely appropriate compensation according to the WIAI contract when insured crops are hit by drought, cold, heat, excessive rains, and other weather- or climate-related damage.

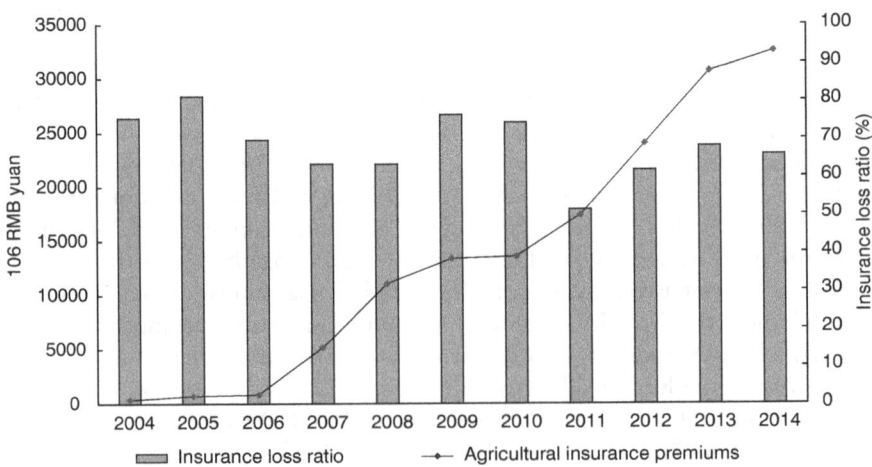

Figure 4.3 Agricultural insurance premiums and simple insurance loss ratio in China

Source: CIRC (2015)

Climate-smart agricultural practices

There are a number of initiatives in China at various levels, from the national to the local level, aimed at developing climate-smart agricultural practices. A few of these are described here.

Efforts to address GHG mitigation

The Chinese government is giving serious attention to climate change, and has included this issue in its long-term socioeconomic development plan. In 2009, China announced that by 2020 it will lower carbon dioxide emissions per unit of GDP by 40 percent to 45 percent from the 2005 level. In this connection, China has enacted and implemented the National Program on Climate Change, the Work Plan for Controlling Greenhouse Gas Emissions during the 12th Five-Year Plan Period, and the National Plan on Climate Change (2014–2020) (National Development and Reform Commission, 2015). China also submitted its intended nationally determined contributions to the Secretariat of UNFCCC and proposed 'to achieve the peaking of CO_2 emissions around 2030 and making best efforts to peak early'. With regard to adaptation, China issued 'China's National Strategy for Climate Change Adaptation' in 2013 (National Development and Reform Commission, 2013).

A series of actions to address climate change have been taken in the agricultural sector. From 2011, China began to implement national grassland compensation incentives for ecological protection to curb further degradation of grassland and increase soil carbon storage. The country also actively promoted conservation tillage technology, increasing soil fertility and agricultural soil carbon storage, and increasing food production stability and sustainability. Since 2005, the Ministry of Agriculture has promoted soil testing and planned to improve fertilizer use efficiency and reduce agricultural nitrous oxide (N_2O) emissions. China has promoted the construction of irrigation and water conservancy projects, enlarged irrigation areas, improved irrigation efficiency, actively promoted water-saving agricultural technology, improved water productivity, and enhanced the ability of farmers to deal with drought. Stress-resistant varieties with high yield, good quality, and resistance to drought, waterlogging, high temperature and pests have also been nurtured and promoted, thereby increasing food production resilience in the face of climate change.

Water management for climate-smart dryland agriculture

Dryland agriculture plays an important role in China's agricultural economy and national food security, producing about 43 percent of total national grain. Water scarcity is the biggest constraint to agricultural production in China. Promoting future dryland agricultural production will rely more on improving rainwater productivity through improved soil and water management, and more attention should be focused on using water efficiently. One successful example of efficient water management has been developed in Gansu province, where dryland agriculture areas account for about 70 percent of total farmland. Water scarcity and frequent drought disasters are major challenges for agricultural production in this province (Shang *et al.*, 2007). Techniques for on-farm rainwater conservation have been practiced there for a long time as a response to water scarcity and climate variability, including rainwater harvesting, conservation tillage and mulching techniques.

Based on these practices, in 2004 local technical experts developed the technology of 'whole plastic-film mulching on double ridges and planting in furrows' through integrating an improved plastic-film mulching technique with a ridge-furrow on-farm water harvesting technique. The improvements of this technology include:

- making one large ridge (70 cm wide and 15–20 cm high) alternating with a small one (70 cm wide and 10–15 cm high), and the furrow between them being used for a planting strip, which could catch rainfall more efficiently, thereby increasing water storage in the soil;
- covering the double ridges with plastic film 120 cm wide instead of covering partially with plastic film 90 cm wide, which could reduce evaporation significantly and increase moisture retention in the field;
- changing the mulching time from the sowing stage to the previous autumn, which could reduce evaporation during the fallow period (Zhao, 2004).

From 2004 to 2007, this technology was tested for maize in different places in Gansu and results indicated that it increased maize yields by an average of 37.1 percent compared with conventional technology. Average rainfall use rate and water use efficiency reached 33 kg/mm and 70 percent, respectively (Li *et al.*, 2009). After farmers saw the benefits in terms of increased water productivity from demonstration projects, this technology was widely adopted in Gansu and application areas reached 7,000 km^2 in 2010. Currently, this technology is being widely spread to other provinces in north China to help them cope with water scarcity. Although there have been problems with fertilization of the next crop if the same film is used for more than one season, development of new technology with a different fertilization technique will solve this problem, allowing the film to be used for two consecutive crops.

Climate-smart soil and nutrient management

CH_4 and N_2O are potent GHGs, with 298 and 25 times higher global warming potential than CO_2 (100 years), and contribute 29 percent and 5 percent to the greenhouse effect, respectively (IPCC, 2007). Rice paddies have been the largest CH_4 emission source, accounting for 10–20 percent of annual global CH_4 emissions to the atmosphere (Wassmann and Aulakh, 2000), and are also an important source of N_2O emissions (Cai *et al.*, 1997). Research on emissions of both CH_4 and N_2O from rice paddies is therefore directly relevant for mitigation of global warming. China is the world's leading rice producer (FAO, 2012) and also the biggest nitrogen fertilizer consumer. Application of nitrogen fertilizer can directly affect rice yield and GHG emissions (Zou *et al.*, 2005; Liang *et al.*, 2013). New fertilizers and related products (i.e. controlled release urea, nitrification inhibitors, and effective microorganisms) adapted to simultaneously promote yield and reduce CH_4 and N_2O emissions from rice paddies are required.

In order to identify the impacts of different fertilizers on CH_4 and N_2O emissions and evaluate the potential of these fertilizers to reduce GHG emissions and increase yields, a field study was conducted to measure CH_4 and N_2O fluxes continuously over two cycles of double rice-winter fallow rotations (2012–2013) under five different nitrogen fertilizer treatments. The treatments were:

- T(1:) Urea (N≥46 percent) used as the control (CK), applied using conventional practices in the study area;
- T(2:) polymer-coated controlled release urea (N≥42 percent, release period 90 d) (CRU);
- T(3:) N-Sever (N≥46 percent) (NU), a synthetic urea mixed with 0.5 percent nitrapyrin in production;
- T(4:) urea added with 1 percent nitrification inhibitor 3, 4–dimethylpyrazole phosphate (DMPP) added; and
- T(5:) urea with a culture solution of effective microorganisms diluted 200 times with same amount of nitrogen added (EM).

Results from the experiments showed that GHG emissions from rice paddies under different treatments averaged over the whole rotation ranged between 77.3–178.23 kg CH_4/ha (Table 4.2) and 4.18–10.11 kg N_2O/ ha (Table 4.3), and significant differences ($p < 0.05$) between the control (urea)and other treatments were found. The study also indicated that, with the application of new nitrogen fertilizers, it is possible to reduce global warming potential while simultaneously increasing yield. Controlled release urea, nitrification inhibitor and effective microorganisms might be effective fertilization options for low-carbon rice production.

Agricultural waste management and CSA: the Hubei Eco-Farming Biogas Project

This project installed one biogas digester with three supplementary renovations for 33,000 households with a reactor size of 8–15m³. Manure produced by 155,684 pigs will be fermented in the biogas digester, with the manure being stored under anaerobic conditions. The biogas will be used as thermal energy to replace the coal currently used to meet the households' daily energy needs. In addition, recovery and use of biogas from digested slurry in a biogas digester will reduce CH_4 emissions from the slurry that would otherwise have been stored in a deep pit. According to the monitoring report for UNFCCC Clean Development Mechanism projects, about 12.17 million m³ biogas were used

Table 4.2 Seasonal CH$_4$ emissions and reduction under different treatments

Treatment	Emissions (kg N$_2$O ha^{-1})				Reduction (%)	
	Early rice	Late rice	Winter fallow	Whole rotation	Rice season	Whole rotation
2012						
CK	0.27abAB	2aA	3.1b	5.37a	–	–
CRU	0.45aA	0.43cBC	3.54ab	4.42ab	61.3	17.7
NU	0.07cB	0.88bcBC	3.23b	4.18b	58.1	22.2
DMPP	0.06cB	0.27cC	3.88a	4.21b	85.3	21.6
EM	0.16bcB	1.36abAB	3.55ab	5.07ab	33.0	5.6
2013						
CK	2.71aA	4.14aA	3.26bB	10.11aA	–	–
CRU	1.47bB	3.12abAB	3.62aA	8.21bAB	33.0	18.8
NU	1.25bB	2.23bB	3.27bB	6.75bB	49.2	33.3
DMPP	1.18bB	1.91bB	3.6aAB	6.69bB	54.9	33.8
EM	1.37bB	2bB	3.34bAB	6.71bB	50.8	33.6

Different lowercase letters indicate significant difference ($p > 0.05$), different capital letters indicate very significant difference ($p > 0.01$)

Source: Authors own compilation from field experiments data at CAAS, Beijing

Table 4.3 Seasonal N$_2$O emissions and reduction under different treatments

Treatment	Emissions (kg N$_2$O ha^{-1})				Reduction (%)	
	Early rice	Late rice	Winter fallow	Whole rotation	Rice season	Whole rotation
2012						
CK	0.27abAB	2aA	3.1b	5.37a	–	–
CRU	0.45aA	0.43cBC	3.54ab	4.42ab	61.3	17.7
NU	0.07cB	0.88bcBC	3.23b	4.18b	58.1	22.2
DMPP	0.06cB	0.27cC	3.88a	4.21b	85.3	21.6
EM	0.16bcB	1.36abAB	3.55ab	5.07ab	33.0	5.6
2013						
CK	2.71aA	4.14aA	3.26bB	10.11aA	–	–
CRU	1.47bB	3.12abAB	3.62aA	8.21bAB	33.0	18.8
NU	1.25bB	2.23bB	3.27bB	6.75bB	49.2	33.3
DMPP	1.18bB	1.91bB	3.6aAB	6.69bB	54.9	33.8
EM	1.37bB	2bB	3.34bAB	6.71bB	50.8	33.6

Different lowercase letters indicate significant difference ($p > 0.05$), different capital letters indicate very significant difference ($p > 0.01$)

Source: Authors own compilation from field experiments data at CAAS, Beijing

each year, and annual emission reduction was about 60,000 tCO_2–e. This project thus supports China's overall sustainable development strategy while at the same time reducing GHG emissions, improving the local environment and public health by replacing coal stoves and reducing coal use as well as improving manure management, and improving local household economies by providing employment for construction, operation, and maintenance of the biogas digesters thereby reducing expenditure on coal.

Conclusions and recommendations

Recognizing the importance and particular characteristics of China's agriculture, climate-smart agriculture in China must focus on improving food productivity and ensuring food security at the same time as building resilience to climate change and reducing GHG emissions from agriculture. This will require both technological and management innovation. In 2013, the Chinese government launched an innovation-driven development strategy, and aimed at building an innovation-oriented country. This would require strategic solutions to promote sustainable agriculture in China that include:

- *Improvement of food productivity and effective supply:* By 2030, gross cereal demand in China is estimated to be 650 million tons, i.e. a 50 million ton increase in cereal productivity is necessary. With limited land and water shortage, as well as climate change, improving crop yield and reducing hazard losses are the main options for meeting the growth requirement. Solutions for meeting such targets should include technological innovation and implementation of a rational crop zonation and cropping system, strengthening of infrastructure for agricultural production, breeding of climate change ready varieties, enhancing water saving and water productivity to combat water scarcity, soil and water management, disaster mitigation and post-harvest protection and processing.
- *Redevelopment of an intensive farming system for income increase and climate resilience:* The increment of cereal production has benefitted from large scale mono-cropping plantations, but higher climatic variability and biodiversity reduction increase the risk of natural and biological disaster, with associated costs. Enhancing the health and sustainability of agro-ecosystems through redeveloping the intensive farming system will however improve climate resilience, reduce losses from disasters and consequently raise incomes. Such agricultural adjustment to climate change should include redeveloping the cereal and cash intercropping system, restructuring crop- or pasture-based livestock systems for recycling, developing agroforestry systems, etc. depending on the local natural resources and socioeconomic conditions.
- *Reduction of GHG emissions from agriculture as much as possible:* Improved management and recycling of livestock manure, other agricultural waste and nitrogen, development and popularization of advanced technologies for water saving and energy-efficient machines, regulated and if possible

reduced use of fertilizers and biological pesticides, and improved soil organic carbon sequestration will reduce GHG emissions from agriculture. Enhancing carbon sequestration, including soil organic pools, through restoration of various agro-ecosystems has the potential for reducing net GHG emissions, but needs to be integrated with productivity improvement, income increment and resilience enhancement.

Due to large regional variations and climate change patterns, a mosaic of approaches is required to address the complex challenges China will encounter. There are no single models that can be replicated. Lessons from past experiences are important to consider while developing new strategies. It is important to look at the knowledge that already exists and where knowledge is not yet available, the need for new studies and practically useable management guidelines may be filled by expert opinion (Clarke *et al.*, 2015).

Notes

1 National standards of 'Parameters for Degradation, Desertification and Salinization of Rangelands' (GB19377–2003) taken from General Administration of Quality Supervision, Inspection and Quarantine of the People's Republic of China (2004).
2 Currently 1 Chinese Yuan Renminbi RMB is equivalent to 0.157 US dollar.

References

Agroclimatic Teaching Group of Beijing Agricultural University (1987) *Agroclimatology*, Beijing: China Agricultural Press.
Bale, J.S., Masters G. J., Hodkinson I. D., Awmack C., Bezemer T. M., and Brown V. K. (2002) Herbivory in global climate change research: direct effects of rising temperature on insect herbivores, *Global Change Biology*, 8 (1): 1–16.
Benzian, B. and Lane, P. (1986) Protein concentration of grain in relation to some weather and soil factors during 17 years of English winter-wheat experiments, *Journal of the Science of Food and Agriculture*, 37(5): 435–444.
Blumenthal, C., Rawson, H. and McKenzie, E. (1996) Changes in wheat grain quality due to doubling the level of atmospheric CO_2, *Cereal Chemistry*, 73 (6): 762–766.
Bütof, A., von Riedmatten, L.R., Dormann, C.F., Scherer-Lorenzen, M., Welk, E. and Bruelheide, H. (2012) The responses of grassland plants to experimentally simulated climate change depend on land use and region, *Global Change Biology*, 18 (1): 127–137.
Cai, Z., Xing, G., Yan, X., Xu, H., Tsuruta, H., Yagi, K. and Minami, K. (1997) Methane and nitrous oxide emissions from rice paddy fields as affected by nitrogen fertilizers and water management, *Plant and Soil*, 196: 7–14.
Campbell, C., Davidson, H. and Winkleman, G. (1981) Effect of nitrogen temperature, growth stage and duration of moisture stress on yield components and protein content of Manitou spring wheat, *Canadian Journal of Plant Science*, 61 (3): 549–563.
Chen, G., Du, S., Jiang, R. and Gao, X. (2013) Fertigation technology application and research status in China, *China Agricultural Technology Extension*, 29 (5): 39–41.
CIRC (China Insurance Regulatory Commission) (2015) China Insurance Regulatory Commission report, http://www.circ.gov.cn/Default.aspx?alias=www.circ.gov.cn/english (accessed 04 November 2015).

Clarke, N., Gundersen, P., Jönsson-Belyazid, U., Kjønaas, O.J., Persson, T., Sigurdsson, B.D., Stupak, I. and Vesterdal, L. (2015) Influence of different tree-harvesting intensities on forest soil carbon stocks in boreal and northern temperate forest ecosystems, *Forest Ecology and Management*, 351: 9–19.

Committee of Second National Assessment Report on Climate Change (2011) *The Second National Assessment Report on Climate Change*, Science Press, Beijing, ISBN 978–7–03–032184–8 (*in Chinese*).

Cooper, K.M., McMahon, C., Fairweather, I. and Elliott, C.T. (2015) Potential impacts of climate change on veterinary medicinal residues in livestock produce: An island of Ireland perspective, *Trends in Food Science & Technology*, 44: 21–35.

Costa A., Chiarello G., Selli E. and Guarino, M. (2012) Effects of TiO$_2$ based photocatalytic paint on concentrations and emissions of pollutants and on animal performance in a swine weaning unit, *Journal of Environmental Management*, 96: 86–90.

Du, J. and Guan, Z. (2007) Climate change impacts on agriculture, *Journal of Anhui Agricultural Sciences*, 35 (16): 4898–4899 *(in Chinese)*.

Fang, H., Cheng, S., Lin, E., Yu, G., Niu, S., Wang, Y., Xu, M., Dang, X., Li, L. and Wang, L. (2015) Elevated atmospheric carbon dioxide concentration stimulates soil microbial activity and impacts water-extractable organic carbon in an agricultural soil, *Biogeochemistry*, 122: 253–267.

Food and Agriculture Organization of the United Nations (FAO) (2012) *Rice Market Monitor*, 15(4), online, http://www.fao.org/docrep/017/ap772e/ap772e.pdf (Accessed 05/10/15).

Fu, G., Yu, J., Yu, X., Ouyang, R., Zhang, Y., Wang, P., Liu W. and Min, L. (2013) Temporal variation of extreme rainfall events in China, 1961–2009, *Journal of Hydrology*, 487 (4): 48–59.

General Office of the State Council (2012) *National Plan for the Development of Modern Agriculture (2012–2020)*, online, http://www.gov.cn/zwgk/2012–12/31/content_2302986.htm (Accessed 05/10/15, *in Chinese*).

Guo, J., Zhang, M., Wang, X. and Zhang, W. (2015) Elevated CO$_2$ facilitates C and N accumulation in a rice paddy ecosystem, *Journal of Environmental Sciences*, 29: 27–33.

Hakala, K. (1998) Growth and yield potential of spring wheat in a simulated changed climate with increased CO$_2$ and higher temperature, *European Journal of Agronomy*, 9 (1): 41–52.

Han G.D., Hao X.Y., Zhao M.L., Wang, M., Ellert, B.H., Willms, W. and Wang, M. (2008) Effect of grazing intensity on carbon and nitrogen in soil and vegetation in a meadow steppe in Inner Mongolia, *Agriculture, Ecosystems and Environment*, 125: 21–32.

Harle, K.J., Howden, S.M., Hunt, L.P. and Dunlop, M. (2007) The potential impact of climate change on the Australian wool industry by 2030, *Agricultural Systems*, 93: 61–89.

He, H. (2010) China gauging station network, *Advances in Water Science*, 4 (005): 460–465.

He, N., Chen, Q., Han, X., Yu, G. and Li, L. (2012) Warming and increased precipitation individually influence soil carbon sequestration of Inner Mongolian grasslands, China, *Agriculture, Ecosystems and Environment*, 158: 184–191.

IPCC (2007) *Climate Change 2007: The Physical Science Basis. Contribution of Working Group I to the Fourth Assessment Report of the Intergovernmental Panel on Climate Change*, Cambridge: Cambridge University Press.

Ju, H., Velde, M., Lin, E., Xiong, W. and Li, Y. (2013) The impacts of climate change on agricultural production systems in China, *Climatic Change*, 120: 313–324.

Ju, W.M., Sun, H. and Tang, Z. (1996) Application of meteorological satellite remote sensing to drought monitoring, *Journal of Catastrophology*, 11 (4): 25–29.

Klumpp, K., Fontaine, S., Attard, E., Le Roux, X., Gleixner, G. and Sousanna, J.F. (2009) Grazing triggers soil carbon loss by altering plant roots and their control on soil microbial community, *Journal of Ecology*, 97: 876–885.

Kuang, Z., Zhu, W., Ding, M., Tan, Z. and Sun, H. (2007) Application of Multi-sources Satellite Data in Monitoring Sugarcane Droughts, *Chinese Journal of Agrometeorology*, 28(1): 93–96.

Lam, S.K., Chen, D., Norton, R. and Armstrong, R. (2012a) Nitrogen demand and the recovery of ^{15}N-labelled fertilizer in wheat grown under elevated carbon dioxide in southern Australia, *Nutrient Cycling in Agroecosystems*, 92: 133–144.

Lam, S.K., Han, X., Lin, E., Norton, R. and Chen, D. (2012b) Does elevated atmospheric carbon dioxide increase wheat nitrogen demand and recovery of nitrogen applied at stem elongation? *Agriculture, Ecosystems and Environment*, 155: 142–146.

Lei, H., Yang, D. and Huang, M. (2014) Impacts of climate change and vegetation dynamics on runoff in the mountainous region of the Haihe River basin in the past five decades, *Journal of Hydrology*, 511 (4): 786–799.

Li, K., Yang, X., Liu, Z., Wang, W. and Chen, F. (2010) Analysis of the potential influence of global climate change on cropping systems in China III. The change characteristics of climatic resources in northern China and its potential influence on cropping systems, *Scientia Agricultura Sinica*, 43 (10): 2088–2097 (*in Chinese*).

Li, L., Liu, G., Yang, Q., Zhao, X. and Zhu, Y. (2009) Research and application development for the techniques of whole plastic-film mulching on double ridges and planting in catchment furrows in dry land, *Agricultural Research in the Arid Areas*, 27 (1): 114–118.

Li, L., Liu, X. and Chen, Z. (1998) Study on the carbon cycle of *Leymus chinensis* steppe in the Xilin River Basin, *Acta Botanica Sinica*, 10: 955–961.

Li, Y., Conway, D., Xiong, W., Gao, Q., Wu, Y., Wan, Y., Li, Y. and Zhang, S. (2011) Effects of climate variability and change on Chinese agriculture: a review, *Climate Research*, 50: 83–102.

Li, Y., Lin E., Han X., Peng, Z., Wang, W., Hao, X. and Ju, H. (2013) Effects of elevated carbon dioxide concentration on nitrous oxide emissions and nitrogen dynamics in a winter-wheat cropping system in northern China, *Mitigation and Adaptation Strategies for Global Change*. doi 10.1007/s11027-013-9513-8.

Li, Y., Xu, G.Q., Huang, G.H. and Shi, Y. (2004) Effects of free-air CO_2 enrichment (FACE) on soil microbial biomass under rice–wheat rotation, *Chinese Journal of Applied Ecology*, 15 (10): 1847–1850.

Liang, X., Li, H., Wang, S., Ye, Y., Ji, Y., Tian, G., Van Kessel, C. and Linquist, B. (2013) Nitrogen management to reduce yield-scaled global warming potential in rice, *Field Crops Research*, 146, 66–74.

Lin, E., Xiong, W., Ju, H., Xu, Y., Li, Y., Bai, L. and Xie, L. (2005) Climate change impacts on crop yield and quality with CO_2 fertilization in China, *Philosophical Transactions of the Royal Society B: Biological Sciences*, 360: 2149–2154.

Liu, Y., Wang, E.L., Yang, X.G. and Wang, J. (2010a) Contributions of climatic and crop varietal changes to crop production in the North China Plain since the 1980s, *Global Change Biology*, 16: 2287–2299.

Liu, Y., Liu, Y. and Guo, L. (2010b) Impact of climatic change on agricultural production and response strategies in China, *Chinese Journal of Eco-Agriculture*, 18(4): 905–910 (*in Chinese*).

Liu, B., Li, M., Guo, Y. and Shan, K. (2010c) Analysis of the demand for weather index agricultural insurance on household level in Anhui, China, *Agriculture and Agricultural Science Procedia*, 1: 179–186.

Liu, Y.N., He, W.L. and Li, Y.L. (2010d) A study on the risk index design of agricultural insurance on apple florescence freezing injury in Shaanxi fruit zone, *Chinese Journal of Agrometeorology*, 31 (1):125–129.

Liu, L., Xu, X., Zhuang, D., Chen, X. and Li, S. (2013) Changes in the potential multiple cropping system in response to climate change in China from 1960–2010, *PLoS ONE*, 8 (12): e80990. doi:10.1371/journal.pone.0080990.

Liu, Y. Li, Y.C., Peng, Z.P., Wang, Y.Q., Ma, S.Y., Guo, L.P., Lin, E.D. and Han, X. (2014) Effects of different nitrogen fertilizer management practices on wheat yields and N_2O emissions from wheat fields in north China, *Journal of Integrative Agriculture*. doi: 10.1016/S2095-3119(14)60867-4

Lou, W.P., Wu, L.H. and Yao, Y.P. (2010) Design of weather-based indemnity indices for paddy rice heavy rain damage insurance, *Scientia Agricultura Sinica*, 43 (3): 632–639.

Mao, Y.D., Wu, L.H., Miao, C.H., Yao, Y.P. and Su, G.L. (2007) A reference design for citrus freeze damage insurance by using meteorological index in Zhejiang Province, *Chinese Journal of Agrometeorology*, 28 (2): 226–230.

Ministry of Agriculture, PRC (1980–2013) *China Agriculture Yearbook*, China Agricultural Press.

Monje, O. and Bugbee, B. (1998) Adaptation to high CO_2 concentration in an optimal environment: radiation capture, canopy quantum yield and carbon use efficiency, *Plant, Cell & Environment*, 21 (3): 315–324.

Mosier, A., Kroeze, C., Nevison, C., Oenema, O., Seitzinger, S. and van Kleemput, O. (1998) Closing the global N_2O budget: Nitrous oxide emissions through the agricultural nitrogen cycle, *Nutrient Cycling in Agroecosystems*, 52: 225–248.

McSherry, M.E. and Ritchie, M.E. (2013) Effects of grazing on grassland soil carbon: a global review, *Global Change Biology*, 19: 1347–1357.

Na, R.H. (2010) *Effects of Diet Composition on Methane and Nitrogen Emissions from Lactating Cattle*, Beijing: Chinese Academy of Agricultural Sciences.

Nardone, A., Ronchi, B., Lacetera, N., Ranieri, M.S. and Bernabucci, U. (2010) Effects of climate changes on animal production and sustainability of livestock systems, *Livestock Science*, 130: 57–69.

National Development and Reform Commission (2013) *China's National Strategy for Climate Change Adaptation*. Government of China, Beijing.

National Development and Reform Commission (2015) Enhanced actions on climate change: China's intended nationally determined contributions, online, http://www4. unfccc.int/submissions/INDC/Published%20Documents/China/1/China's%20 INDC%20-%20on%2030%20June%202015.pdf (Accessed 05/10/15, *in Chinese*).

Parmesan, C. (2007) Ecological and evolutionary responses to recent climate change, *Annual Review of Ecology, Evolution, and Systematics*, 37: 637–669.

Peng, S., Huang, J., Sheehy, J.E., Laza, R.C., Visperas, R.M., Zhong, X., Centeno, G.S., Khush, G.S. and Cassman, K.G. (2004) Rice yields decline with higher night temperature from global warming, *Proceedings of the National Academy of Sciences of the United States of America*, 101: 9971–9975.

Randall, P. and Moss, H. (1990) Some effects of temperature regime during grain filling on wheat quality, *Australian Journal of Agricultural Research*, 41 (4): 603–617.

Rogers, G., Milham, P., Gillings, M. and Conroy, J. (1996) Sink strength may be the key to growth and nitrogen responses in N-deficient wheat at elevated CO_2, *Functional Plant Biology*, 23 (3): 253–264.

Rong, Y., Han, J., Wang, P. and Mao, P. (2001) The effects of grazing intensity on soil physics and chemical properties, *Grassland of China*, 23 (4): 41–47.

Shang, X., Yang, Q. and Liu, G. (2007) The theory and technical system of arid agriculture in Gansu, *Agricultural Research in the Arid Areas*, 25 (Suppl.): 194–196.

Silanikove, N. and Koluman, N. (2015) Impact of climate change on the dairy industry in temperate zones: predications on the overall negative impact and on the positive role of dairy goats in adaptation to earth warming, *Small Ruminant Research*, 123: 27–34.

Tao, F., Yokozawa, M., Hayashi, Y. and Lin, E. (2003a) Future climate change, the agricultural water cycle, and agricultural production in China, *Agriculture, Ecosystems and Environment*, 95 (1): 203–215.

Tao, F., Yokozawa, M., Hayashi, Y. and Lin, E. (2003b) Changes in agricultural water demands and soil moisture in China over the last half-century and their effects on agricultural production, *Agricultural and Forest Meteorology*, 118 (3&4): 251–261.

Tang, Y. and Ren, F. (2005) Reanalysis of surface air temperature over the last 100 years, *Climatic and Environmental Research*, 10 (4): 791–798 (*in Chinese*).

Tester, R., Morrison, W., Ellis, R., Piggo, J., Batts, G., Wheeler, T., Morisson, J. and Hadley, P. (1995) Effects of elevated growth temperature and carbon dioxide levels on some physicochemical properties of wheat starch, *Journal of Cereal Science*, 22 (1): 63–71.

The Group of Chinese Academy of Agricultural Sciences (1984) *Cropping Zonation in China,* Beijing: China Agricultural Press.

Thornton, P., Herrero, M. and Ericksen, P. (2011). Livestock and climate change. ILRI (International Livestock Research Institute), *Livestock Exchange Issue Brief*, 3.

Thu C.T.T., Cuong P.H., Hang L.T., Chao, N.V., Anh, L.X., Trach, N.X. and Sommer, S.G. (2012) Manure management practices on biogas and non-biogas pig farms in developing countries using livestock farms in Vietnam as an example, *Journal of Cleaner Production*, 27: 64–71.

Van Asch, M. and Visser, M.E. (2007) Phenology of forest caterpillars and their host trees: the importance of synchrony, *Annual Review of Entomology*, 52: 37–55.

Wang, F. (2003) *Effects of Global Climate Change on Agriculture Ecosystems*, Beijing: China Meteorological Press (*in Chinese*).

Wang, H. (2012) *Effects of Slow Release Nitrogen Fertilizer on CH_4 and N_2O Emissions in Coastal Fields*, Nanjing: Nanjing Agricultural University.

Wang, S., Lin, E. and She, Z. (2002) *The Impacts of Environmental Evolution on Development in Western China and its Countermeasures, The Assessment of Environmental Evolution in Western China, Vol. 3*, Beijing: Science Press,(*in Chinese*).

Wang, X.L. (1996) CO_2, *Climate Change and Agriculture*, Beijing: China Meteorological Press (*in Chinese*).

Wassmann, R. and Aulakh, M.S. (2000) The role of rice plants in regulating mechanisms of methane missions, *Biology and Fertility of Soils*, 31: 20–29.

Williams, M., Shewry, P., Lawlor, D. and Harwood, J. (1995) The effects of elevated temperature and atmospheric carbon dioxide concentration on the quality of grain lipids in wheat (*Triticum aestivum* L.) grown at two levels of nitrogen application, *Plant Cell and Environment*, 18 (9): 999–1009.

Wrigley, C., Blumenthal, C. and Barlow, P. (1994) Temperature variation during grain filling and changes in wheat-grain quality, *Functional Plant Biology*, 21 (6): 875–885.

Xiong, W., Velde, M., Holman, I., Balkovic, J., Lin, E., Skalsky, R., Porter, C., Jones, J., Khabarov, N. and Obersteiner, M. (2014) Can climate-smart agriculture reverse the recent slowing of rice yield growth in China?, *Agriculture, Ecosystems & Environment*, 196 (14): 125–136.

Yang, J., Xiong, W., Yang, X., Cao, Y. and Feng, L. (2014) Geographic variation of rice yield response to past climate change in China, *Journal of Integrative Agriculture*, 13 (7): 1586–1598.

Yang, T.M., Liu, B.C., Sun, X.B., Li, D. and Xun, S.P. (2013) Design and application of the weather indices of winter wheat planting insurance in Anhui province, *Chinese Journal of Agrometeorology*, 34 (2): 229–235.

Yang, X., Liu, Z. and Chen, F. (2010) The possible effects of global warming on cropping systems in China I. The possible effects of climate warming on northern limits of cropping systems and crop yields in China, *Scientia Agricultura Sinica*, 43 (2): 329–336 (*in Chinese*).

Yang, X., Liu, Z. and Chen, F. (2011) The possible effects of global warming on cropping systems in China VI. Possible effects of future climate change on northern limits of cropping system in China, *Scientia Agricultura Sinica*, 44 (8): 1562–1570 (*in Chinese*).

Yuan, B., Guo, J., Zhao, J. and Ye, M. (2011) Possible impacts of climate change on agricultural production in China and its adaptation countermeasures, *Agricultural Science & Technology*, 12 (3): 420–425.

Zhai, H. (ed.) (2011) *Study on National Strategy of Food Security in China*, Beijing: Chinese Agricultural Science and Technology Press.

Zhang, L., Huo, Z., Wang, L. and Jiang, Y. (2012) Effects of climate change on the occurrence of crop insect pests in China, *Chinese Journal of Ecology*, 31 (6): 1499–1507 (*in Chinese*).

Zhang X. and Cong, Z. (2014) Trends of precipitation intensity and frequency in hydrological regions of China from 1956 to 2005, *Global and Planetary Change*, 117 (6): 40–51.

Zhang, Y., Han, J. and Li, Z. (2002). A study of the effects of different grazing intensities on soil physical properties, *Acta Agrestia Sinica*, 10 (1): 74–78.

Zhang, Y., Shao, X.M., Yin, Z.-Y. and Wang, Y. (2014). Millennial minimum temperature variations in the Qilian Mountains, China: evidence from tree rings, *Climate of the Past*, 10, 1763-1778, doi:10.5194/cp-10-1763-2014, 2014.

Zhao, F. (2004) Technique of whole plastic-film mulching on double ridges and planting in furrow for maize, *Gansu Agricultural Science and Technology*, 11: 22–234.

Zhou, X., Chen, C., Wang, Y., Xu, Z., Han, H., Li, L. and Wan, S.(2013) Warming and increased precipitation have differential effects on soil extracellular enzyme activities in a temperate grassland, *Science of the Total Environment*, 444: 552–558.

Zhou, Y., Qin, Z.H. and Bao, G. (2011) Impacts of climate change on agriculture and its responses, *Chinese Agricultural Science Bulletin*, 27 (32): 299–303.

Zou, J., Huang, Y., Jiang, J., Zheng, X. and Sass, R.L. (2005) A 3-year field measurement of methane and nitrous oxide emissions from rice paddies in China: Effects of water regime, crop residue, and fertilizer application, *Global Biogeochemical Cycles*, 19: GB2021.

5 Climate variability and extremes

Relevance of agroecological-based climate smart farming systems in the sub-Saharan Africa

Sita Ghimire, Monday Ahonsi and Appolinaire Djikeng

Introduction

Climate change refers to any change in climate over time, whether due to natural variability or as a result of human activity. Climate change is a global phenomenon that affects every living being on earth including humans, with resource disadvantaged people being the most affected. Among these are the smallholder farmers of sub-Saharan Africa (SSA) who are dependent on rain-fed agriculture for their livelihoods and have limited access to agricultural inputs, credit and markets. The adverse effects of climate change on African agriculture characterized by frequent crop failures and decreased productivity of crops and livestock, are already impacting on the livelihoods of millions of smallholder farmers on the continent (CDKN, 2014).

SSA is home to about 900 million people, the majority of whom are dependent on agriculture as their main occupation (UN, 2011). Agriculture is thus the backbone of the economy, employing 65 percent of the people and accounting for two thirds of the regional gross domestic product (WB, 2008). Although the majority of farms (80 percent) in SSA are small (less than 2 hectares), these small farms contribute up to 90 percent of total agricultural production in some countries (Wiggins, 2009; Wiggins and Sharada, 2013) and contribute almost 80 percent of food supply on the continent (AGRA, 2014). However, the agricultural production in the region is insufficient to meet the domestic demand due to low yields (Chauvin *et al.,* 2012). Despite high agricultural potential, Africa has remained a net importer of agricultural products over the past four decades (Rakotoarisoa *et al.,* 2012). The reliance on imported foods and agricultural products is of great concern for regional food security, peace, political stability and economic prosperity of the SSA region where population growth rates are high, poverty is rampant and there is increasing dependence on food aid. Recent estimates indicate that 220 million people in SSA are undernourished, accounting for 23.2 percent of the total population (FAO, 2015). In addition, the SSA population is projected to reach 1.5 billion by 2050, thus creating an additional demand for food, and the need to increase food production by almost 360 percent compared to 2006 levels (Sherry, 2013).

The impact of climate change is expected to worsen in SSA given its geographical location, limited adaptive capacity, widespread poverty, political instability and existing low levels of infrastructure development and policy support. This will further aggravate the problem of food and nutritional insecurity to the rapidly growing populations in SSA. One argument is that there is a need to produce more food per unit area using appropriate technologies or farming approaches, including climate-smart agriculture (CSA) and agroecology that have minimum impact on the environment and natural resources, and are beneficial to smallholders. This chapter briefly presents possible climate change scenarios and the associated impacts on agriculture and food and nutritional security in SSA, discusses approaches for adaptation to climate change impacts, and highlights ongoing CSA initiatives and challenges for CSA implementation in SSA. Some of the initiatives may have the potential to contribute to the reduction of greenhouse gases (GHGs). Towards, the end it provides policy recommendations for improving the resilience of African agriculture to the impacts of climate change.

Climate variability and extremes in SSA and impacts

The historical climate data shows that the surface temperature of Africa is warmer today than it was 100 years ago. Africa experienced warming of approximately 0.7°C in most parts of the continent during the twentieth century. The IPCC has reported that land surface temperatures across most of Africa have increased by 0.5°C or more during the last 50–100 years (IPCC 2013, 2014). There is strong evidence that observed temperature increases exceed natural climate variability and have been influenced by greenhouse gas emissions due to human influence. Where rainfall data are available, it indicates a very likely decrease in annual rainfall over the past century in parts of the western and eastern Sahel region in northern Africa, and very likely increases over parts of eastern and southern Africa.

The years 2010 and 2013 were two particularly hot years in Africa (AGRA, 2014). Vioolsdrif in South Africa experienced a temperature of 47.3°C on the 4 March, 2013, which was not only the hottest temperature measured in the world that month but also the hottest March temperature ever measured anywhere on the continent. Similarly, record high temperatures were observed at Navrongo, Ghana on the 6 March (43°C) and at Matam, Senegal on the 21 March (46.2°C). This was the highest temperature ever measured reliably in Ghana and the warmest March temperature on record for Senegal (AGRA, 2014; Burt, 2013). The future climate change scenario for Africa indicates further warming across the continent ranging from 0.2°C per decade to greater than 0.5°C per decade (Desanker and Magadza, 2001; Hulme *et al.,* 2001).

Extreme events

Fluctuations in rainfall in terms of amount, length and intensity are common across the African continent. Following the drought of the 1910s that affected both East and West Africa, all regions received increasing rainfall amounts, but negative trends were observed again from 1950 onwards resulting in widespread famine in West Africa in 1984 (Gommes and Petrassi, 1994). The highest amount of rainfall occurs in central African countries and the highest spatial and temporal variability in rainfall is observed across the countries with a semi-arid climate within western, eastern and southern Africa. Within-country variations in rainfall also occur, for example, the eastern part of Kenya (mostly semi-arid) receives less rainfall than the highlands of Western Kenya (sub-humid). Intra-annual variability in rainfall is observed across regions; Southern Africa receives most of its rainfall during October–March whereas in the Sahel rainfall predominantly falls during July–August, and most countries in the horn of Africa and equatorial eastern Africa have bimodal rainfall patterns with short rains during October-December and long rains during March-May (Masih *et al.,* 2014).

Long-term rainfall data from ten selected locations of nine SSA countries (Botswana, Ghana, Kenya, Malawi, Nigeria, Rwanda, Sudan, Tanzania and Zimbabwe) demonstrated that the inherent variability in seasonal rainfall totals increases disproportionately (as expressed by the coefficient of variation) from humid to sub-humid to the drier semi-arid agroecological zones (AGRA, 2014). Historical rainfall data has shown a decline in precipitation in West Africa ranging from 20 to 40 percent between 1931–1960 and the period 1968–1990 (IPCC, 2007; Sissoko *et al.,* 2011). Rainfall data covering the period 1982–2010 from the semi-humid region of Nigeria has shown an increase in mean monthly rainfall by 65mm per month per decade (AGRA, 2014; Fabusoro *et al.,* 2014). The amount and the patterns of rainfall are associated with drought and flood events. Analysis of drought data covering the period 1900–2013 revealed an intensification of drought events in terms of their frequency, severity and geospatial coverage over the last few decades. These include the intense and widespread droughts during 1972–73, 1983–84 and 1991–92, and severe drought in all regions in the last few decades. The most notable are: the persistent droughts in the Sahel during the 1970s and 1980s; drought in Northern Africa (1999–2002), drought in southern Africa (2001–2003); and drought in the horn of Africa (2010–2011) (Masih *et al.,* 2014). At the same time, some of these regions also in SSA also experienced various flood events. The most serious were the flood of 2007 that affected 1.5 million people in 18 countries in SSA including Ghana, Niger, Mali and Togo in the West to Ethiopia, Uganda and Sudan in the East (WFP, 2007). Most recent is the flood of 2015 in Mozambique, Malawi and Madagascar that affected about 360,000 people (Anonymous, 2015). Although changes in current rainfall patterns in SSA are less clear, it is projected that all the regions except southern Africa are becoming wetter where drying is anticipated, and an increased frequency and severity of extreme weather events such as droughts and floods is most likely across the regions (AGRA, 2014). The anticipated rise in sea levels poses a severe threat of flooding and sea water inundation to low lying coastal regions of SSA that host 320 densely populated cities (Brown *et al.*, 2011). These recurring droughts and floods are devastating to property and livelihoods, and result in loss of lives. Smallholders' resilience in SSA is poor, owing to poor access to information, technology, inputs and their vulnerable locations.

Impacts on food crops

Africa is the most food-insecure continent. The current agricultural productivity in SSA countries is lower than that in other regions of the world (Chauvin *et al.*, 2012). The average production of cereals in SSA is around one ton per hectare, which is one-half of the production in India, less than a quarter of the production in China and less than a fifth of the production in the United States (WB, 2007). Since the majority of agriculture is rain-fed, inadequate and erratic rainfall causes low crop productivity and makes the region vulnerable to climate change (FAO, 2006; IFAD, 2011; Rockström, 2003). Various climate extremes have already been experienced across the region as a major threat to agriculture, making smallholder farmers among the most food-insecure, undernourished and poorest people (Livingston *et al.*, 2011).

The warming in SSA is expected to be greater than the global average, and in parts of the region rainfall will decline (IPCC, 2007). Climate change may further reduce the yield of major cereals in SSA with strong regional variability (Liu *et al.*, 2008; Lobell *et al.*, 2008; Roudier *et al.*, 2011; Walker and Schulze, 2008). Climate change is predicted to reduce crop yields by 8 percent by 2050 (Porter *et al.*, 2014) whereas yield reduction on rain-fed cropland could be as high as 50 percent by 2020 (IPCC, 2007). Maize, the staple food of millions of people in SSA will experience severe yield losses of 18 percent in Southern Africa (Zinyengere *et al.*, 2013) and 22 percent in aggregate across SSA, with yield losses for South Africa and Zimbabwe exceeding 30 percent (Schlenker and Lobell, 2010). The same study reported a possible decrease in sorghum, millet, groundnut and cassava yields by 17, 17, 18 and 8 percent, respectively. Another study showed a decline in cereal production by 3.2 percent in SSA by 2050. The largest decline in yield is projected for wheat (>20 percent) followed by sweet potatoes and yams, cassava, maize and rice (Ringler *et al.*, 2011). In contrast, yields for millet and sorghum are projected to increase slightly, possibly due to their higher tolerance to high temperature and drought.

Expansion of arid land is expected to increase due to changes in temperature and precipitation, most notably in southern Africa. In a 4°C warming scenario, total hyper-arid areas are projected to expand by 10 percent compared to the period 1986–2005 (WB, 2013). The projected changes in temperature and rainfall in a given area will make the place less suitable to cultivate certain crops that are successfully grown at present. A suitability assessment of different geographical regions of SSA for cassava and beans production under projected climate change situations showed no change in the overall suitability of cassava in Sahel, humid West Africa and Central Africa, and significant gains in the suitability in East Africa and Southern Africa. However, there was a marked decline in the suitability of beans in Central, Southern and Eastern Africa with a smaller decline in Sahelian and West Africa. An increased suitability for beans was observed in small areas of the cooler highlands in Eastern and Southern Africa (AGRA, 2014). The study revealed substantial losses of suitable areas for bean production in all regions, in particular East Africa and Central Africa (loss of 28 to 80 percent) where 7 million hectares are currently under beans. Other staples (banana, finger millet, ground nut, pearl millet, sorghum and yam) showed less noticeable but still significant losses of suitable cultivation areas. Banana shows median losses of 8 percent for West Africa and 25 percent for Sahelian Africa, yam with 4 to 6 percent losses in Southern, Eastern and Sahelian Africa and finger millet with 14 percent loss in Sahelian Africa (AGRA, 2014). Pests and diseases outbreaks, emergence of new agricultural pests and diseases (e.g. coffee berry bores in Ethiopia, maize lethal necrosis disease in Kenya), loss of plant biodiversity and shift of plant species to suitable climate zones are other consequences of climate change, some of which are already noticeable. Besides decreasing acreage for suitable cropland, climate change will also affect the suitability of currently available crop varieties.

Impacts on livestock production

Livestock are key assets of poor people in pastoral and agropastoral systems, fulfilling multiple economic, social and risk management functions. Livestock are an important component of SSA agriculture, contributing to the livelihoods of 70 percent of smallholder farmers and providing food, income and major crop production inputs (manure and draft power). However, livestock are vulnerable to climate change, and those in the drylands of Africa are likely to be most affected. Climate change is anticipated to enhance the vulnerability of livestock production systems and increase existing factors that are affecting these systems, such as water and feed availability, biodiversity, changes in ecosystem function and resilience, animal health and overall productivity (Calvosa *et al.*, 2009). As global warming increases from 2 to 4°C, unprecedented heat extremes are projected over an increasing percentage of land areas, resulting in changes in vegetative cover and risk of species extinction. The heat and drought environments will affect water availability for livestock use and at the same time negatively affect livestock feed production systems and pasture yields, causing severe losses of livestock (WB, 2013).

Even without climate change, SSA remains the most food-deprived region worldwide. The impact of climate change on the food security of SSA is expected to be severe because of its geographical location, low agricultural productivity, wide spread poverty, limited adaptive capacity and low levels of infrastructure development and policy support. Even a low warming scenario (2°C) in SSA is expected to have strong repercussions on food security due to significant losses in crop and livestock yields and loss of savannah grasslands, posing a threat to pastoral livelihoods (WB, 2013) and increasing the vulnerability of food systems (food availability, food accessibility, food utilization and food system stability) in the region. It will therefore be difficult with 'business as usual' to produce enough food for rapidly growing populations in SSA countries, and increasing agricultural productivity using technologies/approaches that utilize natural resources better with minimum impacts on the environment will be necessary. The subsequent sections will discuss agroecology-based farming systems and CSA. The discussion will outline their relevance to the current challenges faced by the smallholders in SSA, and their contribution to reduce the risks from climate change.

Sustainable farming systems to address climate change

Agroecology

The Green Revolution of the mid-twentieth century, renowned for increased agricultural production over the past five decades, is often criticized for its negative impacts on the environment and society, including the loss of soil quality and exacerbation of social and income inequalities (Rosset, 2006). The pioneering mission, aimed at combating hunger, was a great success in some parts of the world. It resulted in some unintended consequences leading to biodiversity depletion, high incidence of pests, diseases and weeds, increased erosion and lower soil fertility. In addition, the heavy dependence on external inputs such as fertilizers and other agrochemicals, overuse of natural resources including water for irrigation, pollution of ground water and eutrophication of rivers and lakes resulted in serious impacts on the environment in many regions across the world. (Matson *et al.,* 1997). The negative effects on human and environmental health notwithstanding, meeting the global food demand has remained the priority since the Green Revolution. Most efforts to date focus on improved seeds and inputs that enable a crop to produce maximum yield with little or no consideration given to the sustainability of agroecosystems. Although the Green Revolution largely bypassed Africa, farmers in certain parts of SSA use high yielding crop varieties, keep exotic breeds of animals and use external inputs. Despite yield gains, adequate food is not reaching everyone, and farmers and agricultural workers themselves are the most food insecure people across the globe (Wach, 2015). This could be attributed largely to the inconsistent performance of the Green Revolution technologies in the context of changes in climate variability and extremes. To meeting the global demand for food, fuel, fibre and range of agroecosystem services with minimal pressure on the natural resource base, it has become increasingly important to increase agricultural productivity, ensure food and nutritional security and environmental sustainability. Within this context, an agroecological approach is one potential way forward to meeting future food needs and sustainable agoecosystems while mitigating the effects of climate change.

Agroecology can be defined as the application of ecological concepts and principles to the design and management of sustainable agroecosystems (Altieri,

1995). As a set of agricultural practices, agroecology explores ways to enhance agricultural systems by mimicking natural processes, thus creating beneficial biological interactions and synergies among the components of the agroecosystem (De Schutter, 2010). Agroecology includes integration of crop and livestock production, diversification of species and crops in the agroecosystem over time and space, nutrient and energy recycling on-farm, low external inputs, and a focus on interaction and productivity across agricultural systems. It is based on the idea that farmers should be directly involved in developing new practices by engaging and co-creating with one another, with support from cooperatives, non-governmental organizations, research bodies and government institutions. Studies so far have demonstrated that agroecological production systems can have substantial benefits in conserving biodiversity, increasing resilience, and improving nutrition for farming households (Wach, 2015). It is highly knowledge intensive, founded on the techniques that are developed from the farmers' knowledge and experimentations.

Agroecology is now well supported by a wide range of scientific experts and organizations e.g. the Food and Agriculture Organization of the United Nations (FAO), the United Nations Environment Programme (UNEP) and Biodiversity International as the way to improve the resilience and sustainability of food systems (De Schutter, 2010). Several technologies based on an agroecological perspective have been developed and successfully evaluated in various parts of the world. Some of the practices promote agricultural biodiversity, help to improve soil fertility through fixation of nitrogen, encourage the planting of multifunction trees and advocate the use of intercropping to minimize incidence of pests and diseases. A systemic study of 286 sustainable agricultural projects from 57 resource poor countries covering 37 million hectares of land area revealed that agroecological interventions increased crop productivity on 12.6 million farms by 79 percent, while improving the supply of critical environmental services (Pretty *et al.*, 2006). The average increase in crop yield for projects in African countries in general was 116 percent, and 128 percent for the East African region. It is worth noting that this regional and continental increase in average crop yields in Africa was much above the global average of 79 percent. Another study of 40 projects, based on sustainable agricultural intensification techniques, using participatory plant breeding of neglected orphan crops, integrated pest management and soil conservation and agroforestry, was implemented in 20 African countries. It showed a doubling of the crop yields over a period of 2 to 10 years, with 5.79 million tons per annum increase in food production (Pretty *et al.*, 2011).

There are several promising technologies, which have shown good results in SSA. The push-pull technology developed together by researchers and farmers in Kenya that integrates Desmodium and Napier forages in maize crops helps in doubling the maize yield as well as generating other benefits (Khan *et al.*, 2014; Khan *et al.*, 2011). The legume forage Desmodium pushes stem borers away from maize plants. It reduces the parasitic weed striga, fixes atmospheric nitrogen (N) in soils, and is fed to livestock as nutritious fodder. Napier, the most commonly cultivated fodder grass in the region, attracts stem borer females from maize fields for oviposition but prevents the stem borer larvae to develop to adulthood. About 69,000 farmers from Ethiopia, Kenya, Tanzania, Uganda, Somaliland and Nigeria are currently benefitting from the push-pull technology (Khan *et al.*, 2014). There are other examples such as the Brachiaria Grasses Program that have shown good results and ready for scaling out (Box 5.1).

The Soils, Food and Healthy Communities (SFHC), a farmer organization in Malawi, has successfully implemented agroecological methods among thousands of smallholder farmers in which legume diversification is bringing significant improvements in child nutrition, household food security, improved gender relations and ecosystem services (Bezner Kerr, 2008; Snapp *et al.*, 2013). While some SFHC farmers have tested sorghum as a drought-tolerant climate change adaptation strategy, there are few well-adapted varieties in the north, since this crop declined dramatically in northern Malawi in the past century (Bezner Kerr, 2014). Similarly, the French Agricultural Research Centre for International Development (CIRAD) has worked closely with the farmer organization Association Minim Song Panga (AMSP) in Burkina Faso, to

---

Box 5.1 Brachiaria Grasses Program

The Brachiaria Grasses Program led by the Biosciences eastern and central Africa – International Livestock Research Institute (BecA-ILRI) Hub in partnership with the Kenya Agricultural and Livestock Research Organization (KALRO), Rwanda Agricultural Board (RAB), International Centre for Tropical Agriculture (CIAT) and Grasslanz Technology Limited is an agroecology-based farmer participatory research for development initiative (Djikeng *et al.*, 2014). The program aims to increase livestock production in East Africa by increasing the availability of quality forage through the use of the climate resilient native grass Brachiaria. The program explores suitability of local ecotypes and improved varieties of Brachiaria grass for the semi-arid and sub-humid environments of East Africa. The best bet varieties identified by this program are currently grown by over 1,500 progressive farmers in Kenya and about 525 farmers in Rwanda. These farmers have reported outstanding performance of Brachiaria grass for forage biomass production, livestock productivity and on-farm forage availability especially during the dry seasons (BD, 2015). These grasses not only produce a high amount of nutritious biomass, but also deliver several ecosystem services including carbon sequestration, nitrogen use efficiency enhancement, erosion control, and reduction in greenhouse gases emissions (GHGs) and ground water pollution (Miles *et al.*, 1996). Importantly, they increase livestock productivity and income to smallholders through increased milk and meat production.

Source: BD (2015)

implement participatory plant breeding programs with sorghum. They have developed a range of suitable varieties and training, supporting the farmers groups for seed production, storage/packaging and efficient seed dissemination (vom Brocke *et al.,* 2010; vom Brocke *et al.,* 2014).

The harnessing of indigenous crop and animal genetic resources, farmers' local knowledge and practices, and local inputs and natural resources through the involvement of farmers and relevant stakeholders is a key step for the development of agroecology-based farming systems that can promote sustainable agricultural productivity. Efforts should be placed to build the resilience of the agriculture production systems through enhanced biodiversity and effective ecosystem management (FAO, 2010). Mixed crop-livestock farming with legumes and millets, drought tolerant crop varieties, disease resistant livestock breeds, crops that improve soil fertility, marginal soil adapted crop genotypes, crops with high nutrient and water use efficiency, and flood tolerant crops are all components that can increase the resilience of the system to address climate extremes and variability issues.

Smallholder farmers and climate smart agriculture

The frequency and intensity of extreme weather events such as drought, heavy rains, flooding and high temperatures is likely to increase in SSA (IPCC, 2007; CDKN, 2014). The average and seasonal maximum temperatures are projected to continue rising, with higher average rainfall overall. However, the shorter rainy seasons and inadequate and erratic distribution of rainfall will have severe impacts on rain-fed agriculture, which forms the basis of the livelihoods of 90 percent of people in SSA (Rockström, 2003). The climate change associated climatic extremes in SSA demands an urgent and rapid transformation and reorientation of agricultural systems to ensure food security. Scientists have started to classify these systems under the term Climate Smart Agriculture (CSA) as defined in Chapter 1 of this book.

CSA necessitates building the resilience of smallholder farmers to cope with risk and respond to climate change. It may require incremental adaptation such as better information provision, timely supply of agricultural inputs, changes in agronomic practices, improved market governance, and implementation of insurance and safety net programs. It may also involve system adaptation including major shifts in agricultural production systems, for example, growing new crops or moving away from agricultural to non-agricultural activities. CSA is viewed as one of the best approaches to adapt to and mitigate the negative effects of climate change on agriculture (Lipper *et al.,* 2014). The formulation and effective implementation of CSA programs requires coordination and support from public, private and civil society stakeholders from local to international levels in building evidence and assessment tools, strengthening local and national institutions, developing coordinated and evidence-based policies and increasing financing and its effectiveness (Lipper *et al.,* 2014). One such program is the CGIAR-led CCAFS program operating simultaneously in many countries (Box 5.2).

Box 5.2 CGIAR-led Climate Change, Agriculture and Food Security
The CGIAR-led Climate Change, Agriculture and Food Security
(CCAFS) research program is currently helping farmers in eleven
villages in East and West Africa to adopt practices that will help them
to cope with climate change and enhance food security (Nakweya,
2013). The major interventions in these climate smart villages include
providing timely weather forecasts, resilient water management practices,
initiatives to increase the soil carbon content, rationalised use of nitrogen
fertilizers, use of fuel efficient agromachinery, exchange of knowledge,
and crowdsourcing of seeds of adapted crop varieties (Aggarwal *et al.*,
2013). CCAFS collaborates with national programs while partnering
with rural communities to develop climate smart villages as a model
based on local action that ensures food security, promotes adaptation and
builds resilience to climatic stresses. Researchers, farmers, local agencies
and policy makers need to collaborate to select the most appropriate
technological and institutional interventions based on global knowledge
and local conditions. Scaling up of such CSA programs in many sites
across SSA countries would enable smallholder farmers to better adapt
to and mitigate the effects of climate change to ensure food security in
the region.

Source: Nakweya (2013) and Aggarwal *et al.*, (2013)

Climate-smart agriculture alliance in Africa

Several CSA initiatives funded by government and non-governmental
organizations and international agencies are at the piloting stages in the region.
However, these efforts seem quite fragmented, or at least at a lower scale
and not adequate enough to achieve transformational impact on smallholder
farming communities. A recent initiative, the Africa Climate Smart Agriculture
Alliance (ACSAA) launched on June 2014 at the African Heads of State Summit
in Malabo, Equatorial Guinea, aims to leverage policy, technical and financial
support for local, national and regional level programs and initiatives that
can lead widespread adoption of CSA practices across SSA[1] (NEPAD, 2014).
The ACSAA was convened by the African Union (AU) through NEPAD's
Comprehensive Africa Agriculture Development Program (CAADP). This
AU-NEPAD-INGO research alliance aims to empower six million farming
households in Africa by 2021 and is expected to contribute to the AU's broader
goal of supporting 25 million farming households by 2025. The ACSAA is
developing a platform to promote and support the uptake and use of CSA
practices among smallholder farmers.

ACSAA is positioned to serve as a platform for the engagement of bilateral
and multilateral agencies and private sector actors to support up-scaling of

Box 5.3 Challenges for CSA implementation
(i) Inadequate technological evidence to support effective decision making at local and national levels; (ii) limited number of proven CSA technologies for SSA; (iii) weak national and local institutional support and less organized public, private and civil society stakeholders to support adaptive capacity through enhancing people's access to information, agricultural inputs, markets and other resources; (iv) lack of an enabling policy and regulatory framework to support CSA adoption; (v) lack of dedicated funding to support CSA programs e.g. capacity strengthening, technology development and transfer; and (vi) lack of financing to support the transition to CSA.

Source: Authors' own analysis

CSA. It aims to generate sustainable benefits through farmer-led and inclusive approaches to strengthen the capacity of local systems and resources in agriculture, and to support the efforts of national governments. ACSAA intends to support the scaling-up and practice of CSA across the African continent, beginning with Ethiopia, Niger and Zambia, the countries of greatest need and with the greatest opportunity for impact. According to some, it is a very timely initiative to combat the deteriorating food security situation in SSA, however, successful implementation and out scaling of CSA technologies will have great challenges (Box 5.3).

Policy Recommendations

Two important reports – the IPCC's Fourth Assessment (AR4) and the World Bank's *Turn down the Heat: Climate Extremes, Regional Impacts, and the Case for Resilience* confirm that Africa is a vulnerability hotspot for climate change (Boko et al., 2007; WB, 2013). The effects of climate change on the people living in the hot spots are expected to be severe, and observations to date have shown that most parts of Africa, especially arid and semi-arid lands, are experiencing frequent crop failures and animal deaths due to extreme droughts and rainfall irregularities. Other climate extremes such as floods, wet spells and dry spells are occurring more frequently than in the past. These adverse abiotic stresses coupled with severe biotic stresses such as insect pests and diseases are affecting agricultural productivity across SSA. Moreover, the current methods of using natural resources and agricultural inputs are not appropriate and sustainable to cope with anticipated global warming in Africa of up to a 4°C increase in temperature by the year 2100 (WB, 2013).

Fostering CSA requires coordinated efforts of various stakeholders: farmers, researchers, private sector, civil society and policy makers towards building evidence, increasing local institutional effectiveness, fostering coherence between climate and agricultural policies, and connecting climate and agricultural financing. Integration of the agroecology concept in CSA is crucial to enhance the sustainability of the agricultural production system by reducing overreliance on external production inputs and input-intensive crop varieties and animal breeds. The Global Alliance for Climate Smart Agriculture (GACSA) and ACSAA, at global and continental levels respectively, could serve as good custodians to foster CSA from international to national and local levels.

Africa is the second largest continent with diverse agroecology, necessitating location specific custom-made agricultural technologies for the optimal production and sustainable use of natural resources. Many agricultural technologies in African nations have been introduced from elsewhere and applied on a large scale without thorough evaluation. It is very important to acknowledge the diversity and exclusivity of African agroecology that calls for the development of appropriate agricultural technologies in Africa for Africa. Agriculture in SSA countries needs a significant transformation to address the challenges of food security and climate change. Adoption of an ecosystem approach at landscape scale and ensuring inter-sectorial coordination and cooperation is crucial to effectively respond to the climate change challenges.

Inclusive agricultural and economic growth (i.e. increased agricultural productivity, expansion of markets and trade, and increased economic resilience of vulnerable rural communities) is critical for the food and nutritional security and livelihoods of smallholder farmers worldwide. The efficiency, resilience, adaptive capacity and mitigation potential of the production systems can be enhanced through the improvement of agricultural practices. The AE and CSA approaches have the potential to address aforementioned issues through the combined uses of local knowledge and modern agriculture practices and subsequently increase agricultural productivity and improve food security, livelihoods and environmental sustainability. Therefore, it is important to integrate different approaches while formulating specific technology options for given agroecological conditions in Africa. Various CSA and agroecology technologies have been developed and tested successfully in different regions of the world. Some of these technologies include: maintenance or introduction of agricultural biodiversity; integrated nutrient and soil fertility management; agroforestry; water harvesting; integrated crop-livestock farming; and low use of external inputs (Pretty, 2008). These practices are readily applicable to several rural farming settings to not only increase agricultural production but also improve soil fertility, create more jobs and income, and improve household nutrition and resilience to climate change.

Geographical scaling up of CSA and agroecology and creating an enabling framework for farmers requires the support of national governments in prioritizing spending on public goods (e.g. extension services, storage facilities, roads, electricity, information and communication technologies), investment in knowledge (e.g. agriculture research and extension), strengthening social organizations and gender empowerment and organizing markets (De Shutter, 2010). African governments had committed spending at least 10 percent of their budgets on agriculture by 2015, and 20 nations have pledged to do so under the CAADP rubric (AU, 2003). However, only seven African nations: Ethiopia, Niger, Mali, Malawi, Burkina Faso, Senegal and Guinea have met their commitment as per the Maputo Declaration. Political instability and humanitarian crisis are two main factors responsible for many nations not meeting the commitment. Good governance and effective mobilization of human capital and natural resources are necessary for economic, political and social stability. These efforts should be supplemented with foreign aid to facilitate putting these great ideas into practice. Many SSA countries are not adequately positioned to allocate sufficient national funds to support CSA and agroecology programs. A longer-term engagement between donor and partner countries is necessary for policy and program support, establishment of regional and national knowledge platforms, dissemination and adoption of CSA and agroecology technologies and investment in public goods. Similarly, the CGIAR research community should engage more in conducting CSA and agroecology research and human capacity building.

Many African nations have endorsed the Kyoto Protocol[2] and the United Nations Framework Convention on Climate (UNFCC)[3], both aimed at preventing treacherous human interference with the climate system. African governments coordinate their regional positions and national policies on climate change through the African Ministerial Conference on the Environment (AMCEN). The New Partnership for Africa's Development (NEPAD) is a regional forum in Africa supporting projects and action plans that are relevant to climate change. Since its establishment, the African Climate Policy Centre (ACPC)[4] has been generating knowledge on climate and collecting climate information for Africa to strengthen the use of such information in decision making, by improving analytical capacity, knowledge management and dissemination activities. Despite commitments of African Nations in several ongoing global and regional climate initiatives, only a few countries (e.g. Kenya, Ethiopia and Nigeria) have national climate change policies in place (BNRCC, 2011; César and Ekbom, 2013; Mburia, 2015). The East African community has developed a climate change policy that seeks to mainstream climate change adaptation in national and regional development plans taking a sectoral approach, with emphasis on key socio-economic sectors and sub-sectors that are adversely impacted by climate change and with potential opportunities to contribute to mitigation efforts and sustainable development of the partner states and the region (EAC, 2011). Considering the vulnerability of the region to climate change and its consequences for food security and livelihoods, every SSA nation is urged to have provision for climate change policy, adaptation strategies and appropriate implementation plans.

Conclusions

The impact of climate change on agriculture has already begun to be experienced. Global temperature increases, changes in the amount and irregularity of rainfalls, increases in the atmospheric concentration of GHGs and reductions in crop yields are some major characteristics of climate change. The magnitudes of these changes are anticipated to increase as we continue to produce more food to meet the food demand of the ever-growing global population. Climate change is a global phenomenon affecting everybody on the planet, but people who are comparatively resource disadvantaged, such as smallholder farmers in SSA, are most affected. These farmers, who produce 90 percent of the total food in SSA, are at greatest risk of food insecurity themselves, and the impact of climate change will be severe in this region, which is already highly food insecure. Therefore, a rapid transformation and reorientation of present day traditional agriculture is necessary to ensure food security in the context of climate change – climate smart agriculture (CSA) and agroecology technologies are critical. Adaptation will bring immediate benefits and reduce the impacts of climate change in Africa. Experience from SSA shows that adaptation is growing, but needs strong institutional and policy support, secure land rights and at the same time, investments to support smallholders. This will help the region to benefit from integrated adaptation, mitigation and development approaches.

Several CSA and agroecology approaches have been developed and implemented successfully in SSA and other parts of the world. These technologies significantly increase agricultural productivity. Moreover, farmers are able to use locally available natural resources to substitute or reduce the requirements for major external inputs such as fertilizers, pesticides and improved seeds. Two major alliances, the Global Alliance for Climate Smart Agriculture (GACSA) and the Africa Climate Smart Agriculture Alliance (ACSAA) were established in 2014 as custodians of CSA. These alliances will be instrumental in streamlining CSA activities, advocating for policy support, and garnering the support of donor agencies to develop and up-scale technologies. It is imperative that all SSA nations meet the commitment as per the Maputo Declarations and provision for necessary policy supports to lay the foundation for successful implementation of adaptation measures across the region.

Notes

1 The founding members of ACSAA are AU-NEPAD; INGO partners: CARE International, Catholic Relief Services, Concern Worldwide, Oxfam and World Vision; and research and technical partners: the Consultative Group on International Agricultural Research (CGIAR) – Climate Change, Agriculture and Food Security Research Program (CCAFS), the Food and Agricultural Organization of the United Nations (FAO), the Food, Agriculture and Natural Resources Policy Analysis Network (FANRPAN), and the Forum for Agriculture Research in Africa (FARA).
2 An international agreement linked to the United Nations Framework Convention on Climate Change, which commits its parties by setting internationally binding emission reduction targets. It was adopted in Kyoto, Japan, on 11 December 1997 and entered into force on 16 February 2005.
3 The UNFCC entered into force on 21 March, 1994. A total of 195 countries have ratified the convention.
4 The African Climate Policy Centre (ACPC) is an integral part of the Climate for Development in Africa (ClimDev-Africa) program, which is a joint initiative of the

African Union Commission (AUC), the United Nations Economic Commission for Africa (UNECA) and the African Development Bank (AfDB). ClimDev-Africa has been mandated at regional meetings of African Heads of State and Government, as well as by Africa's Ministers of Finance, Ministers of Planning and Ministers of Environment.

References

African Union (AU) (2003) *Declaration on agricultural and food security in Africa* Assembly/ AU/Decl.7 (II), African Union, Addis Ababa, Ethiopia.

Aggarwal, P., Zougmore, R. and Kinyangi, J. (2013) *Climate-smart villages: A community approach to sustainable agricultural development*, Climate Change Agriculture and Food Security (CCAFS), Copenhagen, Denmark.

Alliance for a Green Revolution in Africa (AGRA) (2014) *The Africa agriculture status report 2014: Climate change and smallholder agriculture in sub-Saharan Africa*, Alliance for a Green Revolution in Africa, Nairobi, Kenya.

Altieri, M. A. (1995) *Agroecology: the science of sustainable agriculture,* Westview Press, Boulder, CO, USA.

Anonymous (2015) Flood death toll across Southern Africa reaches 260, *The Boston Globe*. Bloomberg News.

Bezner Kerr, R. (2014) Lost and found crops: agrobiodiversity, indigenous knowledge, and a feminist political ecology of sorghum and finger millet in northern Malawi, *Annals of the Association of American Geographers,* 104: 577–593.

Bezner Kerr, R. (2008) Gender and agrarian inequality at the local scale. In: Snapp, S.S., and Pound B. (eds). *Agricultural systems: Agroecology and rural innovation*, Elsevier Press, San Diego, USA.

Boko, M., Niang, I., Nyong, A., Vogel, C., Githeko, A., Medany, M., Osman-Elasha, Tabo, R. and Yanda, P. (2007) Africa, climate change 2007: Impacts, adaptation and vulnerability. *Contribution of Working Group II to the Fourth Assessment Report of the Intergovernmental Panel on Climate Change*, Cambridge University Press, Cambridge, UK.

Brown, S., Kebede, A.S. and Nicholls, R.J. (2011) *Sea-level rise and impacts in Africa, 2000 to 2100*, University of Southampton, Southampton, UK.

Building Nigeria's Response to Climate Change (BNRCC) (2011) National adaptation strategy and plan of action on climate change for Nigeria (NASPA-CCN), Federal Ministry of Environment Special Climate Change Unit, Ibadan, Nigeria.

Burt, C.C. (2013) 2013 global weather extremes summary, online, http://beforeitsnews. com/alternative/2013/04/march-2013-global-weather-extremes-summary-2618560. html (accessed 10 October 2015).

Business Daily (BD) (2015) Wonder grass back in Africa, opens new horizon for Kenya's livestock sector, online, http://www.businessdailyafrica.com/Corporate-News/ Wonder-grass-back-in-Africa-opens-new-horizon-for-livestock/-/539550/2880322/-/ item/3/-/uxj23o/-/index.html (accessed 11 October 2015)

Calvosa, C., Chuluunbaatar, D. and Fara, K. (2009) *Livestock and climate change,* International Fund for Agricultural Development, Rome, Italy.

CDKN (2014) The IPPC'S Fifth Assessment Report, What is in it for Africa? Online, http://cdkn.org/wp-content/uploads/2014/04/J1731_CDKN_FifthAssesmentReport_ WEB.pdf (accessed 11 November 2015).

César, E. and Ekbom, A. (2013) *Ethiopia environmental and climate change policy brief*, Sida's Helpdesk for Environment and Climate Change, Göteborg, Sweden.

Chauvin, N.C., Mulangu, F. and Mulangu, G. (2012) *Food production and consumption trends in sub-Saharan Africa: Prospects for the transformation of the agricultural sector,* United Nations Development Programme, Regional Bureau for Africa, Addis Ababa, Ethiopia.

De Schutter, O. (2010) *Report submitted by the Special Rapporteur on the right to food in United Nations General Assembly,* 20 December, 2010. http://www2.ohchr.org/english/issues/food/docs/A-HRC-16-49.pdf (accessed 30 October 2015).

Desanker, P.V. and Magadza, C. (2001) Africa: Climate Change 2001. In: McCarthy, J.J. *et al.* (eds). *Impacts, adaptation and vulnerability, IPCC Working Group II, Third Assessment Report,* Cambridge University Press, Cambridge, UK.

Djikeng, A., Rao, I.M., Njarui, D., Mutimura, M., Caradus, J., Ghimire, S.R., Johnson, L., Cardoso, J.A., Ahonsi, M. and Kelemu S. (2014) Climate-smart Brachiaria grasses for improving livestock production in East Africa, *Tropical Grasslands,* 2: 38–39.

East African Community (EAC) (2011) East African community climate change policy. East African Community Secretariat, Arusha, Tanzania.

Fabusoro, E., Sodiya, C.I., Fasona, M. and Oyedepo, J. (2014) *Vulnerability of settled Fulani agro-pastoralists' livelihoods to climate change and emerging innovations for adaptation and land accessibility in southwest Nigeria,* A Global Environmental Change Research Project, Global System for Analysis, Research and Training (START), Subaward grant 2013-02.

Food and Agriculture Organization (FAO) (2006) *Demands for products of irrigated agriculture in sub-Sahara Africa,* FAO, Rome, Italy.

Food and Agriculture Organization (FAO) (2010) *Climate-smart agriculture policies, practices and financing for food security, adaptation and mitigation,* FAO, Rome, Italy.

Food and Agriculture Organization (FAO) (2015) *The state of food insecurity in the world 2015 meeting the 2015 international hunger targets: taking stock of uneven progress,* FAO, Rome, Italy.

Gommes, R.A. and Petrassi, F. (1994) Rainfall variability and drought in sub-Saharan Africa, online, http://www.fao.org/nr/climpag/pub/EIan0004_en.asp (accessed 10 October 2015).

Hulme, M., Doherty, R., Ngara, T., New, M. and Lister, D. (2001) African climate change: 1900–2100, *Climate Research,* 17: 145-168.

Intergovernmental Panel on Climate Change (IPCC) (2007) Summary for policymakers. In: Solomon, S. *et al.* (eds). *Climate Change 2007: The Physical Science Basis. Contribution of Working Group I to the Fourth Assessment Report of the Intergovernmental Panel on Climate Change,* Cambridge University Press, Cambridge, UK.

Intergovernmental Panel on Climate Change IPCC (2013) Climate change 2013: The physical science basis. In: Stocker, T.F., D. Qin, G.-K. Plattner, M. Tignor, S.K. Allen, J. Boschung, A. Nauels, Y. Xia, V. Bex and P.M. Midgley (eds). *Contribution of Working Group I to the Fifth Assessment Report of the Intergovernmental Panel on Climate Change.* Cambridge University Press, Cambridge, UK and New York, USA.

Intergovernmental Panel on Climate Change IPCC (2014) Climate change 2014: impacts, adaptation, and vulnerability. In: Field, C.B., V.R. Barros, D.J. Dokken, K.J. Mach, M.D. Mastrandrea, T.E. Bilir, M. Chatterjee, K.L. Ebi, Y.O. Estrada, R.C. Genova, B. Girma, E.S. Kissel, A.N. Levy, S. MacCracken, P.R. Mastrandrea, and L.L. White (eds). *Contribution of Working Group II to the Fifth Assessment Report of the Intergovernmental Panel on Climate Change,* Cambridge University Press, Cambridge, UK and New York, USA.

International Fund for Agricultural Development (IFAD) (2011) *New directions for smallholder agriculture,* IFAD, Rome, Italy.

Khan, Z.R., Midega, C.A.O., Pittchar, J.O., Murage, A.W., Birkett, M.A., Bruce, T.J.A. and Pickett, J.A. (2014) Achieving food security for one million sub-Saharan African poor through push–pull innovation by 2020, *Philisophical Transactions of the Royal Society of London: Biology*, 369: 20120284.

Khan, Z.R., Midega, C.A.O., Pittchar, J.O., Pickett, J.A. and Bruce, T. (2011) Push–pull technology: a conservation agriculture approach for integrated management of insect pests, weeds and soil health in Africa, *International Journal of Agricultural Sustainability*, 9: 162–170.

Lipper, L., Thornton, P., Campbell, B.A., Baedeker, T., Braimoh, A., Bwalya, M., Caron, P. Cattaneo, A., Garrity, D., Henry, K., Hottle, R., Jackson, L., Jarvis, A., Kossam, F., Mann, W., McCarthy, N., Meybeck, A., Neufeldt, H., Remington, T., Sen, P.T., Sessa, R., Shula, R., Tibu, A. and Torquebiau, E.F. (2014) Climate-smart agriculture for food security, *Nature Climate Change*, 4: 1068–1072.

Liu, J., Fritz, S., van Wesenbeeck, C.F.A., Fuchs, M., You, L., Obersteiner, M. and Yang, H. (2008) A spatially explicit assessment of current and future hotspots of hunger in Sub-Saharan Africa in the context of global change, *Global and Planetary Change*, 64: 222–235.

Livingston, G., Schonberger, S. and Delaney, S. (2011) Sub-Saharan Africa: The state of smallholders in agriculture, IFAD Conference on New Directions for Smallholder Agriculture, International Fund for Agricultural Development, Rome, Italy.

Lobell, D.B., Burke, M.B., Tebaldi, C., Mastrandrea, M.D., Falcon, W.P. and Naylor, R.L. (2008) Prioritizing climate change adaptation needs for food security in 2030, *Science*, 319: 607–610.

Masih, I., Maskey, S., Mussá, F.E.F. and Trambauer, P. (2014) A review of droughts on the African continent: a geospatial and long-term perspective, *Hydrology and Earth System Sciences*, 18: 3635–3649.

Matson, P.A., Parton, W.J., Power, A.G. and Swift, M.J. (1997) Agricultural intensification and ecosystem properties, *Science*, 277: 504–509.

Mburia, R. (2015) Moving towards climate adaptation policy in Africa, Climate Emergency Institute, online, http://www.climateemergencyinstitute.com/uploads/Moving_Toward_Climate_Change_Adapation_Policy_In_africa.pdf (accessed 10 October, 2015).

Miles, J.W., Maass, B.L. and do Valle, C.B. (1996) *Brachiaria: biology, agronomy, and improvement,* CIAT and CNPGC/EMBRAPA, Cali, Colombia.

Nakweya, G. (2013) Climate-smart villages benefiting African farmers. SciDev. Net's Sub-Saharan Africa desk, online, http://www.scidev.net/sub-saharan-africa/agriculture/news/climate-smart-villages-benefiting-african-farmers.html (accessed 10 October 2015).

New Partnership for Africa's Development (NEPAD) (2014) *Africa climate-smart agriculture alliance launched,* New Partnership for Africa's Development, Addis Ababa, Ethiopia.

Porter, J.R., Xie, L., Challinor, A.J., Cochrane, K., Howden, S.M., Iqbal, M.M., Lobell, D.B. and Travasso, M.I., (2014) Food security and food production systems. In: Field, C.B., *et al.* (eds), *Climate change 2014: impacts, adaptation, and vulnerability. part a: global and sectoral aspects.* Contribution of Working Group II to the Fifth Assessment Report of the Intergovernmental Panel on Climate Change, Cambridge University Press, Cambridge, UK.

Pretty, J. (2008) Agricultural sustainability: concepts, principles and evidence. *Physiological Transactions of the Royal Society B*, 363: 447–465.

Pretty, J.N., Noble, A.D., Bossio, D., Dixon, J., Hine, R.E., Penning De Vries, F.W. and Morrison, J.I. (2006) Resource-conserving agriculture increases yields in developing countries, *Environmental Science & Technology*, 40: 1114–1119.

Pretty, J., Toulmin, C. and Williams, S. (2011) Sustainable intensification in African agriculture, *International Journal of Agricultural Sustainability*, 9: 5–24.

Rakotoarisoa, M.A., Lafrate, M. and Paschali, M. (2012) *Why has Africa become a net food importer: Explaining Africa agricultural and food trade deficits*, FAO, Rome, Italy.

Ringler, C., Zhu, T., Cai, X., Koo, J. and Wang, D. (2011) *Climate change impacts on food security in sub-Saharan Africa: Insights from comprehensive climate change modeling*, International Food Policy Research Institute (IFPRI) Research Brief 15–20, Washington DC, USA.

Rockström, J. (2003) Water for food and nature in drought-prone tropics: vapour shift in rain-fed agriculture, *Philisophical Transactions of the Royal Society of London, Biology* 358: 1997–2009.

Rosset, P.M. (2006) *Food is different: why we must get the WTO out of agriculture*, Zed Books, London, UK.

Roudier, P., Sultan, B., Quirion, P. and Berg, A. (2011) The impact of future climate change on west African crop yields: What does the recent literature say? *Global Environmental Change*, 21: 1073–1083.

Schlenker, W. and Lobell, D.B. (2010) Robust negative impacts of climate change on African agriculture, *Environmental Research Letters 5*, doi:10.1088/1748-9326/5/1/014010.

Sherry, S. (2013) Farming smarter' the key to feed growing world population, BusinessDay BDLive, online, http://www.bdlive.co.za/business/agriculture/2013/12/04/farming-smarter-the-key-to-feed-growing-world-population (accessed 10 October 2015).

Sissoko, K., van Keulen, H., Verhagen, J., Tekken, V. and Battaglini, A. (2011) Agriculture, livelihoods and climate change in the West African Sahel, *Regional Environmental Change,* 11: 119–125.

Snapp, S.S., Bezner Kerr, R., Smith, A., Ollenburger, M, Mhango, W., Shumba, L.; Gondwe, T. and Kanyama-Phiri, G.Y. (2013) Modeling and participatory, farmer-led approaches to food security in a changing world: a case study from Malawi, *Science et changements planétaires/Sécheresse,* 24: 350–358.

United Nations (UN) (2011) *World population prospects: The 2010 revision,* United Nations Population Division, New York, USA.

vom Brocke, K., Trouche, G., Weltzien, E., Kondombo-Barro, C.P., Sidibe, A., Zougmoré, R. and Gozé, E. (2014) Helping farmers adapt to climate and cropping system change through increased access to sorghum genetic resources adapted to prevalent sorghum cropping systems in Burkina Faso, *Experimental Agriculture*, 50: 284–305.

vom Brocke, K., Trouche, G., Weltzien, E., Barro-Kondombo, C.P., Gozé, E. and Chantereau, J. (2010) Participatory variety development for sorghum in Burkina Faso: Farmers' selection and farmers' criteria, *Field Crops Research*, 119: 183–194.

Wach, E. (2015) Should agribusiness or agroecology be the future of Africa? Institute of Development Studies, online, http://www.ids.ac.uk/opinion/should-agribusiness-or-agroecology-be-the-future-of-africa (accessed 10 October 2015).

Walker, N.J. and Schulze, R.E. (2008) Climate change impacts on agro-ecosystem sustainability across three climate regions in the maize belt of South Africa, *Agriculture, Ecosystems and Environment*, 124: 114–124.

Wiggins, S. (2009) *Can the smallholder model deliver poverty reduction and food security for a rapidly growing population in Africa?*, FAO, Rome, Italy.

Wiggins, S. and Sharada, K. (2013) *Looking back, peering forward: what has been learned from the food-price spike of 2007–2008?* Overseas Development Institute, London, UK.

World Bank (WB) (2007) *World development report 2008*, World Bank, Washington DC, USA.

World Bank (WB) (2008) *The growth report: strategies for sustained growth and inclusive development*, World Bank, Washington DC, USA.

World Bank (WB) (2013) *Turn down the heat: climate extremes, regional impacts and the case for resilience*, World Bank, Washington DC, USA.

World Food Program (WFP) (2007) Floods across sub-Saharan Africa hit 1.5 million people. World Food Program, online, https://www.wfp.org/stories/floods-across-sub-saharan-africa-hit-15-million-people (accessed 10 October 2015).

Zinyengere, N., Crespo, O. and Hachigonta, S. (2013) Crop response to climate change in southern Africa: A comprehensive review, *Global and Planetary Change*, 111: 118–126.

6 Building resilience in African smallholder farming communities through farmer-led agroecological methods

Rachel Bezner Kerr, Hanson Nyantakyi-Frimpong, Esther Lupafya, Laifolo Dakishoni, Lizzie Shumba and Isaac Luginaah

Introduction

In Africa, much of the concern surrounding climate change stems from potential negative effects on smallholder agriculture (Dumenu and Obeng, 2016; Müller *et al.*, 2011), food security (Olsson *et al.*, 2014; Porter *et al.*, 2014), dietary quality (Tirado *et al.*, 2015), and the gendered consequences of the adoption of technologies that mitigate climate risks (Fisher and Carr, 2015). According to international climate scientists, African agriculture is highly sensitive to climate change and will become more so in the next fifty years (Niang *et al.*, 2014). Under current climate change models, it has been shown that arid and semi-arid Africa will experience truncated growing seasons and the increased occurrence of extreme weather events such as erratic rainfalls, droughts, and floods (Niang *et al.*, 2014; Li *et al.*, 2009). Although impacts will vary by sub-region, there is a broader consensus that climate change will have adverse consequences for Africa as a whole, with grave impacts on cereal yields, nutrition, and health (Müller *et al.*, 2011; Olsson *et al.*, 2014; Porter *et al.*, 2014; Tirado *et al.*, 2015). Particularly in the West African Sahel, scientific evidence shows that the magnitude and speed of projected climatic changes will outstrip farmers' ability to manage these changes (Elagib, 2015; Niang *et al.*, 2014). Given these concerns, there is an emerging body of scholarship examining the everyday experiences of farming households, and how to build climate change resilience by harnessing existing local knowledge, farming innovations, and indigenous-based adaptation practices (Jiri *et al.*, 2015; Mapfumo *et al.*, 2015; Mortimore and Adams, 2001; Nyong *et al.*, 2007).

Our primary goal in this chapter is to contribute to this emerging literature. We summarize findings from a case study illustrating how local knowledge and participatory farmer-led agroecological research could be used to enhance the resilience of smallholder agriculture in Africa. We draw our empirical findings from a broader participatory action research based in Malawi. The research had two key objectives: (1) to document the ways in which northern and central Malawian farmers experience and perceive climate change, and how these experiences vary by age, gender, HIV status, and household structure; and (2) to conduct participatory action research with farmers to assess different adaptation strategies for addressing climate change, health, and food security. We present part of our empirical findings from this project by bringing together the literature on resilience, agroecology, participatory action research, and climate-smart agriculture (Altieri *et al.*, 2015; Berkes *et al.*, 2003; Folke, 2006; Garrity *et al.*, 2010).

We argue that although climate change is imposing significant constraints on African smallholder agriculture, it is possible to use agroecological farming methods to manage these impacts and build resilience into traditional farming systems. We demonstrate that under climate variability, smallholders who use a diverse range of agroecological farming practices can manage drought stress, and its effects on food security and poor nutrition. Some critics have argued that due to the nature of the current capitalist world economy, this farming approach may be unsuitable for smallholders who are using agroecological options not out of choice but of necessity due to limited alternatives (Bernstein, 2014). Others suggest that agroecology takes a longer time to yield benefits, and that given projected climate dynamics, farmers in the Global South, who are more vulnerable to climatic changes and have more degraded lands need a much quicker change in food production (e.g. Tomich *et al.*, 2011; Foley *et al.*, 2011; Godfray and Garnett, 2014; Mueller *et al.*, 2012). Based upon our empirical evidence, we argue that even for the labor-stressed, poor smallholder households, agroecology can be used to build farmer resilience to climate change, and this farming approach can yield food security and nutritional benefits not only in the long term but also in the short term.

The chapter is organized as follows. We begin by situating our study within the broader literature on the concept of resilience. We proceed by reviewing past studies that have shown climate-smart and agroecological strategies used to address climate change. Next, we introduce our case study setting by discussing environmental and social factors affecting food systems resilience under climate change. We then outline our experimental research methods, before presenting our empirical findings. In the last section of the chapter, we discuss how our work contributes to the social context of climate change impacts, and how to build farmers' climate change resilience using a bottom-up approach.

Background literature

Resilience

Resilience is a concept that stems from the Latin word *resilire*, which means 'bouncing back,' 'rebounding,' or 'recoiling' (Alexander, 2013). The concept has its roots in complex systems theories, development psychology, and ecology (Berkes *et al.*, 2003; Folke, 2006). However, in the existing literature on environmental change and food systems, resilience ideas are drawn from the fields of ecology and agroecosystems (Folke, 2006). In this section, we briefly highlight the state of knowledge on the concept, before shifting our attention to different agricultural strategies used to address resilience.

There are multiple definitions of resilience, and these definitions hinge on disciplinary tradition, the unit of analysis, and the research context. Irrespective of how resilience is defined, however, there are certain crosscutting themes, including flexibility, innovation, adaptability, capacity, connectedness, and feedback (Brown, 2013; Folke, 2006). A widely used definition of resilience is the capacity of a system to experience shocks while retaining essentially the same structure and function, the ability to self-organize, and the capacity to learn and adapt (Brown, 2013; Folke, 2006; Walker *et al.*, 2006). Three fundamental attributes are inherent in this definition: (1) retaining structure and function, (2) self-organization, and (3) learning and adaptation.

A system's ability to retain the same 'structure and function' means being flexible or maintaining the same identity in response to changing socio-ecological contexts. The attribute of 'self-organization' refers to response or recovery following a socio-ecological stress. For example, after a disaster, a resilient system should be able to recover without external help. The attribute of 'learning and adaptation' describes lessons from past experiences, and putting this knowledge into practice without repeating past mistakes. This attribute also emphasizes building individual or community capacity to learn and adapt through formal and indigenous knowledge.

There is an immense literature that has examined some of the inherent weaknesses of resilience and its application in practical contexts (Folke, 2006; Lebel *et al.*, 2006; Berkes *et al.*, 2003). A common criticism is that resilience fails to take account of politics and power relations (Lebel *et al.*, 2006), and it reinforces existing inequalities brought about by neoliberal governance (Welsh, 2014). It has been argued that as a concept, resilience downplays social inequalities, gender, and power asymmetries that shape a system's ability to maintain the same structure and function, to self-organize, and to learn and adapt whilst undergoing socio-ecological stresses (Nelson and Stathers, 2009). Further, much of the literature on resilience tends to focus on a system disturbed by exogenous forces. In so doing, the concept often overlooks internal, endogenous, and social dynamics that can also cause stress (Folke, 2006). Criticisms have also been raised that as a concept, resilience rarely addresses the questions: 'who decides what should be resilient, when, and to what purpose?' (Lebel *et al.*, 2006).

Climate-smart strategies for resilience

The term 'Climate-Smart Agriculture' was initially launched at an FAO meeting on Agriculture, Food Security and Climate Change in The Hague in 2010. While the definition is in flux, the most recent definition (Lipper *et al.*, 2014) includes three major objectives: (1) sustainably increasing agricultural productivity to support equitable increases in incomes, food security, and development; (2)

adapting and building resilience to climate change from the farm to national levels; and (3) developing opportunities to reduce greenhouse gas (GHG) emissions from agriculture compared with past trends. A relatively recent concept developed to help policy-makers and governments address the climate impacts of agriculture, CSA does not have as developed a set of approaches as agroecology. A range of agricultural methods have fallen under the designation of 'climate-smart approaches, ranging from the *zäi* stone bunds, an indigenous soil conservation strategy developed in West Africa, to conservation agriculture methods, to use of improved hybrid seeds and more efficient fertilizer and irrigation methods (Harvey *et al.,* 2014). The role of corporations, governments, and scientists in deciding what is CSA is currently under political contestation, with some authors suggesting that there is a need to be attentive to ecological limits and social inequities as well as clarifying the specific practices associated with the term and recognizing the tradeoffs and limitations (Harvey *et al.,* 2014; Neufeldt *et al.,* 2013).

Agroecological strategies to address resilience

Agroecology is a much more developed set of practices, as well as being a scientific discipline and a broader social movement (Wezel *et al.,* 2009). Broadly defined as 'the integrative study of the ecology of the entire food system, encompassing ecological, economic and social dimensions' (Francis *et al.,* 2003), agroecological approaches include attention to social, economic, and political dynamics that shape food production, local knowledge, and building farmer capacity to innovate, as well as specific technical practices that draw on ecological principles. A recent review of agroecological methods noted the value of agrobiodiversity to both production and resilience of food systems, alongside greater substitution of organic sources for nitrogen, such as cover crops, legumes, composting, and manure application (Tomich *et al.,* 2011). More research is needed on the potential of livestock integration into smallholder agriculture, and whether agroecological approaches can effectively address food needs in low-income countries (Tomich *et al.*, 2011). While there is increasing political mobilization related to agroecology, particularly in Latin America, there remains limited empirical evidence of the efficacy of these methods to address food security needs under smallholder conditions, particularly in Sub-Saharan Africa (Ponisio *et al.,* 2015).

Alexander *et al.* (2011) called for greater integration of indigenous ecological knowledge, defined as 'a cumulative body of knowledge, practice, and belief, evolving by adaptive processes and handed down through generations by cultural transmission, about the relationship of living beings (including humans) with one another and with their environment' (Berkes, 2012) and climate impacts. Recognizing that there are many unequal power relationships inherent in scientific and indigenous communities, nonetheless these authors and others advocate for respectful collaboration that recognizes both the heightened vulnerability of many indigenous communities to climate change, as well as the detailed observation and experience of a given environment (Nyong *et al.,* 2007).

While Alexander *et al.* (2011) were focused on impact studies, traditional ecological knowledge could also contribute to more resilient agricultural approaches.

Sub-Saharan Africa has long been characterized by inter- and intra-annual natural climate variability (Niang *et al.*, 2014). A number of studies have examined different agricultural strategies used to address resilience to these climatic changes. Common strategies include diversification of agroecosystems, including the use of agroforestry, crop-livestock integration, and polycultures (Garrity *et al.*, 2010; Mbow *et al.*, 2014). The general enhancement of agrobiodiversity, the use of organic soil management strategies, and water conservation and harvesting have also been identified in many agrarian settings. Agroforestry has been used to improve soil fertility, mainly through biological nitrogen fixation by leguminous trees (Duguma and Hager, 2011). In semi-arid Africa, trees have been integrated into farming systems to ensure tighter nutrient recycling, while improving soil structural properties (Bayala *et al.*, 2008; Mbow *et al.*, 2014; Lott *et al.*, 2009). In African agrarian settings, agroforestry systems for farm resilience comprise different land management practices, including boundary plantings, perennial crops, hedgerow intercropping, live fences, and mixed strata agroforestry (Garrity *et al.*, 2010; Mbow *et al.*, 2014).

Crop-livestock integration has also been identified as a major strategy to address agricultural resilience to climate change (Ickowicz *et al.*, 2012). At the farm level, mixed crop livestock systems serve as insurance against droughts, and they are essential to the security and reproduction of farming systems. They provide alternative food sources in times of severe droughts and crop failures (Toutain *et al.*, 2010). They also augment incomes, contribute to biodiversity maintenance, and enhance soil fertility through organic matter transfer (Ickowicz *et al.*, 2012). Other agricultural strategies to address resilience include intercropping, and the introduction of new crop varieties that can withstand drought stress. For example, as a risk management mechanism, some African farmers are increasingly intercropping maize with beans, cassava, cowpeas, groundnuts, pigeonpeas, pumpkins, sorghum, and sweet potatoes (Vincent *et al.*, 2013). There is now an emerging understanding that instead of focusing on technological solutions to address climate impacts, it is important to build resilience, taking into account local knowledge and practices (Nyong *et al.*, 2007; Ifejika Speranza, 2010) although such an approach contrasts to the dominant rhetoric of more technically oriented strategies.

Participatory action research

This study used a Participatory Action Research approach, an engaged research epistemology and methodology, which involves people in dialogue, critical reflection, and analysis of their circumstances, followed by methodical, applied research aimed to transform society, address their problems and eliminate oppression (Fals-Borda and Rahman, 1991; Chevalier and Buckles, 2013). Part of the action research approach is one that relies on 'complexity thinking' – non-linear, contingent and context-specific approaches rather than reductionist models (Rogers *et al.,* 2013). Complexity approaches in action research mean that, in place of 'case studies', researchers and stakeholders are cooperatively working to address a problem, through iterative research processes that foster reflection and shared learning (Rogers *et al.,* 2013).

The research setting

This study took place in two sites: Ekwendeni and Kasungu in northern and central Malawi (Figure 6.1). These two areas are mid-altitude (1000–1200 m a.s.l.), and have sub-tropical ecosystems, with unimodal rainfall during the months of December to April (700–1300 mm/yr).[1] The typical cropping pattern of smallholders in both sites has been maize (*Zea mays*) as the dominant staple crop, and other crops grown at low density, including tobacco (*Nicotiana tabacum*), sweet potatoes (*Ipomoea batatas),* and groundnut (*Arachis hypogaea).* We selected these two sites as a focus for research partly due to the evidence of frequent drought occurrence in the historical records (Mandala, 1995), and the limited agricultural and social research in Malawi's north and central regions compared to the south. Another major reason for choosing these sites is our longer term partnership with farmer groups in each area.

The two sites have some differences that allowed for useful comparison. In Kasungu, research was carried out in the Simlemba Traditional Authority (TA), about 140 km north of Lilongwe, the capital city of Malawi (Figure 6.1). This region is a drier, more challenging ecosystem, with sandy soils that have a lower pH (Snapp, 1999). Kasungu District as a whole has been characterized by the cultivation of tobacco and maize in estate farming, with the highest concentration of tobacco estates in the country, influencing both food security and social differentiation. Many rural farmers of Kasungu have worked as tenants on tobacco estates, which have some of the lowest wages in the world (De Schutter, 2013). Kasungu was one the most food insecure districts in Malawi during the 2005–2006 food crisis, and low rainfall was a critical factor in worsening the food situation (MVAC, 2006). With a population of 29,241, the Simlemba TA is an isolated and remote area, with limited access to public services, and one of the most disadvantaged areas in Malawi's central region. The most recent Demographic and Housing Survey for Malawi indicated an HIV prevalence rate of 7.6% for the Central Region, compared to 6.6% in the Northern Region (NSO and ORC Macro, 2010).

The other site, Ekwendeni, is located further north, about 20 km north of Mzuzu. Ekwendeni has more clay-rich soils that are generally low in nitrogen (Snapp, 1999). The population of Ekwendeni catchment area is approximately 70,000 people, and the majority of farmers are smallholders, with only a few tobacco estates in the region. This area has also experienced a high level of food insecurity in the last two decades (Bezner Kerr, 2005). Due to the distance from Lilongwe, Malawi's capital city, there are fewer organizations and government activities in Ekwendeni compared to the central and southern regions. The town has relatively high rates of HIV/AIDS, attributed to the high numbers of

Figure 6.1 The study sites

Produced by: The Cartographic Section, Dept. of Geography, Western, 2006. #37-06

truckers stopping in the town enroute to Tanzania. There is a patrilineal system of inheritance, and thus women do not own land, but gain access through their husbands.

Methods

We designed the study as a longitudinal research project comparing households in Ekwendeni and Kasungu over a four-year period. Overall, we relied on multiple data collection procedures, including in-depth interviews, focus groups, a baseline survey, adaptation experiments by farmers, and an endline survey.

In 2009, we conducted six focus group discussions and twenty-five in-depth interviews on farmers' perceptions and ideas about climate change. This qualitative part of the research was meant to understand the social context of farmer perceptions of environmental change, and possible ways to improve adaptation and resilience. The six focus groups were conducted at six different sites considered to have maximum variation in rainfall patterns. Participants were selected using purposive sampling. Each focus group involved between 10 and 15 mixed groups of men and women. Interview participants were selected using purposive, non-probability sampling based on variation in agroecosystems, knowledge of indigenous crops, level of food security, age, gender, and health status. We used an interview guide, which was modified as questions and new issues arose during the interview. The interview guide contained questions about rainfall patterns, experiences during periods of drought, main worries (to get a sense of the importance of changing climate compared to other concerns) and crops grown in the past. Both the interviews and focus groups took between 1 and 1.5 hours to complete, and were conducted in local languages. All responses were audio recorded, with permission, and then transcribed and translated into English by hired research assistants. Data analysis was done by reading through the data and identifying key themes and contradictions (Miles and Huberman, 1994).

We further carried out a cross-sectional baseline survey in 2010 with 1,213 randomly selected households, 800 in Ekwendeni and 413 in Kasungu. Only one respondent aged 18 years or above was interviewed per household. Questions were asked about perceptions of rainfall patterns, drought occurrence, flooding, temperature changes, and general demographic and farming information. We also assessed food security and dietary diversity status. We measured food security by using the Household Food Insecurity Access Scale (Coates *et al.,* 2007). This scale contains a set of nine standardized questions that measure different dimensions of food insecurity, including anxiety about food, the quality of food supply, and the quantity of food intake. Based upon affirmative responses to these nine questions, we classified households into four different food security categories: food secure, mildly food insecure, moderately food insecure, and severely food insecure (Coates *et al.*, 2007). We also measured dietary diversity by adopting a standardized set of questions for a 24-hour dietary recall (Swindale and Bilinsky, 2006). These questions sought to establish whether a household member had eaten food items from 11 food groups over a 24-hour period. From the 24-hour dietary recall, a Household Dietary Diversity Score (HDDS) was calculated by counting the number of food groups consumed by each household. The HDDS ranged from a minimum of 0 to a maximum of 11. Following Labadarios *et al.* (2011), we classified the households into three levels of HDDS, namely Low HDDS (if households consumed 0 to 2 food groups); Medium HDDS (if households consumed 3 to 4 food groups); and High HDDS (if households consumed 5 to 11 food groups). A team of 10 enumerators pre-tested and carried out the survey. The survey results were analyzed using descriptive, bivariate, and multivariate analyses.

After the baseline survey in 2010, we encouraged and assisted 425 interested farmers to conduct different experiments as a means to improving food security and adapting to climate change. Using an agroecological approach, each farmer selected several strategies to test, including some of the following elements: (1) integration of trees (reforestation, use of fruit trees and/or agroforestry); (2) soil fertility and/or conservation strategy; (3) crop diversification, including legume intercrops and (4) livelihood diversification, such as dimba gardens with small-scale irrigation, livestock, and bee-keeping. These experiments were not imposed on farmers; rather, farmers selected their own preferred experiments based upon labor availability, landholding, land tenure security, and other social factors that influence smallholder farming. The experiments were conducted in 31 villages, which were selected using random sampling. Purposive, non-probability sampling methods were used to select participating households within villages. Farmers were selected by village members based on their level of household food security (i.e. low), HIV status (if known), and age (to ensure inclusion of youth).

A final survey of all participating households was conducted in February 2011 and 2013 to assess changes over time in agricultural practices, food security, dietary diversity, farming knowledge, and other social and environmental variables. We also conducted additional focus groups and in-depth interviews. In the final survey, we measured food security and dietary diversity by using the same methods adopted in the baseline survey. We entered the data into SPSS 18 and STATA 11, and conducted descriptive, bivariate, and multivariate analyses. In all interviews and surveys, the study was explained in full, and informed consent was requested, including addressing issues of privacy and confidentiality. The University of Western Ontario Research Ethics Board approved all research activities of this study.

Findings

The research findings are organized into four parts. We first present farmer perceptions and ideas about climate change, before discussing the participatory experiments, and their impacts on food security, dietary diversity, gender, and community relations. These findings are presented using both quantitative and qualitative evidence, with the later used to contextualize and give depth to the former.

Farmers' perceptions and ideas about climate change

Farmer perceptions and ideas about climatic changes are important because such views frame and form a basis for on-farm adaptation strategies (Gbetibouo,

2009; Mortimore and Adams, 2001; Nyong *et al.*, 2007). In the interviews we conducted in 2009, respondents consistently indicated that planting rains are beginning later in the season and finishing earlier. They also said that rains that used to come in June, September and October, that have different names from the planting rains, no longer come. Older people indicated that it has been over a decade since these rains came. We verified this initial finding by asking young people if they remember these rains, and found consistently that they did not know or remember these rains. Some people also said that rainfall intensity has decreased, and that the rains are also stopping in the middle of the rainy season.

In the baseline survey, deforestation was the most frequently named cause of climate change, mentioned by 46.4% of respondents compared to greenhouse gas emissions (2.6%). One in every 5 respondents (20.1%) linked climate change to God's will. This linkage was a more common response among participants with primary or no formal education, than those with secondary or higher education (although the sample size for these categories was low) in both men and women. Despite these fatalistic views of climate change, more than half (56%) stated that climate change can nonetheless be prevented, with men more likely to state that climate change can be prevented (61%), compared with women (49%).

In the final survey, about half of all respondents noted that there have been droughts in the last 3 years preceding the survey. Only 12% of farmers mentioned that there has been incidence of floods over the 3 years preceding the survey. We also asked farmers whether they think the climate is changing very rapidly, rapidly, slowly, or whether the climate is not changing at all. Slightly less than half of the farmers (47%) indicated that the climate is changing, but the pace of change was slow, while 30% described rapid change in climate and less than 1% of farmers (0.3%) indicated that the climate was not changing at all.

Enhancing the resilience of smallholder agriculture under climate change in northern and central Malawi

After documenting farmers' experiences and understanding of climate change, we used these as a basis to help build resilience adaptation. Using participatory action research and an agroecological approach, farmers selected a range of experiments as climate adaptation strategies, including the integration of trees (reforestation, use of fruit trees and/or agroforestry); soil fertility and conservation strategy; crop diversification, including legume intercrops; and livelihood diversification, such as dimba gardens with small-scale irrigation, livestock and bee-keeping. All these options were left open to farmers, and they selected a range of combinations based on particular needs.

Crop diversification was by far the most common adaptation experiment selected by the majority of farmers. We measured crop diversification by counting the total *number* and *types* of crops cultivated per hectare of farm field. We found 2.5 different crops per hectare in 2011, which increased to 3 crops per hectare in the 2013 final survey. Additionally, in the endline survey, a significant percentage of households had put their fields under legume cultivation (46%), compared to 31% in the baseline survey. As well, only 29% of households in the 2011 survey were using intercropping, compared to a statistically significant increase to 93% of households in 2013.

The most common intercrops were maize-cowpea, maize-groundnut, and maize-cowpea-groundnuts. Maize was the most commonly cultivated cereal, whilst soya bean was the most commonly cultivated legume, followed by groundnuts, with a statistically significant increase in the number of fields under soya beans and groundnuts, compared to the baseline data. In addition, the final survey showed a statistically significant increase in the number of fields under pigeonpea compared to the baseline survey. We found an increased adoption of 'lost indigenous crops' (Bezner Kerr, 2014), including sorghum, cowpea, and finger millet, although the increase was not statistically significant for millet. The survey findings also illustrate a significant decrease in the percentage of households who cultivated cash crops such as tobacco. There were no statistically significant differences between cropping patterns in the Ekwendeni and Kasungu sites. All the key differences were found between the 2011 and 2013 survey periods.

In terms of building soil health, the final survey showed that households have expanded the range of agroecological methods used to improve soil fertility. The majority of household use manure (59%), or leave crop residue on the land and incorporate it into soils to increase organic matter (54%). We found that both the *range* of agroecological practices had increased, and a significantly higher percentage of farmers were using these strategies in 2013, compared to 2011. As well, we found that the use of 'dimba', dry season vegetable gardens, had also increased significantly from 54% of households in 2011, to 61% of households in 2013. In 2011, the average number of crops in the 'dimba' was 1.2, compared to a significant increase of 2.5 in 2013.

We asked farmers whether there have been any changes in the community as a result of the climate change project. The majority mentioned that since the initiation of the project, tree planting has increased. Farmers often noted that more trees could serve as a windbreak, and could also reverse increasing climate variability. Furthermore, the focus groups revealed that non-project farmers were more eager to adopt legume intercropping and other agroecological practices, having seen the benefits being derived by participating farmers. As one participant explained, 'nowadays, people are eager to try planting sweet potato, which has vitamin A and is being tried by the climate change project farmers. Other farmers also want to grow most of the legumes being grown by members of the climate change project.'

Food security and nutrition implications of smallholder climate-resilient farming

We found a significant change in food security over the 2 years, with 27% of households shifting from severe food insecurity to either mildly food insecure or food secure in 2013 (Figure 6.2).

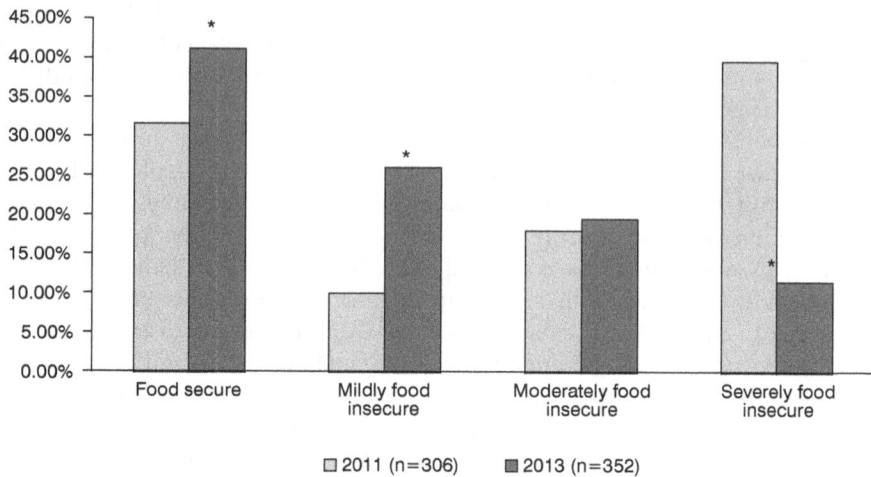

Figure 6.2 Percentage of households at different food security levels, 2011 & 2013

Source: Authors own compilation based on field survey data in Malawi

While we anticipated a high number of severely food insecure households in 2011, since this was one of the criteria for selecting participating households, we did not anticipate such a precipitous rise in food security in two years. This quantitative finding was confirmed in our qualitative in-depth interviews and focus groups. As one participating farmer said in a focus group held in April 2014:

> I was going to forget a very important thing. Hunger has gone down. We used to have many malnourished children. Some even dying. But because of the legumes and other crops, and using recipes, children are not dying.[2]
> (Oliet Zgambo, Bwabwa climate change
> group discussion, April 13 2014)

We also found a significant difference in food security status between male-headed and female-headed households. Overall, households that were headed by women reported greater food insecurity in the 2011 survey period, compared to those headed by men. After the participatory climate change experiments, however, we found a significantly positive change in household food security over time for all participating farmers. More significantly, we found that there was a greater change for female-headed households compared to male-headed households who were severely food insecure in 2011 compared to 2013.

In order to better understand the determinants of these significant changes in household food security, we conducted bivariate and multivariate analysis of the 2013 data set (Bezner Kerr *et al.*, under review). We explored the relationship between the dependent variable food security, and independent variables relating to farming practices, socio-economic factors, and the participatory climate change experiments. The results showed that the use of crop residue is a significant predictor of food security – which concurs with Mr. Zgambo's quote above, since he explicitly linked legume production to improved food security. In addition, a unit increase in the total number of crops planted by farmers (farm production diversity) resulted in a greater likelihood of food security. One female farmer, Rhoda Jere, described how farm production diversity is linked to food security:

> We are food secure because we grow a lot of soya, [finger] millet, sweet potatoes and groundnuts. We have all sorts of food and we sell the surplus. After selling, the money helps us with other needs.
> (Chilida climate change discussion group, April 9, 2014)

Similarly, our bivariate results showed that farmers who adopted maize and double legume intercrop were more likely to be food secure relative to those who planted sole maize (Bezner Kerr *et al.* unpublished data).

Intra-household gender relations also played a significant role in the changes in food security over time. Relative to farmers who did not discuss farming practices with their spouses, those who frequently chatted about farming were 2.37 times more likely to be food secure. Furthermore, farmers were significantly likely to be food secure if they reported using information learnt from our climate change experiments. Health status was also a strong determinant of food security. We found that farming households with no members experiencing chronic illness (like HIV/AIDS) were more likely to be food secure relative to those who were caring for a sick family member. Farmers from the Kasungu site were significantly less likely to be food secure, compared to those from the Ekwendeni site.

In our multivariate analysis, we examined the effects of crop residue incorporation into soil, the total number of crops cultivated, intercropping, crop diversity, and storage of harvested crops. We also examined discussion of farming with spouses, use of information learnt from the climate change experiments, and the use of animal manure for soil fertility, in addition to level of education, level of income, gender of the household head, and the location of the research site (Bezner Kerr *et al*, under review). The results indicated that at an alpha level of 0.05, farmers who grew more crops, stored crops in the past year, intercropped 2 or more crops, discussed farming practices with their spouses, and used manure to improve soil fertility were significantly more likely to be food secure than those who did not, while holding all other variables constant.

The use of agroecological farming not only resulted in food security, but also dietary diversity, which is one of the measures of nutrition. We found a statistically significant increase in the consumption of different food groups in 2013, compared to 2011. Food groups for which average consumption increased significantly include white tubers and roots, dark green leafy vegetables, vitamin A rich fruits, organ meat, eggs, fish, legumes, nuts and seeds, milk and milk products, oils and fats, and sweets. In the in-depth interviews, many farmers described eating a greater range of food due to participating in the project. As described by one widow:

There is a great change in my family since joining the climate change project, like now we are enjoying soya, we are also enjoying other recipes like sweet beer from sorghum. Children are healthier these days because they enjoy different recipes from crops provided by the project and we have enough food after harvest because our soils have improved.

(Mvula, female farmer, widow, age 42 – interviewed July 29, 2013)

The majority of respondents also reported a general improvement in health, which they attributed to crop diversification (using legumes), improvements in soil health, and crop yields.

Among farmers who reported significant and unprecedented benefits, the majority were widows, divorced women, or other types of female household heads. This finding was consistent with results from our final experimental survey, which revealed a significantly higher positive impact on food security for female-headed households compared to male-headed households.

Participatory agroecology and social relations at the household and community scales

The in-depth interviews further revealed significant improvements in intra-household social relations. In describing general improvements in soil health and dietary diversity, for example, farmers often interlaced these accounts with comments about improved decision-making between spouses:

My family is very healthy because we eat a variety of foods – groundnuts, soya beans, sorghum, cassava. My soils have improved looking at the soil colour, dark. Soya beans [and] groundnut when incorporated [into soils] they improve soils. My wife and I make decisions on what type of crops to grow, how to use the money after crop sales.

(Peter, male farmer, married, age 28 – interviewed July 26, 2013)

Project benefits were reported not only at the household level, but also at the community scale. In the final survey, over half of the participants reported that their community had taken action about climate change in the past 3 years. During interviews and focus groups, many project participants noted that they have seen significant changes in the community as a result of the climate change project. There were several different types of collaboration. Tree planting was one of the most significant community-level changes reported by the project participants. As one male farmer mentioned:

We have learned so much from the climate change project. I have learned to grow different crops, like pigeonpea, soya and sorghum, and even agroforestry trees. We've planted fruit trees, like orange and guava. We have learned how to plant trees, because the trees have all been cut down. So we have learned how to have a tree nursery and care for trees.

(Luhumero discussion group, Siuli Tembo village, April 2014)

Community tree nurseries were established in 20 villages, and over 20,000 tree seedlings were grown and distributed through these tree nurseries, with approximately a 20% survival rate. Over 80% of respondents in the endline survey named tree planting as an activity they undertook in response to climate change, after having joined the project.

There were several other notable examples of community organizing evident. In two villages, there was increased production of finger millet, an indigenous grain that is more drought tolerant. While finger millet had traditionally been grown by clearing and burning forest land, participating farmers experimented with growing millet in ridges – a practice which had been encouraged unsuccessfully by colonial and post-colonial governments in Malawi (Bezner Kerr, 2014). Initial experimentation in one village was shared with another pair of participating villages that lay over 25 km away, and that had previously never interacted. This second set of villages enthusiastically took up the 'ridge' millet growing, producing enough seed to distribute to an additional 40 farming households in 4 neighbouring villages. The climate change adaptation experimenting villages met with the village leaders of the neighbouring villages and asked for 'serious' farmers who were willing to try these new growing techniques.

A third type of community organizing was around fuel-efficient stoves. Some villages, after learning how to build kilns and make fuel-efficient stoves from clay, built community kilns and mobilized community members to mold clay stoves. These stoves were initially distributed freely to village members, but eventually became a source of income generation, by selling to neighbouring villages.

Another type of community organizing was around the distribution of small livestock, namely pigs and goats. Livestock were managed at a community level. The participating farmers constructed a village corral for the pigs, contributed feed (e.g. maize stalks), cared for the pigs and even purchased medicine if the pigs became sick. Villages received 2 pigs, and once the pigs reproduced, the first group of offspring were distributed to other participating farmers using the pass-on model. Over 100 pigs and goats were distributed within the participating villages using this system. The livestock provided both manure for soil improvement as well as a source of income to cash-strapped farming households.

Related to the pass-on program is another community-level impact. As a diverse group, including those who were HIV positive, highly food insecure or the young, the livestock care fostered cohesion and helped the participating farmers to work as a team. The community tree nurseries had a similar effect. AIDS-affected farmers talked about the support that they received from their fellow participating farmers if they became ill. This informal support system, which came out through community collaboration that emphasized inclusive, farmer-led research, is an important long-term outcome of the project.

The majority of farmers reported learning new information about climate change as a result of the project, and implementing changes in their lives because of what they had learned. Over 70% of women and 40% of men had no interaction with the agricultural extension service over the past year. A small percentage of farmers had more than 3 visits a year. Many respondents reported improved information sharing on farming practices, knowledge about climate change, crop diversification, and the cultivation of drought-resistance crops. In one of the focus group discussions, participants revealed that non-project farmers, both in and around the research villages, were eager to adopt legumes or crop diversification practices.

Discussion and conclusion

This chapter set out to show how diversified agroecological farming practices can be used to build resilience into agricultural systems under environmental changes. We have examined part of the empirical results from a broader study that sought to examine farmer experiences of climate change, and different adaptation strategies. A number of important issues emerged from the preceding analyses. Our main findings showed that farmers are aware of climate variations and its impacts on smallholder agriculture. These findings are similar to climate change perceptions reported among smallholder farmers in other parts of southern Africa (e.g. Bryan *et al.,* 2009; Gbetibouo, 2009), and in East, Central and West Africa (e.g. D'haen *et al.,* 2014; Silvestri *et al.,* 2012).

Our work also shows that when presented with climate change adaptation options, farmers tend to select practices that increase the diversity of the farming system. This finding could be explained by the fact that the cultivation of a diverse range of crops helps to optimize or spread risks. Crop diversification was not selected in isolation, but in combination with the use of mulching and agroforestry. All these practices help to build some resilience into farming systems because farmers using cover crops, intercropping, mulching and agroforestry can experience less crop failure and lower economic loss in the event of climate extremes.

Our data sets consistently show greater improvements in food security among farmers who chose to maintain farm production diversity under climatic variations. Similarly, most households had a dietary diversity score between 4 and 5, which is categorized as 'medium' dietary diversity (Labadarios *et al.,* 2011; Swindale and Bilinsky, 2006). The average household dietary diversity score was 3.48 in 2011, which significantly increased to 5.16 in 2013, suggesting a better nutrient intake after the participatory climate change experiments. These findings could be explained by the increased crop diversity, which is a significant indicator of dietary diversity (Fanzo *et al.,* 2013). Our survey results are also fairly similar to nationally representative data, which shows that farm production diversity is strongly associated with household dietary diversity and food security in Malawi (Jones *et al.,* 2014).

Through an agroecological approach, we have also shown not only how communities and farming systems could be made resilient, but also how to address some of the inequalities in todays' food systems. Indeed, a significant finding of our work is how greater crop diversity could be used to quickly improve the food security of female-headed households, often considered to be at greater risk of poor nutrition in Malawi. Previous research shows that in comparison to men, women in Malawi tend to have lower decision-making power, limited control over household resources, and higher workloads (Bezner Kerr, 2005). The voluminous literature on climate change also shows how marginalized women will be severely affected by projected climatic variations in Africa (Niang *et al.*, 2014). Our findings are therefore quite illuminating in demonstrating that smallholder agroecology can lead to significant improvements in food security among women.

In sum, our work shows that agroecological practices such as crop diversity, cereal-legume intercrops, and the use of organic material to build soil health, are very effective strategies to improve food security and address social inequalities in the context of environmental change. From these findings, several suggestions for policy and practice can be reached. Given the looming climate change threats on African agriculture (Elagib, 2015; Li *et al.*, 2009; Niang *et al.*, 2014), considerable attention is now being given to building resilient farming systems (Ifejika Speranza, 2010). As our introductory section makes clear, however, the mainstream resilience literature is often limited in many ways. Resilience suffers from a lack of attention to power dynamics within households, communities, and the broader international community (Nelson and Stathers, 2009). It is a somewhat 'steady-state' concept and does not foster transformational change, both environmentally and socially, of a system. If the transformational challenges presented by climate change are to be addressed, resilience thinking needs to be combined with other concepts that address distributive power issues and the unsustainable nature of current food systems. Food sovereignty and agroecological approaches can help to build community resilience in environmentally and socially marginal areas, if done along with co-learning from farmers, and modification and adaptation of existing innovations. In using food sovereignty and agroecological approaches to build resilience, however, these should be implemented as a 'basket of options,' but not a 'one-size-fits-all' approach. They should recognize and address existing inequalities, and also value social, environmental, and political factors influencing agricultural systems. Under severe climatic variations, a food sovereignty approach looks for holistic farming methods that take seriously economic, social, environmental and cultural dimensions of food systems. It looks for the rights of small-scale farmers to decide about what, how, when and for whom to produce.

The case study presented in this chapter falls within these alternative frameworks. Rather than distributing ready-made solutions, our ongoing research in Malawi pays greater attention to local alternatives that are based on the knowledge and practices of communities themselves. Many participating farmers have now designed local food production systems that are more adapted to climate change and in tune with local contexts and needs. Our research findings demonstrate that highly food insecure households, including both youth and HIV-affected households, are able to carry out climate change adaptation research using an agroecological approach, and that this research can lead to positive changes in food security and nutritional outcomes.

A key aspect of this approach is to ensure opportunities for farmer exchange, support, sharing and experimentation, in ways that emphasize equity alongside other important aspects of well-being such as nutrition. At the same time, there is a need to move to landscape, regional and international level strategies in order to build long-term resilience. Some communities in Malawi, for example, including several participating in this research project, have been subject to foreign land acquisition, and the land is now being used for high-input, mechanized export crop production, hiring only a handful of people for security, and leaving these rural communities with few options. There is an urgent need to take stances against those policies and practices that work against building up resilience in rural communities. An example is the Global Alliance for Food Security and Nutrition, which commits Malawi (and other Sub-Saharan African countries) to 'freeing up' thousands of hectares of land for large-scale agricultural investments, which are unlikely to benefit smallholder farmers (Patel *et al.*, 2015). In these advocacy efforts, the process is also critical, in that marginalized, poor farming communities' perspectives must be brought into the forefront while addressing inequalities within rural communities. A more resilient rural community will be one that can address these political and social inequalities alongside the environmental challenges that climate change now poses.

Notes

1 www.worldweatheronline.com/Ekwendeni-weather-averages/Rumphi/MW.aspx.
2 Paraphrased quotation.

References

Alexander, C., Bynum, N., Johnson, E., King, U., Mustonen, T., Neofotis, P., and Weeks, B. (2011) Linking indigenous and scientific knowledge of climate change. *BioScience*, 61(6): 477–484.

Alexander, D. (2013) Resilience and disaster risk reduction: An etymological journey. *Natural Hazards and Earth System Sciences,* 13: 2707–2716.

Altieri, M. A., Nicholls, C. I., Henao, A., and Lana, M. A. (2015) Agroecology and the design of climate change-resilient farming systems. *Agronomy for Sustainable Development*, 35 (3), 869–890.

Bayala, J., Heng, L.K., Noordwijk, M. V., and Ouedraogo, S. J. (2008) Hydraulic redistribution study in two native tree species of agroforestry parklands of West African dry savanna. *Acta Oecolo* 34(3) 370–378.

Berkes, F. (2012) *Sacred Ecology: Traditional Ecological Knowledge and Resource Management.* 3rd edition. New York: Routledge.

Berkes, F., Colding, J., and Folke, C., eds. (2003) *Navigating Social-Ecological Systems: Building Resilience for Complexity and Change.* Cambridge, UK: Cambridge University Press.

Bernstein, H. (2014) Food sovereignty via the 'peasant way': a sceptical view. *The Journal of Peasant Studies*, 41(6): 1031–1063. http://doi.org/10.1080/03066150.2013.852082

Bezner Kerr, R. (2005) Food security, entitlements and gender relations in northern Malawi. *Journal of Southern African Studies,* 31(1): 53–74.

Bezner Kerr, R. (2014) Lost and found crops: agrobiodiversity, indigenous knowledge, and a feminist political ecology of sorghum and finger millet in Northern Malawi. *Annals of the Association of American Geographers*, 104(3): 577–593. DOI: 10.1080/00045608.2014.892346.

Brown, K. (2013) Global environmental change: A social turn for resilience? *Progress in Human Geography*, 38(1): 107–117.

Bryan, E., Deressa, T. T., Gbetibouo, G. A., and Ringler, C. (2009) Adaptation to climate change in Ethiopia and South Africa: options and constraints. *Environmental Science & Policy*, 12(4): 413–426.

Chevalier, J. M. and Buckles, D. J. (2013) *Participatory Action Research: Theory and Methods for Engaged Inquiry*. New York: Routledge.

Coates, J., Swindale, A., and Blinsky, P. (2007) *Household Food Insecurity Access Scale (HFIAS) for Measurement of Household Food Access: Indicator Guide (Volume 3)*. Washington, DC: United States Agency for International Development.

De Schutter, O. (2013) Report on the Special Rapporteur on the Right to Food, Mission to Malawi. United Nations, New York., online: http://www.srfood.org/images/stories/pdf/officialreports/20140310_Malawi_en.pdf (accessed 19 November 2015).

Duguma, L. A., and Hager, H. (2011) Farmers' assessment of the social and ecological values of land uses in central Highland Ethiopia. *Environmental Management*, 47(5): 969–982.

Dumenu, W. K., and Obeng, E. A. (2016) Climate change and rural communities in Ghana: Social vulnerability, impacts, adaptations and policy implications. *Environmental Science & Policy*, 55: 208–217.

D'haen, S. A. L., Nielsen, J. Ø., and Lambin, E. F. (2014) Beyond local climate: rainfall variability as a determinant of household nonfarm activities in contemporary rural Burkina Faso. *Climate and Development*, 6(2): 144–165.

Elagib, N. A. (2015) Drought risk during the early growing season in Sahelian Sudan. *Natural Hazards*, 79(3): 1549–1566.

Fals-Borda, O. and M. A. Rahman (1991) *Action and Knowledge: Breaking the Monopoly with Participatory Action Research*. New York: Apex.

Fanzo, J., Hunter, D., Borelli, T., and Mattei, F., eds. (2013) *Diversifying Food and Diets: Using Agricultural Diversity to Improve Nutrition and Health*. Earthscan/Routledge, London.

Fisher, M., and Carr, E. R. (2015) The influence of gendered roles and responsibilities on the adoption of technologies that mitigate drought risk: The case of drought-tolerant maize seed in eastern Uganda. *Global Environmental Change*, 35, 82–92.

Foley, J.A., Ramankutty, N., Brauman, K.A., Cassidy, E.S., Gerber, J.S., Johnston, M., Mueller, N.D., O'Connell, C., Ray, D.K., West, P.C. and Balzer, C. (2011) Solutions for a cultivated planet. *Nature,* 478 (7369): 337–342.

Folke C. (2006) Resilience: the emergence of a perspective for social-ecological systems analyses. *Global Environmental Change,* 16: 253–67.

Francis, C., Lieblein, G., Gliessman, S., Breland, T.A., Creamer, N., Harwood, R., Salomonsson, L., Helenius, J., Rickerl, D., Salvador, R. and Wiedenhoeft, M. (2003). Agroecology: the ecology of food systems. *Journal of Sustainable Agriculture*, 22 (3), pp .99-118.

Garrity, D. P., Akinnifesi, F. K., Ajayi, O. C., Weldesemayat, S. G., Mowo, J. G., Kalinganire, A., and Bayala, J. (2010) Evergreen agriculture: a robust approach to sustainable food security in Africa. *Food security*, 2(3): 197–214.

Gbetibouo, G. A. (2009) *Understanding Farmers' Perceptions and Adaptations to Climate Change and Variability: The Case of the Limpopo Basin, South Africa* (Vol. 849). Washington DC: International Food Policy Research Institute.

Godfray, H. C. J. and Garnett, T. (2014) Food security and sustainable intensification. *Philosophical Transactions of the Royal Society of Biology,* 369: pp. 20120273.

Harvey, C. A., Chacón, M., Donatti, C. I., Garen, E., Hannah, L., Andrade, A., and Wollenberg, E. (2014) Climate-Smart Landscapes: opportunities and challenges for integrating adaptation and mitigation in tropical agriculture. *Conservation Letters,* 7(2): 77–90. http://doi.org/10.1111/conl.12066.

Ickowicz, A., Ancey, V., Corniaux, C., Duteurtre, G., Poccard-Chappuis, R., Touré, I., Vall, E., and Wane, A. (2012) Crop–livestock production systems in the Sahel – increasing resilience for adaptation to climate change and preserving food security. In: *Proceedings of FAO/OECD Workshop on Building Resilience for Adaptation to Climate Change in the Agriculture Sector.* Rome, FAO-OCDE. pp. 261–294.

Ifejika Speranza, C. (2010) *Resilient Adaptation to Climate Change in African Agriculture.* Deutsches Institut fur Entwicklungspolitik: Bonn.

Jiri, O., Mafongoya, P. L., and Chivenge, P. (2015) Indigenous knowledge systems, seasonal 'quality' and climate change adaptation in Zimbabwe. *Climate Research,* 66, 103–111.

Jones, A. D., Shrinivas, A., and Bezner Kerr, R. (2014) Farm production diversity is associated with greater household dietary diversity in Malawi: Findings from nationally representative data. *Food Policy,* 46, 1–12.

Labadarios, D., Steyn, N. P., and Nel, J. (2011) How diverse is the diet of adult South Africans. *Nutrition Journal,* 10: 33.

Lebel, L., Anderies, J. M., Campbell, B., Folke, C., Hatfield-Dodds, S., Hughes, T.P. and Wilson, J. (2006) Governance and the capacity to manage resilience in regional social-ecological systems. *Ecology and Society,* 11 (1): 19. www.ecologyandsociety.org/vol11/iss1/art19/

Li, Y., Ye, W., Wang, M., and Yan, X. (2009) Climate change and drought: a risk assessment of crop-yield impacts. *Climate research,* 39(1): 31.

Lipper, L., Thornton, P., Campbell, B. M., Baedeker, T., Braimoh, A., Bwalya, M., and Torquebiau, E. F. (2014) Climate-smart agriculture for food security. *Nature Climate Change,* 4(12): 1068–1072. http://doi.org/10.1038/nclimate2437

Lott, J. E., Ong, C. K., and Black, C. R. (2009) Understorey microclimate and crop performance in a Grevillea robusta-based agroforestry system in semi-arid Kenya. *Agricultural and Forest Meteorology,* 149:1140–1151.

Mandala, E. C. (1995) *The End of Chidyerano: A History of Food and Everyday Life in Malawi, 1860–2004.* Portsmouth: Heinemann.

Mapfumo, P., Mtambanengwe, F., and Chikowo, R. (2015) Building on indigenous knowledge to strengthen the capacity of smallholder farming communities to adapt to climate change and variability in southern Africa. *Climate and Development,* 8 (1): 72–82.

Mbow, C., Smith, P., Skole, D., Duguma, L., and Bustamante, M. (2014) Achieving mitigation and adaptation to climate change through sustainable agroforestry practices in Africa. *Current Opinion in Environmental Sustainability,* 6: 8–14.

Miles, M. B. and Huberman, A. M. (1994) *Qualitative data analysis: An expanded sourcebook.* Thousand Oaks, CA: Sage Publication.

Mortimore M., and Adams W. M. (2001) Farmer adaptation, change and 'crisis' in the Sahel. *Global Environmental Change,* 11: 49–57.

Mueller, N.D. Gerber, J. S., Johnston, M. Ray, D.K., Ramankutty, N. and Foley, J.A. (2012) Closing yield gaps through nutrient and water management. *Nature,* 490 (7419): 254–57.

Müller, C., Cramer, W., Hare, W.L., and Lotze-Campen, H. (2011) Climate change risks for African agriculture. *Proceedings of the National Academy of Sciences,* 108(11): 4313–4315.

National Statistics Office (NSO), Malawi and ORC Macro. (2010) *Malawi Demographic and Health Survey 2010*. Zomba, Malawi and Calverton, MD: National Statistical Office of Malawi (NSO) and ORC Macro.

Nelson, V., and Stathers, T. (2009) Resilience, power, culture, and climate: a case study from semi-arid Tanzania, and new research directions. *Gender & Development*, 17(1): 81–94.

Neufeldt, H., Jahn, M., Campbell, B.M., Beddington, J.R., DeClerck, F., Pinto, A.D., and Zougmoré, R. (2013) Beyond climate-smart agriculture: toward safe operating spaces for global food systems. *Agriculture & Food Security*, 2(1): 1–6. http://doi.org/10.1186/2048-7010-2-12.

Niang, O.C. Ruppel, M. Abdrabo, A. Essel, C. Lennard, J. Padgham, P. and Urguhart, P. (2014) Africa. In: *Climate Change 2014: impacts, adaptation, and vulnerability Contribution of Working Group II to the Fifth Assessment Report of the Intergovernmental Panel on Climate Change*. Cambridge University Press: Cambridge/New York, pp. 1199–1265.

Nyong, A., Adesina, F., and Osman Elasha, B. (2007) The value of indigenous knowledge in climate change mitigation and adaptation strategies in the African Sahel. *Mitigation and Adaptation Strategies for Global Change*, 12(5): 787–797.

Olsson, L., M. Opondo, P. Tschakert, A. Agrawal, S.H. Eriksen, S. Ma, L.N. Perch, and S.A. Zakieldeen (2014) Livelihoods and poverty. In: *Climate Change 2014: impacts, adaptation, and vulnerability. Contribution of Working Group II to the Fifth Assessment Report of the Intergovernmental Panel on Climate Change*. Cambridge University Press: Cambridge/New York, pp. 793–832.

Patel, R., R. Bezner Kerr, L. Shumba and L. Dakishoni. (2015) Cook, eat, man, woman: understanding the new alliance for food security and nutrition in Malawi, and its alternatives. *Journal of Peasant Studies*, 42 (1): 21–44. November 2014, online, DOI: 10.1080/03066150.2014.971767.

Ponisio, L. C., M'Gonigle, L. K., Mace, K. C., Palomino, J., de Valpine, P., and Kremen, C. (2015) Diversification practices reduce organic to conventional yield gap. *Proceedings of the Royal Society B: Biological Sciences* 282. DOI: 10.1098/rspb.2014.1396.

Porter, J.R., L. Xie, A.J. Challinor, K. Cochrane, S.M. Howden, M.M. Iqbal, D.B. Lobell, and M.I. Travasso (2014) Food security and food production systems. In: *Climate Change 2014: impacts, adaptation, and vulnerability. Contribution of Working Group II to the Fifth Assessment Report of the Intergovernmental Panel on Climate Change*. Cambridge University Press: Cambridge/New York, pp. 485–533.

Rogers, K. H., Luton, R., Biggs, H., Biggs, R., Blignaut, S., Choles, A. G., Palmer, C. G., and Tangwe, P. (2013) Fostering complexity thinking in action research for change in social–ecological systems. *Ecology and Society* 18(2): 31.

Silvestri, S., Bryan, E., Ringler, C., Herrero, M., and Okoba, B. (2012) Climate change perception and adaptation of agro-pastoral communities in Kenya. *Regional Environmental Change*, 12(4): 791–802.

Snapp, S. (1999) Soil nutrient status of smallholder farms in Malawi. *Communications in Soil Science and Plant Analysis*, 29(17 & 18): 2571–2588.

Swindale, A. and Bilinsky, P. (2006) Household Dietary Diversity Score (HDDS) for measurement of food access: indicator guide, Version 3. Washington, D.C.: Food and Nutrition Technical Assistance Project, Academy for Educational Development.

Tirado, M. C., Hunnes, D., Cohen, M. J., and Lartey, A. (2015) Climate change and nutrition in Africa. *Journal of Hunger & Environmental Nutrition*, 10(1): 22–46.

Tomich, T. P., Brodt, S., Ferris, H., Galt, R., Horwath, W. R., Kebreab, E., and Yang, L. (2011) Agroecology: a review from a global-change perspective. *Annual Review of Environment and Resources*, 36(1): 193–222. http://doi.org/10.1146/annurev-environ-012110-121302.

Toutain, B., Ickowicz, A., Dutilly-Diane, C., Reid, R., Diop, A. T., Taneja, V. K., Gibon, A., Genin, D., Ibrahim, M., Behnke, R. and Ash, A. (2010) Impacts of extensive livestock systems on terrestrial ecosystems. In: H. Steinfeld, H. Mooney, F. Schneider and L. Neville, eds. *Livestock in a changing landscape. Volume I: Drivers, consequences, and responses.* SCOPE. Washington, DC: Island Press, pp 165–195.

Vincent, K., Cull, T., Chanika, D., Hamazakaza, P., Joubert, A., Macome, E., and Mutonhodza-Davies, C. (2013) Farmers' responses to climate variability and change in southern Africa–is it coping or adaptation?, *Climate and Development*, 5(3): 194–205.

Walker, B., L. Gunderson, A. Kinzig, C. Folke, S. Carpenter, and L. Schultz. (2006) A handful of heuristics and some propositions for understanding resilience in social-ecological systems. *Ecology and Society* 11 (1): 13.

Welsh, M. (2014) Resilience and responsibility: governing uncertainty in a complex world. *The Geographical Journal*, 180(1): 15–26.

Wezel, A., Bellon, S., Doré, T., Francis, C., Vallod, D., David, C. (2009) Agroecology as a science, a movement or a practice. A review. *Agronomy Sustainable Development* 29: 503–515.

7 Precision nutrient management under conservation agriculture-based cereal systems in South Asia

Tek B. Sapkota, Kaushik Majumdar,
Ritika Khurana, Raj K. Jat, Clare Maeve Stirling
and Jat M. L.

Overview

Cereal crops such as rice, wheat and maize are the major food sources in South Asian (SA) countries. The second half of twentieth century marked a significant achievement in production of these major cereals with impressive gains mainly due to the introduction of improved germplasm, increased use of fertilizer inputs and an expansion of irrigated areas. However, agriculture in the region is now facing a new set of challenges due to degradation of land and natural resources and changing climatic conditions. At the same time, production of these cereals will have to increase to meet the food demand of an increasing population. Given the circumstances, agricultural systems need a paradigm shift so as to increase food production and adapt to climatic variability whilst at the same time minimizing the effects on the environment.

Conservation agriculture (CA), defined as minimum soil disturbance, permanent soil cover and appropriate crop rotation is promoted as resource conservation and sustainable production system. Each principle of CA involves a set of practices to affect the nutrient dynamics of the soil. Conventional fertilizer recommendations, which have been calibrated mainly based on tillage-based systems, are not necessarily appropriate for CA systems and 4R nutrient stewardship (applying Right source of fertilizer at the Right time in Right place using Right method) must be formulated taking into account the specific nutrient dynamics of CA systems. Research suggests that retention of crop residues as mulch under CA immobilizes some of the applied N during initial years but supplies additional N through mineralization in subsequent years. This needs to be factored into the nutrient management (NM) system in CA. Broadcast application of fertilizer N results in more volatilization loss under CA than under conventional systems. Sub-surface drilling of fertilizer during planting as well as in the standing crops has been found to be effective in improving nutrient use efficiency and increasing crop yield in CA systems.

Various tools, techniques and decision support systems have been developed and used for soil-based and plant-based precision nutrient management. There is still a large knowledge gap in understanding of nutrient dynamics and NM in CA systems, particularly in SA where fertilizer recommendation is largely based on the response trials conducted over a wide geographical area. In this chapter, we describe and discuss various aspects of precision nutrient management under CA-based cereal systems in SA and implications for the future of food security in the region.

Production and productivity trends of major cereals in South Asia

The Green Revolution (GR) has helped many SA countries to reduce their dependency on cereal imports. This remarkable achievement of the GR was largely due to expansion of crop area, and increased cropping intensity by the use of high-yielding varieties, chemical fertilizer, pesticides, irrigation and mechanization. Although the productivity of rice, wheat and maize has increased after the GR, the rate of yield increase has reduced recently. For example, the annual production increase of rice, wheat and maize in India was much higher during 1961–1990 than during 1991–2013 (Figure 7.1). This is mainly because the natural resource base has deteriorated over time due to exploitative farming. On the other hand, projections indicate that production of rice, wheat, and maize will have to increase by about 1.1 percent, 1.7 percent and 2.9 percent per annum, respectively, over the next four decades to meet the increasing food demand in the region (Jat *et al.,* 2011). Horizontal expansion of agriculture to increase food production is an unlikely and costly solution now because of the competition for land from other human activities and the social cost associated with using forest land or wetlands for crop cultivation. Therefore, most of the needed increase in production has to result from increased crop yield from the existing cultivable land. Chemical fertilizers play an important role to increase agricultural production. Some people argue that fertilizer was as important as seed in the Green Revolution (Tomich *et al.,* 1995), contributing as much as 50 percent of the yield growth in Asia (Hopper, 1993). Increasing food demand in this region means use of a correspondingly larger quantity of fertilizer in agriculture. If used inappropriately, this may have negative impacts on soil, water and the environment that could undermine future food production. Two major negative externalities of inappropriate fertilizer use are the release of greenhouse gas, and environmental pollution due to nutrient run-off and leaching.

Nutrient management strategies and challenges for major cereals in South Asia

Rice, wheat and maize show a wide range of yield gaps in SA (Table 7.1; Lobell *et al.,* 2009). Various factors are responsible for those yield gaps. Hengsdijk and Langeveld (2009) analysed yield and yield gaps of major crops in the world,

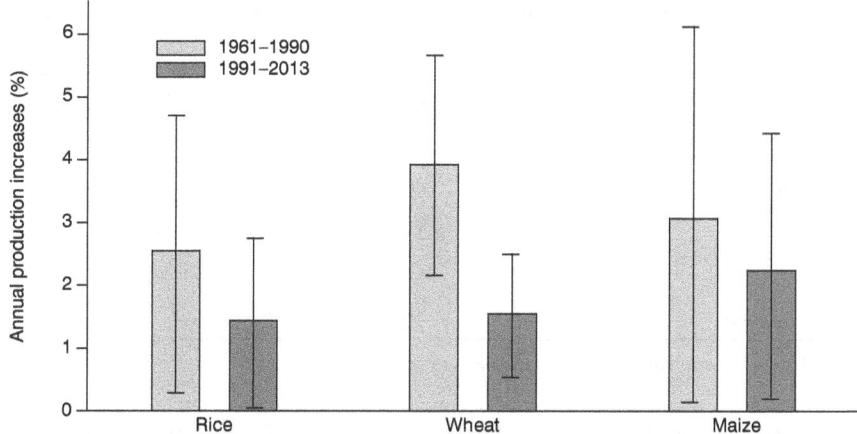

Figure 7.1 Annual increase in rice, wheat and maize production of India before and after 1990. Individual bars represent mean annual increase (%) for the period. Vertical line represents the standard error of the means

Source: Authors' analysis based on FAOSTAT (2015)

Table 7.1 Average yield and yield gap of rice, wheat and maize in India, Nepal and Bangladesh

Crop	Country	Average yield (ton/ha)	Potential yield (ton/ha)	Yield gap (ton/ha)
Rice	India	3.6	6.5	2.9
	Nepal	4.2	5.1	0.9
	Bangladesh	4.6	5.4	0.8
Wheat	India	3.2	4.5	1.3
	Bangladesh	2.9	4.2	1.3
Maize	South Asia	1.4	4.5	3.1

Source: Adapted from Lobell *et al.* (2009)

and concluded that about 20 percent of maize yield gap was explained by suboptimal NM.

In SA, high disparity between nutrient applied and nutrient uptake by harvested products is a serious threat to long-term fertility and soil health. This is mainly because nutrient recommendation for crops in SA is based upon crop response data averaged over large geographic areas and do not take into account the spatial variability in the inherent nutrient supplying capacity of soils (Majumdar *et al.*, 2013a). In general, nitrogen, phosphorus and potassium are the nutrient elements most commonly applied by farmers for cereal production.

Table 7.2 General fertilizer recommendation for rice, wheat and maize in South Asia

Crop	Fertilizer recommendation (kg/ha)		
	N	P_2O_5	K_2O
Rice	60–80	30	30
Rice (hybrid)	100	60	60
Wheat	80–100	15–30	25–50
Maize (OPV)	90–120	50	30–50
Maize (hybrid)	150	70	50

Source: Authors' compilation from various sources

The general recommendation for fertilizer N, P and K for rice, wheat and maize in South Asia is given in Table 7.2.

In general, 33 percent of N and all of P and K are applied as a basal dose at the time of planting and a remaining dose of N is applied in two equal splits at specific growth stages, which differ from crop to crop. In the conventional approach to cultivation, basal fertilizer is applied by broadcasting the fertilizer followed by some form of tillage to incorporate fertilizer into the soil. In case of rice, basal application is done by broadcasting the fertilizer over the puddled soil. The remaining dose of N is almost always applied by broadcasting the urea over the crop. In the CA system, on the other hand, the basal dose of fertilizer is drilled just below the seed row by using seed-cum-fertilizer drill. By and large, under the CA system the remaining dose of N is also applied by broadcasting urea. Surface application of urea fertilizer can lead to substantial loss of N by means of ammonia (NH_3) volatilization.

Blanket fertilizer application, therefore, results in under-fertilization in some cases and over-fertilization in others. Inappropriate and imbalanced use of nutrients has led to multiple nutrient deficiencies and low nutrient use efficiency. On average, efficiency of fertilizer N is only 30–40 percent in rice and 50–60 percent in other cereals (Brar *et al.*, 2011; Dobermann, 2006). The efficiency of fertilizer P is 15–20 percent in most crops and that of K is 60–80 percent. The fertilizer response ratio of cereals in irrigated areas in India is continuously declining over time (Figure 7.2; Biswas and Sharma, 2008). Increased nutrient use efficiency through precision nutrient management can substantially reduce production cost thereby increasing economic benefit and reducing the environmental burden from farming. For example, an increase in nitrogen use efficiency by one percent globally is estimated to be worth USD 234 million (Magen and Nosov, 2008). Given the situation, a renewed focus is needed on NM based on local pedoclimatic conditions, not only to achieve full benefit of good genetic materials but also to overcome crop yield barriers. A site-specific approach of nutrient management (SSNM) captures the spatial and temporal variability in soil fertility in smallholder production system and provides an approach to 'feeding' crops with all the required nutrients based on crop's needs and thus improves the crop yield (Das *et al.*, 2009).

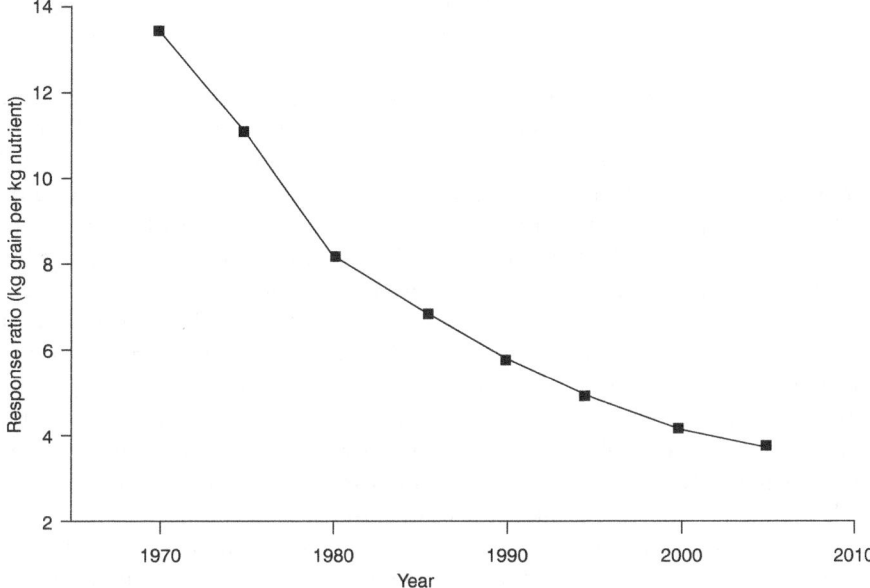

Figure 7.2 Fertilizer response of cereals over time in irrigated areas in India

Source: Biswas and Sharma (2008)

Conservation Agriculture (CA) in the quest of sustainability

The term CA was coined in the late 1990s in an attempt to integrate several crop management systems based on zero, reduced, minimum or conservation tillage under one single umbrella (Wall *et al.*, 2013). CA practices are designed to achieve sustainability by conserving resources while maintaining high yield and minimizing environmental degradation. CA is characterized by three interlinked principles: minimum physical soil disturbance, permanent soil cover with live or dead plant materials (e.g. crop residue mulches) and crop diversification in space and time (e.g. crop rotation, cover crops or intercrops). Each of the three principles of CA is linked to a number of desirable functions when they are translated into practice. Minimum tillage (MT) reduces the mechanical impact of cultivation on the soil's physical structure and soil life. Soil cover through surface residue retention protects the soil from the impacts of rain, thereby increasing water infiltration, and reducing evaporation, water run-off and soil erosion (Sommer *et al.*, 2014). Surface residue retention also reduces the soil temperature, thereby reducing the rate of organic matter decomposition and increasing concentration of organic matter on surface soil layer (Dordas, 2015). Crop diversification through rotations, cover crops and intercrops contributes to recycling nutrients. Biological N fixation improves when legumes are included, and pest and diseases cycles are broken. Therefore, CA is a paradigm shift from conventional agriculture with regard to management of crops, soil, water, nutrients, weeds and farm machinery.

CA helps build soil organic matter, conserve soil moisture and buffers against drought, as well as temperature extremes. Experimental evidence from various production environments suggests that CA-based management can have both immediate, e.g. reduced production costs, reduced erosion, stabilized crop yield, improved water productivity, adaptation to climatic variability (Hobbs, 2007; Jat *et al.,* 2014); and long-term benefits, e.g. higher soil organic matter contents and improved soil quality (Gathala *et al.,* 2011; Kienzler *et al.,* 2012). From a long-term CA experiment in Rice-Wheat (RW) cropping system of eastern Indo-Gangetic Plains (IGP), Jat *et al.* (2014) reported that there is immediate yield advantage in a No-till (NT) wheat crop, but the potential advantage of CA in NT direct seeded rice was realized only after 3–4 years of experimentation. In this long-term trial, the RW system grain yield was always higher in the CA-based system than in the CT-based system except for the first year (Figure 7.3). Through its positive effects on soil and water conservation, CA often increases average yields in the long-term and helps stabilize the yield against weather extremes. Improvement in grain and straw production encourages farmers to leave crop residues in their fields, thereby ensuring sustainability of the production system.

The evidence suggests that CA has emerged as an important management strategy to address many of the pressing challenges confronting the cereal systems in South Asia. NT has been widely adopted by farmers in wheat production systems, particularly in North Western IGP, primarily to facilitate early planting of wheat in areas where rice is harvested late, to lower production

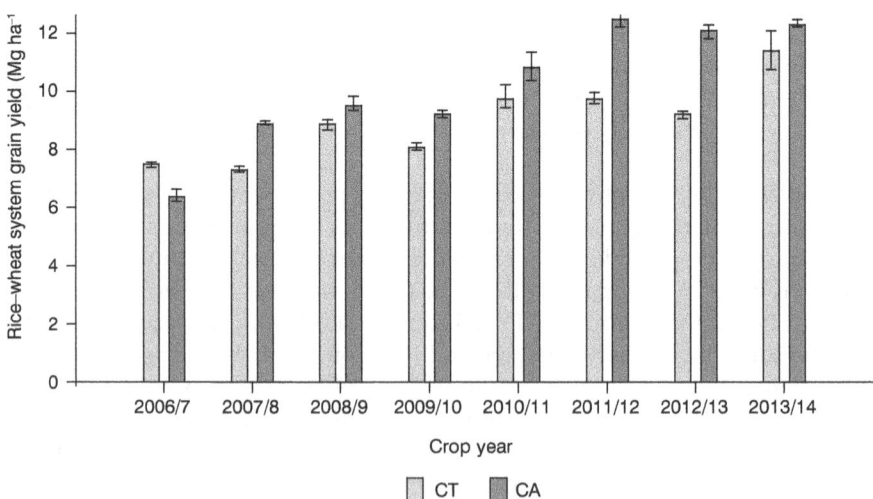

Figure 7.3 Long-term yield advantage of CA in rice-wheat cropping system
CT is an intensive tillage-based production system with all crop residues taken off the field. CA is ZT-based production of both rice and wheat with 50 percent rice and 25 percent wheat residue retention

Source: Adapted from Jat *et al.* (2014)

cost and increase yield and profitability (Chhokar *et al.*, 2007; Jat *et al.*, 2009; Saharawat *et al.*, 2010). With the development of planting equipment that can handle loose straw left in the field after combine harvesting of rice, and drill seed and fertilizer directly through the residues at appropriate depth (e.g. Turbo happy seeder), farmers are also retaining previous crop residue and moving towards full CA-based wheat system (Sharma *et al.*, 2012; Sidhu *et al.*, 2007).

Nutrient dynamics in CA-based cropping systems

In general, mineralization-immobilization, sorption-desorption, dissolution-precipitation and oxidation-reduction determine the dynamics of nutrient in the soil systems. CA, through its three key principles, influences the above-mentioned chemical and biochemical processes considerably. The changes in physical and biological properties of the soil associated with CA practices are expected to modify the direction and kinetics of the chemical and biochemical processes, leading to altered nutrient dynamics in the soil. In the following sections, we describe the effect of tillage and residue management (components of CA) on nutrient dynamics in cereal systems, particularly for the macronutrients.

Influence of tillage practices on soil nutrient dynamics

Mineralization is the transformation of nutrients from an organic to an inorganic state while immobilization is the reverse process. These two processes significantly influence the dynamics of several nutrients, namely nitrogen (N), phosphorus (P), sulphur (S) and the micronutrients. There is potentially a greater likelihood of more immobilization, denitrification or leaching of applied nitrogen in NT systems because of the increased microbial activity at the residue-soil interface in NT systems. The total amount of organic N mineralized in NT systems during the growing season is less than conventional systems, even though there is a potentially larger source of mineralizable N and greater microbial activity in NT soils. This is due to less surface area of organic residues exposed to microbial action when the residue exists as undisturbed mulch as compared to incorporating the residue through ploughing. The above discussion generally suggests that N recommendations should be higher in NT systems than conventional tillage (CT) systems, at least at the initial phases of establishment of a continuous NT system, until a new steady state equilibrium between immobilization and mineralization is reached, and supply of N from the labile organic pool increases. However, Dordas (2015) suggested that the net immobilization phase in the initial years of NT adoption is transitory, and immobilization of N under NT systems reduces the possibility of leaching and denitrification losses of soil mineral N. Yadvinder-Singh *et al.* (2015) suggested that although short-term soil N mineralization is lower under a NT system, total soil N mineralization may be similar in NT and tilled soil over the crop season.

Tillage practices, on the other hand, increase the exposure of soil organic carbon to microbial decomposition thereby increasing the possibility of loss of mineral N from active and physically protected N pool (Six *et al.,* 2002). Other researchers argue that 15–25 percent extra available moisture (Lafond *et al.,* 1992) during the growing season with NT as compared to CT also favours N losses from the system through leaching and gaseous emission (Thomas *et al.,* 1973; McMahon and Thomas, 1976; Tyler and Thomas, 1977). This happens also due to higher number of aerobic and anaerobic microorganisms in NT than in CT soil. Doran (1980) reported that populations of nitrifying organisms increased up to 20–fold while the population of denitrifiers increased up to 44–fold in the surface layer of soil under NT corn as compared to CT. Besides, a higher amount of organic substrate, nitrate and soil moisture together with existence of larger aggregates under NT than in a CT system may contribute to higher N loss through denitrification. However, lesser mineralization in NT soils and common practices of deep placement of N at crop establishment when water content is high in soils and splitting of N application to match crop demand can considerably decrease the denitrification potential in NT systems. Erosion control and subsurface placement of fertilizers such as N has the potential to significantly improve N availability and nutrient use efficiency (NUE) (Raun and Johnson, 1999), and are generally representative of NT systems.

The grain yield response and N uptake curve (Figure 7.4) shows that whilst yields are less at lower N levels, they are higher as N levels increase under an NT system. Lower grain yields and N uptake observed with NT at

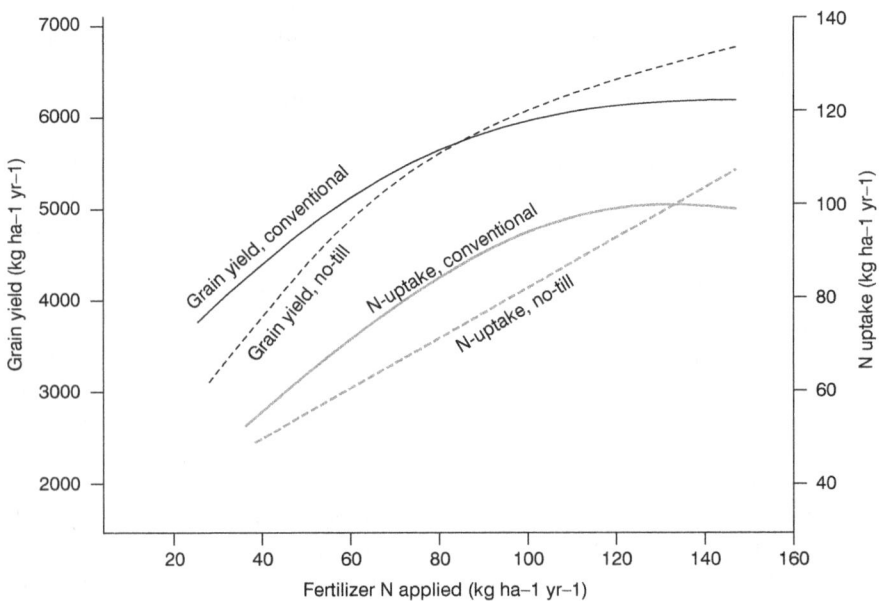

Figure 7.4 N uptake and grain yield of corn at various N fertilizer rates under conventional and no-tillage systems

Source: Adapted from Moschler and Martens (1975)

suboptimal rates of fertilizer N application probably resulted from either greater immobilization of fertilizer N, losses of N from denitrification and leaching, lower mineralization of soil organic N, or some combination of these factors. However, Majumdar *et al.* (2012), through their on-farm trials in India, showed that the yield of NT corn was higher than CT crop even at similar N application rates.

Influence of crop residue management on soil nutrient dynamics

When residues are surface applied or incorporated into the soil, the impacts of crop residues on nutrient availability differ. Since organic matter is a key factor in soil quality and nutrient dynamics, the management of previous crop residues has a profound effect on soil nutrient dynamics. The amount of residues left in the field, the composition of the residues and its placement (retained or incorporated) influence the decomposition rates in the soil. Further, whether N is mineralized or immobilized depends on the C:N ratio of the organic matter being decomposed by soil microorganisms (Hadas *et al.*, 2004). The progress of N mineralization and immobilization following residue addition is illustrated in Figure 7.5. There is a

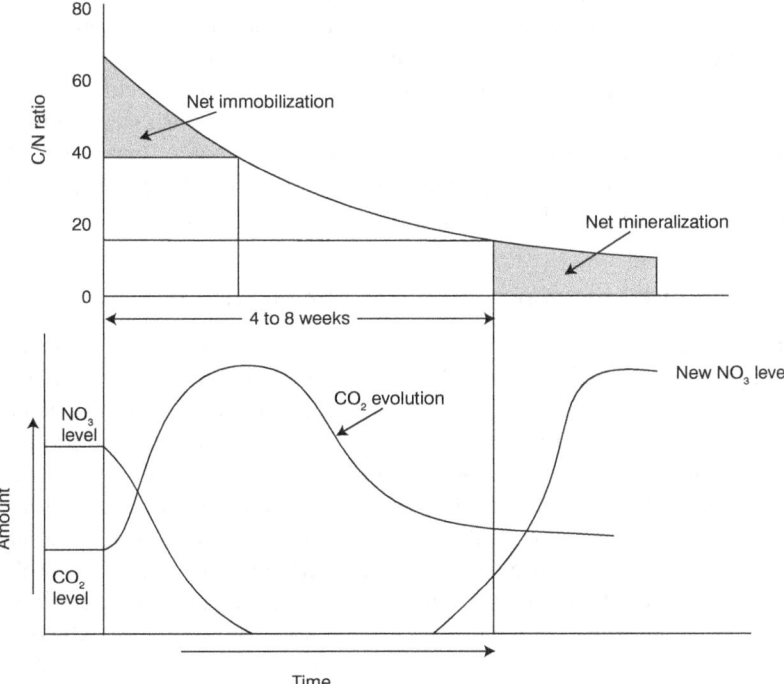

Figure 7.5 General description of N mineralization and immobilization following addition of crop residues in the soil

Source: Adapted from Havlin *et al.* (2005)

rapid increase in the number of heterotrophic organisms during the initial stages of fresh organic matter decomposition as indicated by elevated CO_2 evolution. If the C:N ratio of the residue is > 20:1, net immobilization will occur as shown in the hatched area under the top curve (Figure 7.5). The insufficient nitrogen in the substrate will induce the organisms to draw on the mineral nitrogen in the soil leading to immobilization of N (Zagal and Persson, 1994). The residue C:N ratio will, however, decrease as the decay proceeds because of decreasing C (respiration as CO_2) and increasing N (N immobilized from soil solution) and a new equilibrium will be reached, accompanied by mineralization of N (Figure 7.5) (Havlin *et al.*, 2005). Evidence suggests that retention of plant residue having high C:N ratio combined with NT reduce N availability to plants at least at the initial phases of crop growth, particularly during the initial years of implementation of CA (Verhulst *et al.*, 2010). High C:N ratio of cereal straw i.e. 60:1 to 100:1 (Havlin *et al.*, 2005) and generally low available N in soils of SA is expected to prolong the stage of N immobilization in this region.

Surface placement of the straw reduces N immobilization as compared to straw incorporated into the soil. In addition, Schnürer *et al.* (1985) demonstrated that residue added to soil with manure or nitrogen fertilizer led to residue decomposition rates that were two times greater than when no amendments were added. Rasmussen *et al.* (1997) found that standing straw residue had a strong adverse effect on yield, lowering the grain production by 13 percent compared with chopped straw. Additionally, where the surface temperature during the spring corn seeding period was reduced by 2–6 °C, lower yield was reported where stubble was retained on the surface, as compared with treatments where stubble was removed or incorporated (Cai and Wang, 1999). Rennie and Heimo (1984) reported that incorporation of straw into the soil led to significantly lower barley yields than when the straw was left on the soil surface.

Precision nutrient management in the context of CA

Nutrient management is an important aspect of CA for crop productivity and for the adoption of CA by farmers (Vanlauwe *et al.*, 2014). It has received little attention in CA despite its direct effect on crop yield. Experiences show that current fertilization management in CA needs revision to improve soil, crop productivity and environmental quality. CA-based NT and MT systems often result in greater stratification of soil nutrients than CT systems (Grant and Bailey, 1994; Lupwayi *et al.*, 2006; Jones *et al.*, 2007). NT conserves and increases the availability of K and other nutrients near the soil surface where crop roots often proliferate (Franzluebbers and Hons, 1996). Ismail *et al.* (1994) showed that extractable P was 42 percent greater at 0–5 cm, but 8–18 percent lower at 5–30 cm depth compared with CT treatments in silt loam soil under 20 years of NT. Similarly, N and K levels have also been found to be higher in surface 15 cm under NT than CT, gradually decreasing to the level of CT at lower soil depths (Govaerts *et al.*, 2007).

In a fully established CA system, the aim of NM is to maintain soil nutrient levels, replacing the losses resulting from the nutrients exported by the crops.

Because CA systems have a diverse crop mix, including legumes, and nutrients are stored in the soil organic matter, nutrients and their cycles must be managed more at the system or crop mix level. Thus, fertilization would not anymore be strictly crop specific. Additionally, undisturbed soils are habitats for free-living nitrogen-fixing bacteria and there is rhizospheric fixation of nitrogen (Sprent and Sprent, 1990). Furthermore, CA systems are based on building and breaking down organic matter to maintain soil health and productive capacity. As microorganisms decompose soil and organic matter, organic acids are continuously being formed. If these acids are not neutralized by free bases, then soil acidity will increase.

4R Nutrient Stewardship in the context of CA

4R Nutrient Stewardship is a new innovative approach for fertilizer best management practices that considers the economic, social and environmental dimensions of fertilizer management and is essential to sustainability of agricultural systems. The concept is simple: apply the Right source of nutrient, at the Right rate, at the Right time and in the Right place. All farmers, irrespective of the size of farm, knowledge and awareness levels, consider what fertilizer to apply, how much, when and how before making a fertilizer application decision in any crop. The 4R Nutrient Stewardship principle connects these fertilizer application decisions to scientific principles and guides the application decisions specific to crops, soils and the local site. The key scientific principles and examples of practices for application of the right source of nutrient, at the right rate, at the right time and in the right place are given in Figure 7.6. The four 'rights' provide a simple checklist to assess whether a given crop has been fertilized properly. Asking 'Was the crop given the right source of nutrients at the right rate, time and place?' helps farmers and advisors to identify opportunities for improvement in fertilizing each specific crop in each specific field (IPNI, 2012). The following sections briefly illustrate the 4Rs as they guide users to improved nutrient management strategies for higher crop productivity and profitability in an environmentally sustainable manner.

Right source

The idea of selecting the most appropriate nutrient source seems simple in concept, but many factors need to be considered when making this choice. An appropriate source will supply the nutrients in a plant-available form, so that nutrients are ready for uptake when the plants need them. The appropriate source of fertilizer should also suit the physical and chemical properties of the soil, so that the nutrients remain in available forms and are not held strongly by the soil matrix or lost from the soil. Typical examples would include nitrate application to flooded soils, or surface applications of urea on high pH soils where the potential of nutrient loss through leaching or through NH_3 volatilization is high. Surface application of urea results in severe loss of N under NT system (Bandel

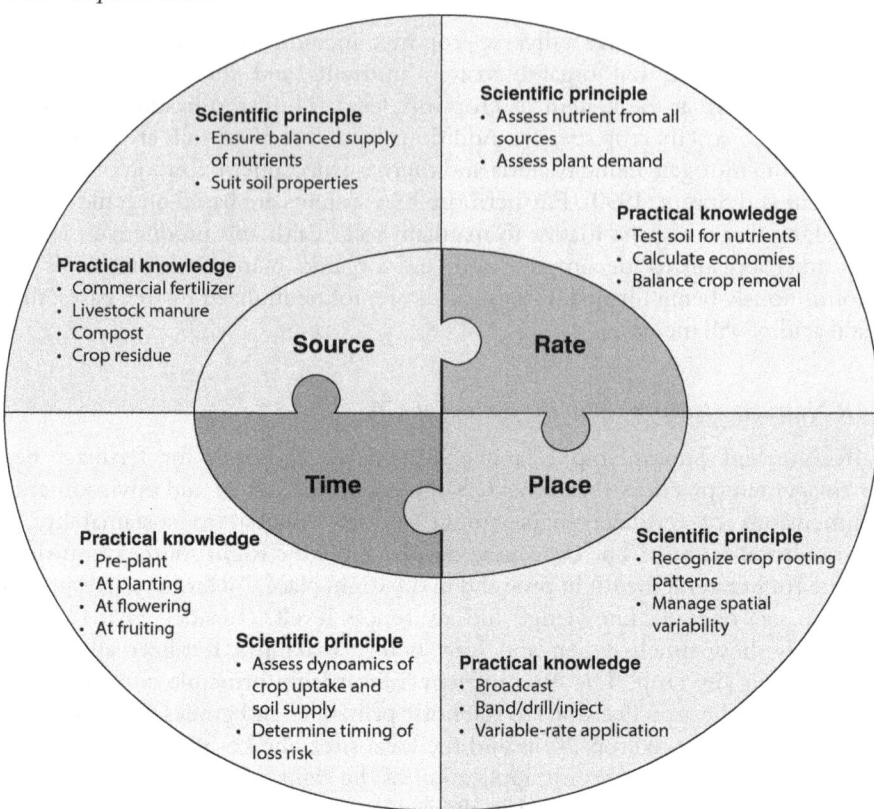

Figure 7.6 Examples of key scientific principles and associated practices of 4R nutrient stewardship

Source: Adapted from IPNI (2012)

et al., 1980) particularly at the early phases of crop establishment when there is ample moisture and a substantial amount of undecomposed organic substrate at the surface of the soil. Therefore, a different form of N (e.g. Ammonium nitrogen) should be used to minimize the loss. NT with residue retention creates continuous pores between the surface and sub-surface soil (Kay, 1990), providing more rapid passage of soluble nutrients deeper into the soil profile than when the soil is tilled (Franzluebbers and Hons, 1996). Choice of nutrient source, for example NH_4, may provide opportunities to reduce nutrient losses under a CA system. Halvorson and Del Grosso (2012) compared the growing season soil nitrous oxide (N_2O) emissions as a function of N source averaged over strip-till and NT-irrigated corn at Fort Collins, Colorado in 2009 and 2010. In this experiment, they observed significant reduction in N_2O emissions with controlled release N fertilizers whereas grain yield was not different among the N sources (Figure 7.7).

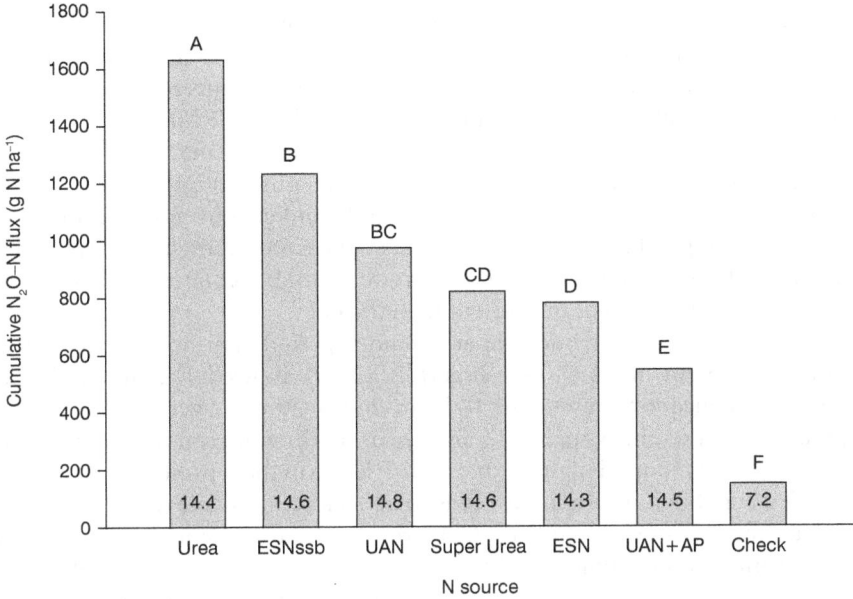

Figure 7.7 Effect of fertilizer N sources on N₂O emission in irrigated corn
Average grain yields (ton/ha) are shown at the bottom of each bar. The bars bearing different letters differ significantly from each other at α=0.05 probability level. ESN=polymer-coated urea, ESNssb=ESN subsurface band, UAN=Urea-ammonium nitrate, AP=AgrotainPlus

Source: Redrawn from Halvorson and Del Grosso (2012)

Right rate

A key scientific principle of selecting the right fertilizer rate is matching nutrient supply with plant nutrient demand. The selection of a meaningful yield target attainable with optimal crop and nutrient management is the first step for determining the right nutrient rate. An assessment of the quantity of nutrients already present in the soil through soil testing or nutrient omission trials, and estimation of the amount and plant availability of nutrients from other sources namely manure, composts, bio-solids, crop residues, atmospheric deposition and irrigation water helps finalizing the amount of nutrients that needs to be applied through external sources.

Loss or unavailability of a fraction of the applied nutrients for plant uptake is expected and fertilizer use efficiency is reduced. The expected efficiency of the applied nutrient must be factored in while calculating the required amount of nutrients to meet plant demand. The aim of any fertilizer rate decision is not applying too much fertilizer, that may lead to leaching and other losses to the environment, or too little that may result in lower yields and crop quality (Roberts, 2007). Economic considerations are major drivers of deciding appropriate fertilizer rates (Murrell and Bruulsema, 2008). For nutrients unlikely to be retained in the soil, the most economic rate of application is where the last unit of nutrient applied is equal in value to the increase in crop yield it generates (law of diminishing returns). For nutrients retained in the soil, their value to future crops should also be considered (IPNI, 2012). Recently, Singh *et al.,* (2014) showed significant yield improvement in RW systems through optimizing nutrient application based on nutrient input-output balance (Figure 7.8). The authors critically assessed the indigenous nutrient supply and externally supplied nutrients, based on the uptake requirement of crops and the efficiency factors that led to improved productivity, better farm profitability, and significant improvement in nutrient use efficiency.

NT with residue retention for an extended period may increase the available nutrient content in a soil. Govaerts *et al.* (2007) showed 1.65 and 1.43 times higher K concentrations in the 0–5 cm and 5–20 cm layers respectively, on permanent raised beds than CT raised beds, both with crop residue retention. Unger (1991) reported higher extractable P levels in the topsoil in NT compared to tilled soil. Such information needs to be factored in while deciding fertilizer application rates in zero tillage (ZT) residue management scenarios. At the same time, the possibility of higher losses of mobile nutrients in CA systems may also alter the rate requirement. Retention of crop residues in a CA system may result in a soil deficient in N, and will require external N application to compensate for this to meet the needs of the microorganisms and the growing crop, particularly during initial years of conversion.

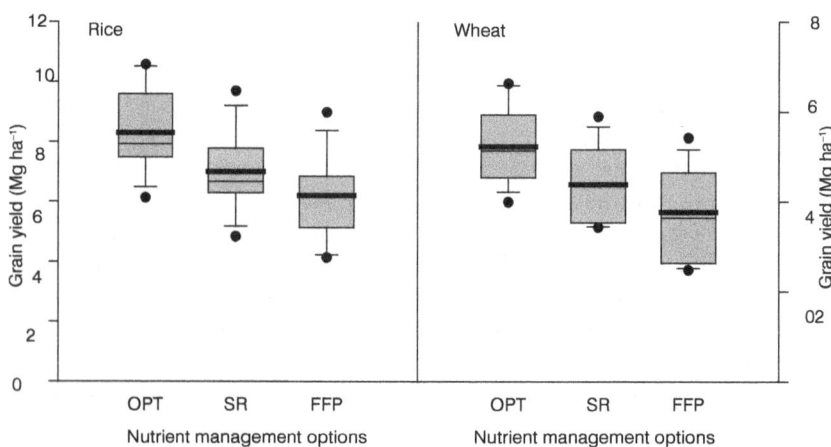

Figure 7.8 Rice and wheat yield under different nutrient management options
The error bars represent 10th to 90th percentile of the data. The thin and the thick line represents the median and mean, respectively. OPT=optimum application rate, SR=state recommendation and FFP=farmers fertilizer practice

Source: Adapted from Singh *et al.* (2014)

Right time

Nutrients are used most efficiently when their availability is synchronized with crop demand. The uptake of major nutrients and dry matter accumulation patterns are similar for most crops, and usually follow a sigmoid or 'S' shaped curve. This is characterized by rather slow early uptake, increase to a maximum during the rapid growth phase and decline as the crop matures (IPNI, 2012). Applications timed and targeted at specific growth stages may be beneficial to crop yield and/or quality (Majumdar *et al.,* 2013b). Tuning fertilizer application with crop demand minimizes the fertilizer loss, thereby reducing environmental impact. In a recent study, Bijay-Singh *et al.* (2015) showed significant improvement in rice yield, agronomic efficiency and recovery efficiency of N through the GreenSeeker optical sensor-based N application in rice (Table 7.3). The GreenSeeker optical sensor helped in-season estimation of the right time and rate of N application matching the uptake requirement of rice in a site-specific manner.

The synergy and antagonistic effect of different fertilizer nutrients should also be considered while timing the fertilizer application. For instance, if N and K are not applied together, nutrient imbalances result in pest attacks, as is well

Table 7.3 Evaluation of GreenSeeker optical sensor-based fertilizer N management in rice

Cultivar	N management Strategy	Total N applied (kg/ha)	Rice grain yield (kg/ha)	Total N uptake (kg/ha)	AE_N (kg/kg)	RE_N (percent)
PAU 201	No N	0	3.99	57.8	–	–
	Three equal splits	120	6.96	131.7	24.7	61.6
	GreenSeeker-based	102	7.16	130.8	31.0	71.5
PUSA 44	No N	0	3.94	63.1	–	–
	Three equal splits	120	6.38	121.6	20.3	48.7
	GreenSeeker-based	97	6.37	117.0	25.1	55.6
HKR 127	No N	0	3.75	57.9	–	–
	Three equal splits	120	6.04	120.4	19.1	52.1
	GreenSeeker-based	102	6.19	117.7	23.8	58.6

AE_N agronomic efficiency of applied N [kg grain (kg N applied)-1]; RE_N recovery efficiency of applied N (percent)

Source: Adapted from Bijay-Singh *et al.* (2015)

documented with oil palm leaf eaters that benefit from high N, low K foliage (Majumdar *et al.,* 2013b). The logistics of fertilizer distribution, field operations and application equipment are important factors affecting timing decisions. Slow release and other enhanced efficiency fertilizer technologies may be useful tools where logistics demand a single application at what might normally be an inopportune time. These speciality sources could significantly reduce the total number of applications, saving money and labour.

Nitrogen application timing in CA systems needs to be managed carefully to avoid N deficiency due to slow mineralization, immobilization and volatilization, and to avoid excess N fertilization. There are several options that allow sufficient time for soil organic matter to decompose before sowing the crop. Application of N fertilizer (25–70 kg/ha) before sowing will speed up mineralization. During sowing, N can be applied to prevent immobilization and provide young seedlings with adequate N (Dordas, 2015).

Right place

Right place means positioning needed nutrient supplies strategically so that a plant has access to them. Plant genetics, placement technologies, tillage practices, plant spacing, crop rotation or intercropping, weather variability, and a host of other factors can affect which placement is appropriate (IPNI, 2012). However, one of the primary objectives of right nutrient placement is to ensure that roots access nutrients immediately after application, thus reducing the possibility of loss. Root architecture differs greatly among plant species, changes with age and interacts strongly with soil conditions. All these factors should be considered while deciding the right place for fertilizer application.

In CA systems, broadcast and seed-placed P and K fertilizer applications lead to the accumulation of these nutrients in the surface and depletion in the deeper soil profile. Therefore, sub-surface banding of P and K with the seed, or ideally about 6–10 cm below the seed, is highly recommended to promote deeper root growth and avoid stranding these nutrients near the soil surface under the CA system (Randall and Hoeft, 1988). Further, phosphorus is strongly held by the soil. Therefore, band application of phosphatic fertilizer confines its interaction with smaller soil volume. Similarly, surface application of urea and urea-containing fertilizers results into severe loss of N under NT system (Bandel *et al.,* 1980), particularly at the early phases of crop establishment when there is ample moisture and a substantial amount of undecomposed organic substrate at the surface of the soil. Incorporating urea-containing fertilizer deep into the soil can reduce N loses through NH_3 volatilization. Therefore, such urea-containing fertilizers can be used in the NT system through injection or banded placement. Malhi and Nyborg (1992) suggested that precise placement of N fertilizer in a NT system with side-banding can reduce the immobilization effects as the NT drill separates the fertilizer and residue. Verhulst *et al.* (2010), however, suggested that the efficiency of chemical fertilizers can be increased by applying them to mulch rather than to soil. Yadvinder-Singh *et al.* (2015), in their N management systems evaluation under NT wheat in northwest India, found that up to 75 percent of the recommended N fertilizer can be drilled at sowing without loss of yield. Similarly, Hobbs and Gupta (2004) reported 10–15 percent increase in fertilizer use efficiency in a rice-wheat system through better placement of fertilizer with the seed drill in CA systems compared with broadcasting in the traditional system.

Site-specific nutrient management (SSNM)

SSNM is an approach of supplying plants with nutrients to optimally match their inherent spatial and temporal needs for supplemental nutrients (Buresh and Witt, 2007). The SSNM approach aims to enable farmers to dynamically adjust their fertilizer use to optimally fill the deficit between the nutrient needs of the crop and the supply from naturally occurring indigenous sources such as soil, crop residues, organic inputs and irrigation water. The SSNM approach does not necessarily aim to either reduce or increase fertilizer use. Instead, it aims to recommend nutrients at optimal rates and times to achieve high profit for farmers, with high efficiency of nutrient use by crops across spatial and temporal scale, thereby preventing leakage of excess nutrients to the environment.

The basic steps to be followed in SSNM are: i) set an attainable yield target, ii) estimate indigenous supply of nutrients through nutrient omission trials, iii) estimate crop response to nutrients (difference between target yield and yield in nutrient omission plot) and iv) estimate the nutrient rate based on response and agronomic use efficiency of nutrient (Shaobing *et al.*, 2010). In case of P and K, the yield target provides an estimate of the total nutrient needed by a crop. A portion of this requirement can be fulfilled from non-fertilizer sources such as soil, crop residues, organic inputs and irrigation water known as the indigenous nutrient supply. SSNM uses a nutrient balance approach, in which fertilizer P and K are recommended in amounts sufficient to close the gap between the indigenous supply and plant needs to achieve the yield target so that soil fertility is maintained without mining the soil. In case of N, within season nutrient estimation is used to determine the amount of N to be applied at the time of crop establishment. Subsequent application can dynamically be varied to match the spatial and temporal needs of crop through periodic monitoring (Buresh and Witt, 2007). Sensor-based, site-specific application of fertilizer has been reported to improve fertilizer use efficiency, and to increase the grain yield of many crops around the world. For example, Shaobing *et al.* (2010), through their field experiments and demonstration trials in rice, reported that SSNM reduced N fertilizer by 32 percent and increased grain yield by 5 percent compared with farmers' N fertilization practices. Similarly Bijay-Singh *et al.* (2015), through their seven field experiments in IGP, obtained similar rice yields in SSNM as the blanket fertilizer practice, but with reduced N rate. This increased recovery efficiency as well as the agronomic efficiency of N. Khurana *et al.* (2008) evaluated SSNM in 56 on-farm experiments with transplanted rice and irrigated wheat crops in Northwest India from 2002–03 to 2004–05. In this

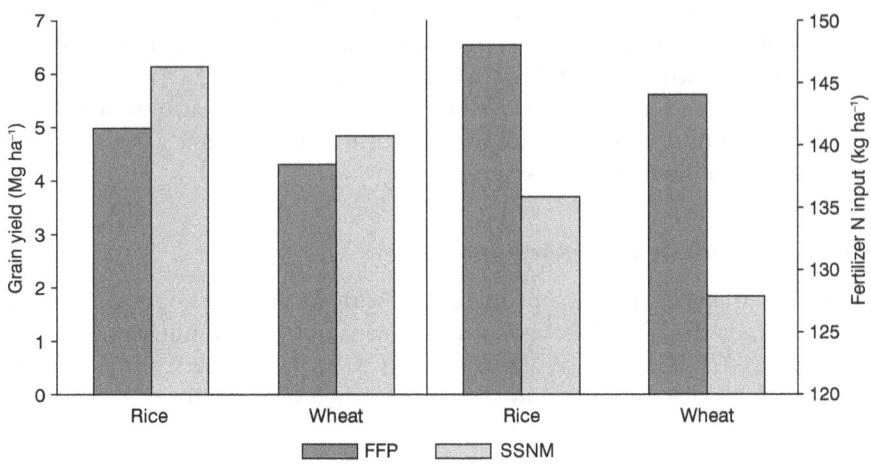

Figure 7.9 Grain yield (left panel) and fertilizer N input (right panel) in rice and wheat crops under farmers' fertilization practice (FFP) and site-specific nutrient management (SSNM). Means of 56 on-farm experiments averaged over two years

Source: Redrawn from Khurana *et al.* (2008)

study, they recorded higher grain yield of both rice and wheat with application of a lower amount of fertilizer N under SSNM system as compared to farmers' fertilization practice (Figure 7.9).

Tools and techniques for precision nutrient management

Managing the right source of fertilizer at the right rate, the right time and in the right place is best accomplished with the right tools. Various technologies are available to help make decisions related to nutrient management. These tools enhance the ability to fine-tune nutrient management decisions and develop the SSNM plan for each field. In the soil test-based SSNM approach, the fertilizer rates are established based on the concept of crop removal adjusting the soil residual nutrients. A plant-based approach, on the other hand, focuses on managing field-specific spatial variation in the indigenous nutrient supply and temporal variability in plant N status occurring within a growing season. The following paragraphs describe the tools and techniques available to implement precision nutrient management, particularly under smallholder production systems.

Leaf chlorophyll content can be linked with leaf N content because the majority of leaf N is contained in chlorophyll molecules. Therefore, measurement of leaf greenness by a chlorophyll meter such as SPAD (Soil Plant Analysis Development) throughout the growing season can signal potential N deficiency early enough to correct it without reducing yields (Shapiro *et al.*, 2013). However, luxury consumption of N does not increase leaf chlorophyll content, which causes SPAD meter readings to reach a plateau when N availability is adequate, regardless of how much extra N is taken up by the plant.

As a simple alternative to SPAD meter, International Rice Research Institute (IRRI) and Philippines Rice Research Institute (Philrice) jointly developed a Leaf Colour Chart (LCC) to measure the relative greenness of the crop leaf. The LCC was intensively validated across South Asia by several universities and research organization. The leaf greenness can be used as an indicator of the plant N status to determine in-season N demand. LCC is a plastic ruler-shaped strip containing four to six panels that range in colour from yellowish green to dark green. It is an innovative cost effective tool for real-time or crop-need based N management in rice, maize and wheat. It is used to rapidly monitor leaf N status at tillering to panicle initiation stage, guiding the application of fertilizer N accordingly. LCC can be employed to estimate fertilizer requirement during regular top-dressing at active tillering and panicle initiation, determined by measuring leave greenness. Leaf colour can be regularly monitored and fertilizer N applied when leaves become more yellowish-green than the critical threshold value indicated on the LCC. In both the cases, amount of N applied will be less if the crop leaf colour is greener and more if it is yellower (LCC, 2015). Bijay-Singh *et al.* (2002) reported 12–25 percent less N requirement through use of SPAD meter or LCC with no loss in yield in a rice-wheat system. SPAD meter-based SSNM increased the partial factor productivity of N in rice by 48 percent over the farmer's practice in China (Yao *et al.*, 2012).

GreenSeeker (GS) is a variable rate application and mapping equipment designed for use throughout a growing season. Here, crop vigour, measured as normalized difference vegetative index (NDVI), is used as the basis for N prescription rates. In this approach, NDVI is calculated as the difference between the reflectance in the near infrared and visible red divided by sum of both (Eq. 7.1):

$$NDVI = \frac{\rho NIR - \rho red}{\rho NIR + \rho red} \tag{7.1}$$

where ρNIR represents the fraction of emitted near-infrared radiation returned from the sensed area and ρred represents the fraction of emitted red radiation from the sensed area. It has active sensors (has its own light source), which allows measurements to be taken during day or night, nullifying the effect of atmospheric interference (Crain *et al.*, 2012). The device uses the software to calculate NDVI and ratio vegetation index {(RVI), NIR/Red} directly and generates sensor readings. However, adoption of the original GS was patchy, mainly because of its size (about 5 kg) and price. As a cost-effective alternative to original GS, a handheld version has been developed by Trimble's Agriculture Division. The hand-held GS works with same principle as original GS. The greatest advantages of hand-held GS are its small size, light weight, ease of use, and lower cost, which makes it affordable for small farms in both developing and developed countries. For South Asia, the algorithm for GreenSeeker-based precision nitrogen application in rice and wheat were developed and validated through collaborative research of Punjab Agriculture University (PAU),

International Maize and Wheat Improvement Centre (CIMMYT) and Indian Council of Agriculture Research (ICAR) (Bijay-Singh *et al.*, 2011, 2015). In these validation trials, GreenSeeker sensor-based N management resulted in similar (in rice) to higher yield (in wheat) with reduced N rates thereby increasing N use efficiency. In China, GS-based precision nutrient management increased partial factor productivity of N in rice by 65 percent over the farmer's practice (Yao *et al.*, 2012).

SPAD meter and LCC guide fertilizer N management based only on colour of the leaves, without considering plant biomass. An optical sensor has the distinct advantage that it can work out fertilizer N requirement based on expected yields as well as achievable greenness of the leaves.

Decision support systems (DSS) for precision nutrient management

As stated earlier, SSNM optimizes the supply of nutrients over space and time to match the crop requirement. However, SSNM is a knowledge intensive concept, which limits its on-farm application at a large scale. Farmers and their advisors need a decision support tool that allows easy implementation of SSNM at farm level. Computer-based systems can be useful for analysing complexity of nutrient dynamics in soil and providing recommendation for nutrient management specific for the site. Decision support systems are nowadays progressively being used to facilitate application of improved nutrient management practices in farmers' fields. Such tools have become increasingly important in geographies where blanket fertilizer recommendations prevail. These tools provide small-scale rice, wheat and maize farmers with crop and nutrient management advice customized to their farming conditions and needs.

For example, International Plant Nutrition Institute (IPNI) in collaboration with CIMMYT and other partners have developed the *Nutrient Expert®* (NE) for wheat and hybrid maize for smallholder farmers of South Asia (http://software. ipni.net). These recently developed decision support tools synthesized the on-farm research data into a simple delivery system that enables farmers to rapidly implement SSNM for their individual fields (Pampolino *et al.*, 2012). The tool estimates the attainable yield for a farmer's field based on growing conditions; determines the nutrient balance in the cropping system based on yield and fertilizer/manure applied in the previous crop; and combines such information with the expected N, P and K response in the concerned field to generate a location specific nutrient recommendation for wheat. It utilizes information provided by a farmer or a local expert to suggest a meaningful yield goal for a given location and formulates a fertilizer management strategy required to attain the yield goal. The software also does a simple profit analysis, comparing costs and benefits between the farmer's current practice and the recommended alternative improved practice.

Satyanarayana *et al.* (2012) evaluated NE in CA and CT maize during the summer and winter seasons. Nutrient recommendations from NE maize were tested against farmers' practice (FP) and state recommendation (SR) during

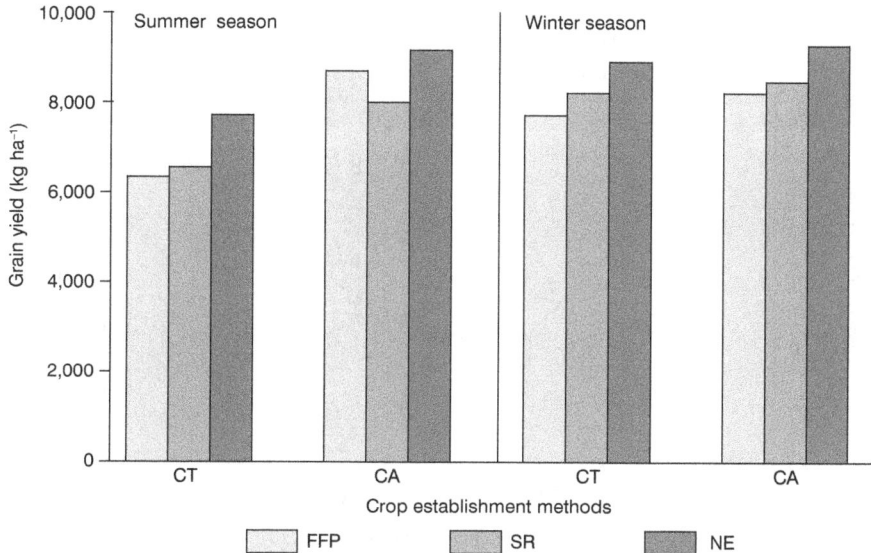

Figure 7.10 Nutrient Expert® improved grain yield in CA and CT maize in Southern India. FFP = farmers fertilizer practice, SR = state recommendation, NE = nutrient expert, CA = conservation agriculture, CT = conventional tillage

both the growing seasons. Across the seasons, NE recorded higher grain yield in CA (9.3 ton/ha) in comparison to CT (8.4 ton/ha). The NE maize tool was able to capture the inherent differences between conventional and CA practices of crop management, and NE-based fertilizer recommendations performed better than FP and SR for maize (Figure 7.10).

Based on two years of study to evaluate NE over farmers' fertilization practice (FFP), Bhende and Kumar (2014) recorded significantly higher wheat yields from NE-based nutrient management (5.2 and 3.8 ton/ha in NE and FFP, respectively). In this study, the gross return over fertilizer cost was higher by about USD 300 per ha in NE than in FFP.

Combined use of Nutrient Expert® and GreenSeeker™ minimized the environmental footprint of wheat production under a NT system in north-western India. The estimated total global warming potential (GWP) per tonne of wheat grain production and per dollar of net return, was lower for NE-based strategies than other nutrient management strategies (Sapkota *et al.* 2014, Figure 7.11).

Similarly, the International Rice Research Institute (IRRI), together with its partners, has developed another DSS called Crop Manager for Rice (released in Philippines and Bangladesh and under evaluation in different states of India), Maize and Rice-Wheat System (http://cropmanager.irri.org/home), which can also be used by extension workers, crop advisors and service providers to give advice to farmers specific to their growing condition. Besides nutrient application, crop manager also provides general crop management based on local growing conditions.

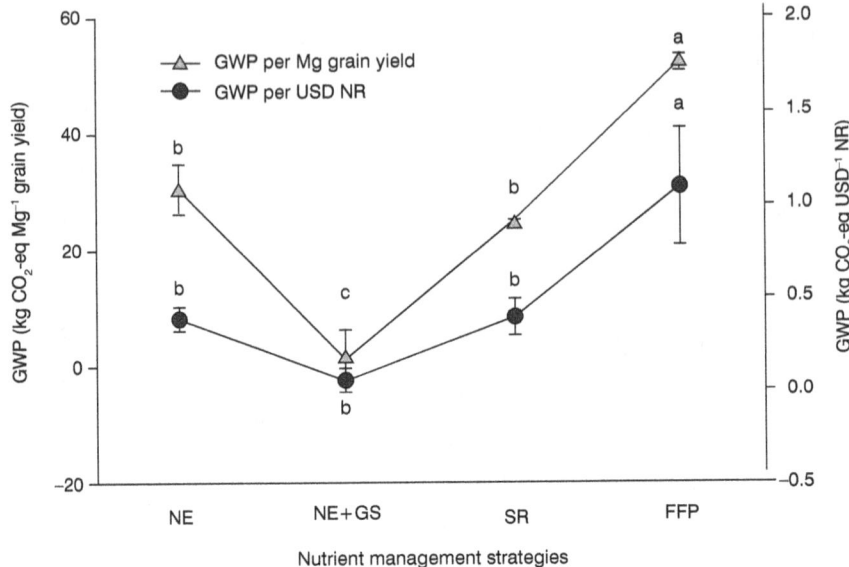

Figure 7.11 Total global warming potential (GWP) per Mg grain yield and per US$ net return (NR) under different nutrient management strategies in NT wheat production systems in Haryana, India. Means within category bearing same lowercase letters are significantly different from each other based on LSD ($\alpha=0.05$). Vertical bars are the standard error of the means
NE=nutrient expert, NE+GS=nutrient expert followed by GreenSeeker, SR=state recommendation, FFP=farmers fertilizer practice

Source: Authors' own analysis

Mechanization for precision N Application

Applying nutrients at the right place (both in horizontal and vertical dimension) in the soil ensures that plant roots can absorb enough of each nutrient at all times during the growing season. Placement near the seed-row may increase access of crops to the nutrient early in the growing season and provide a 'starter' effect that improves early-season growth. This is particularly important for less mobile nutrients such as P and K. Surface application of fertilizer under CA leads to more loss of nutrients resulting in poor nutrient use efficiency and environmental pollution. Therefore, proper placement of fertilizer is very crucial to increase NUE in a CA system. With development of Turbo-happy seeder, farmers are now able to drill fertilizer P and K directly through the residues at appropriate depth (Sidhu *et al.*, 2007). In this method, the fertilizer placement is done in the same row of the crop and just 1–2 cm below the seed placement. Generally 30 percent N is drilled at the time of planting and remaining recommended N is applied through broadcasting urea in moist soil, which is vulnerable to volatilization loss. NH_3 volatilization from urea can be as high as 42 percent (Katyal *et al.*, 1987) and could be more under NT (Patra *et al.*, 2004). Application of urea just before irrigation reduced its losses from 42 percent to 15 percent (Katyal *et al.*, 1987). Drilling of N in standing crop can precisely place the N near root zone thereby reducing the volatilization losses. CIMMYT-India and Borlaug Institute for South Asia (BISA) in close collaboration with Indian NARS and manufacturers have developed the machinery that can be used for drilling fertilizer N in NT and surface residue conditions either during planting or post-emergence application in standing crops. With these developments, proper placement of fertilizer N has been made possible also during first and second top dressing. Application in a localized band allows for more efficient use of N and as a result, lower application rates can be used than would be needed if the urea was broadcast on the surface and not incorporated. This is because subsurface drilling reduces the potential for ammonia gas volatilization.

It was found that in a RW system, fertilizer efficiency is increased by 10–15 percent due to better placement of fertilizer with the seed drill compared with broadcasting in the traditional system (Hobbs and Gupta, 2004). In a research trial conducted in loamy soils of Bihar and black cotton soils of central India (Madhya Pradesh) on N drilling in standing wheat crop, it was observed that the urea drilling before irrigation is better under black cotton soils and after irrigation in loamy soils of Bihar. In the experiment to compare different rate and method of N application in maize-wheat system in Bihar, we recorded higher grain yield and partial factor productivity of N in the side-drill method than in the broadcast application method (Figure 7.12). Drilling of half N at planting of wheat has been shown to perform better than broadcasting (Sarkar *et al.*, 1991).

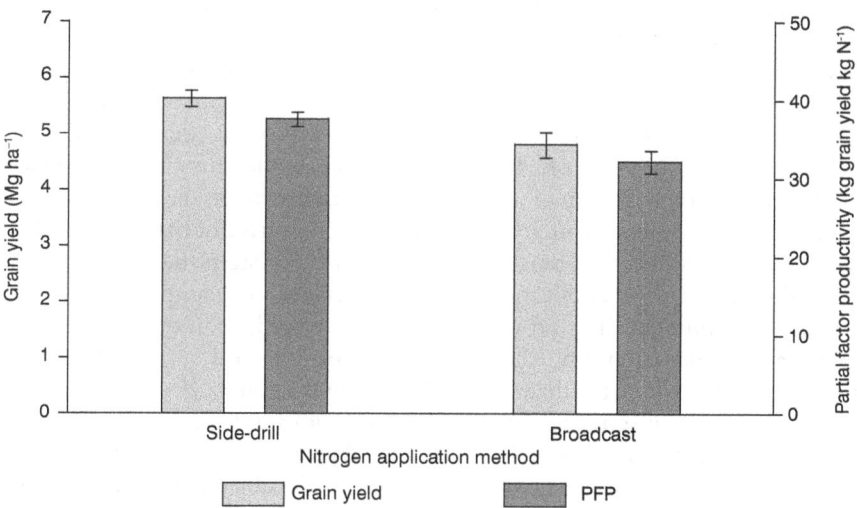

Figure 7.12 Wheat grain yield and partial factor productivity (PFP) of N as affected by two fertilizer application methods in the maize-wheat systems

Source: Authors' own analysis

Kumar *et al.* (2010) found that placement of N in between rows of wheat *vis-à-vis* broadcast resulted in significant yield increase. In general, the broadcasting of fertilizer nutrients offered the plant roots to become surface feeder, whereas drilling facilitate roots to grow deeper. Deeper root systems efficiently forage nutrients available in the deeper layer, which reduce the leaching losses of the nutrients and deeper root system reduced crop lodging.

Conclusions and way forward

In the post-GR era, the natural resource base has been under progressive pressure and deteriorated over time due to exploitative farming. This has led to a significant slowdown in the rate of yield growth of major cereals in SA, leading to concerns for food security and economic growth. In response, CA is increasingly being advocated as a management strategy for saving production resources, improving farm profitability and soil health on a sustainable basis. All components of CA i.e. no tillage, residue retention/organic soil cover and diverse crop rotations substantially influence nutrient dynamics in soil, thereby requiring paradigm shift in nutrient management. For example, increased soil organic matter content and soil porosity, increased biological nitrogen fixation by legumes in rotation, or exploitation of the deeper soil layers through crops with deep and dense root systems, have a significant bearing on nutrient dynamics across the soil profile. Efficient nutrient management under CA systems involves maintaining the nutrient levels in soil, considering the plant extraction and losses/addition resulting from various soil processes induced by CA systems. Because CA systems involve a diverse crop mixture including legumes, nutrients and their cycles must be managed at cropping system level rather than at single crop level.

Research findings show that in CA systems, nutrient requirements are lower, nutrient efficiencies are higher and risks of environment pollution are lower. Few researchers claim that CA enhances immobilization, denitrification or leaching of applied N, requiring higher rate. However, the immobilization phase in CA is transitory. Build-up of a larger pool of readily mineralizable organic N eventually requires less N in a CA system than in conventional systems. The presence of undisturbed system and residue in a CA system means the timing and method of fertilizer application is very important. For example, sub-surface banding of P and K in the seed row, but below the seed, promotes deeper root growth and avoids stranding these nutrients near the soil surface. Similarly, precise placement of N fertilizer through side banding in a CA system reduces immobilization (as it separates fertilizer and residue) and volatilization loss.

The yield and profitability of CA-based cereal production can be increased by adopting precision nutrient management systems. The objective is to capture the spatial and temporal variability in soil fertility and to provide recommendations to feed the crops with all the required nutrients based on the crop's need. The nutrients recommendation should be at optimal rates and times to achieve high profit for farmers, with high efficiency of nutrient use by crops across spatial and temporal scales, thereby preventing leakage of excess nutrient to the environment. Various tools, techniques and DSS are available to develop site-specific precision nutrient management plans for each field and dynamically fine-tune in-season nutrient management. Preliminary findings suggest that application of such tools help increase crop yield, use nutrients efficiently and profitably, and reduces the environmental footprint. More research is needed to fine-tune precision nutrient management in CA systems under different agroecological conditions to fill the knowledge gap. Such research findings should be used to create enabling policy and institutional frameworks for wide-scale adoption of CA as a sustainable agricultural practice.

Acknowledgement

The authors acknowledge the support of International Maize and Wheat Improvement Center (CIMMYT), International Plant Nutrition Institute (IPNI) and CGIAR research programme on Climate Change Agriculture and Food Security (CCAFS) for the review and synthesis of this chapter.

References

Bandel, V.A., Dzienia, S. and Stanford, G. (1980) Comparison of N fertilizers for no-till corn, *Agronomy Journal,* 72: 337–341.

Bhende, S.N. and Kumar, A., (2014) Nutrient Expert®-based fertiliser recommendation improved wheat yield and farm profitability in the Mewat. Better Crops, *South Asia,* 8: 21–23.

Bijay-Singh, Yadvinder-Singh, Ladha, J.K., Bronson, K.F., Balasubramanian, V., Singh, J. and Khind, C.S. (2002) Chlorophyll meter– and leaf color chart–based nitrogen management for rice and wheat in Northwestern India, *Agronomy Journal,* 94: 821–829.

Bijay-Singh, Sharma, R.K., Jaspreet-Kaur, Jat, M.L., Martin, K.L., Yadvinder-Singh, Varinderpal-Singh, Chandna, P., Chaudhary, O.P., Gupta, R.K., Thind, H.S., Jagmohan-Singh, Uppal, H.S., Khurana, H.S., Ajay-Kumar, Uppal, R.K., Vashistha, M., Raun, W.R. and Gupta, R. (2011) Assessment of the nitrogen management strategy using an optical sensor for irrigated wheat, *Agronomy for Sustainable Development,* 31: 589–603.

Bijay-Singh, Varinderpal-Singh, Purba, J., Sharma, R.K., Jat, M.L., Yadvinder-Singh, Thind, H.S., Gupta, R.K., Chaudhary, O.P., Chandna, P., Khurana, H.S., Kumar, A., Jagmohan-Singh, Uppal, H.S., Uppal, R.K., Vashistha, M. and Gupta, R. (2015) Site-specific fertilizer nitrogen management in irrigated transplanted rice (Oryza sativa) using an optical sensor, *Precision Agriculture,* 1: 1–21.

Biswas, P.P. and Sharma, P.D. (2008) A new approach for estimating fertiliser response ratio-the Indian scenario, *Indian Journal of Fertilisers,* 4: 59–62.

Brar, M.S., Bijay-Singh, Bansal, S.K. and Srinivasarao, C. (2011) Role of potassium nutrition in nitrogen use efficiency in cereals, *Optimizing Crop Nutrition,* International Potash Institute, E-ifc No. 29.

Buresh, R.J. and Witt, C. (2007) Site-specific nutrient management. In: IFIA (ed.), *Fertilizer Best Management Practices: General Principles, Strategy for Their Adoption and Voluntary Initiatives vs Regulations. Paper Presented at IFA International Workshop on Fertilizer*

Best Management Practices, 7–9 March 2007, Brussels, Belgium. International Fertilizer Industry Association.

Cai, D.X. and Wang, X.B. (1999) Conservation tillage systems for spring maize in the semi-humid to arid areas of China. In: Stott, D. E. *et al.* (eds.), *Simulating the Global farm*—Selected Papers from the 10th International Soil Conservation Organization Meeting, Purdue University. pp. 24–29.

Chhokar, R.S., Sharma, R.K., Jat, G.R., Pundir, A.K. and Gathala, M.K. (2007) Effect of tillage and herbicides on weeds and productivity of wheat under rice–wheat growing system, *Crop Protection*, 26: 1689–1696.

Crain, J., Ortiz-Monasterio, I. and Raun, B. (2012) Evaluation of a reduced cost active NDVI sensor for crop nutrient management, *Journal of Sensors*, 2012: 1–10.

Das, D.K., Maiti, D. and Pathak, H. (2009) Site-specific nutrient management in rice in Eastern India using a modeling approach, *Nutrient Cycling in Agroecosystems*, 83: 85–94.

Dobermann, A. (2006) Nitrogen use efficiency in cereal systems. In: 13th ASA Conference, 10–14 September, Perth, Western Australia.

Doran, J.W. (1980) Soil microbial and biochemical changes associated with reduced tillage, *Soil Science Society of America Journal*, 44: 765–771.

Dordas, C. (2015) Nutrient management perspectives in conservation agriculture. In: Farooq, M. and Siddique, K.. (eds.), *Conservation Agriculture*, Springer International Publishing, Cham.

FAOSTAT (2015) Faostat Common Database.Food and Agriculture Organization of the United Nations. Rome, Italy Available at: http://faostat3.fao.org/home/E[Accessed 30 July 2015]

Franzluebbers, A.J. and Hons, F.M. (1996) Soil-profile distribution of primary and secondary plant-available nutrients under conventional and no tillage, *Soil Tillage Research*, 39: 229–239.

Gathala, M.K., Ladha, J.K., Saharawat, Y.S., Kumar, V., Kumar, V. and Sharma, P.K. (2011) Effect of tillage and crop establishment methods on physical properties of a medium-textured soil under a seven-year rice-wheat rotation, *Soil Science Society of America Journal*, 75: 1851–1862.

Govaerts, B., Sayre, K.D., Lichter, K., Dendooven, L. and Deckers, J. (2007) Influence of permanent raised bed planting and residue management on physical and chemical soil quality in rain fed maize/wheat systems, *Plant Soil*, 291: 39–54.

Grant, C.A. and Bailey, L.D. (1994) The effect of tillage and KCl addition on pH, conductance, NO_3-N, P, K and Cl distribution in the soil profile, *Canadian Journal of Soil Science*, 74: 307–314.

Hadas, A., Kautsky, L., Goek, M. and Kara, E.E. (2004) Rates of decomposition of plant residues and available nitrogen in soil, related to residue composition through simulation of carbon and nitrogen turnover, *Soil Biology and Biochemistry*, 36: 255–266.

Halvorson, A.D. and Del Grosso, S.J. (2012) Nitrogen source and placement effects on soil nitrous oxide emissions from no-till corn, *Journal of Environmental Quality*, 41: 1349–1360.

Havlin, J., Beaton, J.D., Tisdale, S.L. and Nelson, W.L. (2005) *Soil Fertility and Fertilizers: An Introduction to Nutrient Management*, Pearson Prentice Hall, Upper Saddle River, New Jersey, USA.

Hengsdijk, H. and Langeveld, J.W.A. (2009) *Yield Trends and Yield Gap Analysis of Major Crops in the World*, WOT Working Document 170, Wageningen, The Netherlands. WOT Working Document 170, Wegeningen, The Netherlands.

Hobbs, P.R. (2007) Conservation agriculture: what is it and why is it important for future sustainable food production?, *Journal of Agricultural Science,* 145(2): 127.

Hobbs, P.R. and Gupta, R. (2004) Problems and challenges of no-till farming for the rice-wheat systems of the Indo-gangetic plains in South Asia. In: Lal, R. *et al* (eds.), *Sustainable Agriculture and the International Rice-Wheat System,* Boca Raton, FL, CRC Press.

Hopper, W. (1993) Indian agriculture and fertilizer: An outsider's observations, Keynote Address to the FAI Seminar on Emerging Scenario in Fertilizer and Agriculture: Global Dimensions. Fertilizer Association of India, New Delhi.

International Plant Nutrition Institute (IPNI) (2012) *4R Plant Nutrition: A Manual for Improving the Management of Plant Nutrition,* North American version. International Plant Nutrition Institute, Norcross, GA, USA.

Ismail, I., Blevins, R.L. and Frye, W.W. (1994) Long-term no-tillage effects on soil properties and continuous corn yields, *Soil Science Society of America Journal,* 58: 193–198.

Jat, M.L., Gathala, M.K., Ladha, J.K., Saharawat, Y.S., Jat, A.S., Kumar, V., Sharma, S.K., Kumar, V. and Gupta, R. (2009) Evaluation of precision land leveling and double zero-till systems in the rice–wheat rotation: Water use, productivity, profitability and soil physical properties, *Soil Tillage Research,* 105: 112–121.

Jat, M.L., Saharawat, Y.S. and Gupta, R.K. (2011) Conservation agriculture in cereal systems of south Asia : Nutrient management perspectives, *Karnataka Journal of Agricultural Science,* 24: 100–105.

Jat, R.K., Sapkota, T.B., Singh, R.G., Jat, M.L., Kumar, M. and Gupta, R.K. (2014) Seven years of conservation agriculture in a rice–wheat rotation of Eastern Gangetic Plains of South Asia: Yield trends and economic profitability, *Field Crops Research,* 164: 199–210.

Jones, C., Chen, C., Allison, E. and Neill, K. (2007) Tillage effects on phosphorus availability, Western Nutrient Management Conference, Salt Lake City, UT.

Katyal, J.C., Bijay-Singh, Vlek, P.L.G. and Buresh, R.J. (1987) Efficient nitrogen use as affected by urea application and irrigation sequence, *Soil Science Society of America Journal,* 51: 366–370.

Kay, B.D. (1990) Rates of change of soil structure under different cropping systems, *Advances in Soil Science,* 12: 1–52.

Khurana, H.S., Bijay-Singh, Dobermann, A., Phillips, S.B., Sidhu, A.S. and Yadvinder-Singh, (2008) Site-Specific nutrient management performance in a rice-wheat cropping system, *Better Crops South Asia,* 92: 26–28.

Kienzler, K.M., Lamers, J.P.A., McDonald, A., Mirzabaev, A., Ibragimov, N., Egamberdiev, O., Ruzibaev, E. and Akramkhanov, A. (2012) Conservation agriculture in Central Asia—What do we know and where do we go from here?, *Field Crops Research,* 132: 95–105.

Kumar, M., Sheoran, P. and Yadav, A. (2010) Productivity potential of wheat (*Triticum aestivum*) in relation to different planting methods and nitrogen management strategies, *Indian Journal of Agricultural Science,* 80: 427–429.

Lafond, G.P., Loeppky, H. and Derksen, D.A. (1992) The effects of tillage systems and crop rotations on soil water conservation, seedling establishment and crop yield, *Canadian Journal of Plant Science,* 72: 103–115.

LCC (2015). Leaf Colour Chart (LCC) [WWW document] Online www. nitrogenparameters.com/lcc.html

Lobell, D.B., Cassman, K.G. and Field, C.B. (2009) Crop yield gaps: their importance, magnitudes, and causes, *Annual Reviews Environment and Resources,* 34: 179–204.

Lupwayi, N.Z., Clayton, G.W., O'Donovan, J.T., Harker, K.N., Turkington, T.K. and Soon, Y.K. (2006) Soil nutrient stratification and uptake by wheat after seven years of conventional and zero tillage in the Northern Grain belt of Canada, *Canadian Journal of Soil Science,* 86: 767–778.

Magen, H. and Nosov, V. (2008) Putting potassium in the picutre: achieving improved nitrogen use efficiency, *Bangladesh Journal of Agricutlure Environment,* 4: 115–127.

Majumdar, K., Jat, M.L., Pampolino, M., Satyanarayana, T., Dutta, S. and Kumar, A. (2013a) Nutrient management in wheat : current scenario, improved strategies and future research needs in India, *Journal of Wheat Research,* 4: 1–10.

Majumdar, K., Jat, M.L. and Shahi, V.B. (2012) Effect of spatial and temporal variability in cropping seasons and tillage practices on maize yield responses in eastern India, *Better Crops South Asia,* 6: 4–6.

Majumdar, K., Johnston, A.M., Dutta, S., Satyanarayana, T. and Roberts, T.L. (2013b) Fertilizer best management practices: Concept, global perspectives and application, *Indian Journal of Fertilisers,* 9: 14–31.

Malhi, S.S. and Nyborg, M. (1992) Placement of urea fertilizer under zero and conventional tillage for barley, *Soil Tillage Research,* 23: 193–197.

McMahon, M.A. and Thomas, G.W. (1976) Anion leaching in two Kentucky soils under conventional tillage and a killed-sod mulch, *Agronomy Journal,* 68: 437–442.

Moschler, W.W. and Martens, D.C. (1975) Nitrogen, phosphorus, and potassium requirements in no-tillage and conventionally tilled corn, *Soil Science Socociety of America Journal,* 39: 886–891.

Murrell, T.S. and Bruulsema, T.W. (2008) Principles of allocating funds across nutrients, *Better Crop,* 92: 3–5.

Pampolino, M.F., Witt, C., Pasuquin, J.M., Johnston, A. and Fisher, M.J. (2012) Development approach and evaluation of the Nutrient Expert software for nutrient management in cereal crops, *Computers and Electronics in Agriculture,* 88: 103–110.

Patra, A.K., Chhonkar, P.K. and Khan, M.A. (2004) Nitrogen loss and wheat (*Triticum aestivum* L.) yields in response to zero-tillage and sowing time in a semi-arid tropical environment, *Journal of Agronomy and Crop Science,* 190: 324–331.

Randall, G.W. and Hoeft, R.G. (1988) Placement methods for improved efficiency of P and K fertilizers: A review, *Journal of Production Agriculture,* 1: 70–79.

Rasmussen, P.E., Rickman, R.W. and Kleeper, B.L. (1997) Residue and fertility effects on yield of no-till wheat, *Agronomy Journal,* 89: 563–567.

Raun, W.R. and Johnson, G.V. (1999) Improving nitrogen use efficiency for cereal production, *Agronomy Journal,* 91: 357–363.

Rennie, D.A. and Heimo, M. (1984) Soil and fertilizer-N transformations under simulated zero till: Effect of temperature regimes, *Canadian Journal of Soil Science,* 64: 1–8.

Roberts, T.L. (2007) Right product, right rate, right time and right place... the foundation of best management practices for fertilizer. In: *IFA International Workshop on Fertilizer Best Management Practices (FBMPs),* 7–9 March 2007, Brussels, Belgium, International Fertilizer Industry Association (IFA), Paris, France.

Saharawat, Y.S., Singh, B., Malik, R.K., Ladha, J.K., Gathala, M., Jat, M.L. and Kumar, V. (2010) Evaluation of alternative tillage and crop establishment methods in a rice–wheat rotation in North Western IGP, *Field Crops Research,* 116: 260–267.

Sapkota, T.B., Majumdar, K., Jat, M.L., Kumar, A., Bishnoi, D.K., Mcdonald, A.J. and Pampolino, M. (2014) Precision nutrient management in conservation agriculture based wheat production of Northwest India : Profitability, nutrient use efficiency and environmental footprint, *Field Crops Research,* 155: 233–244.

Sarkar, M.C., Banerjee, N.K., Rana, D.S. and Uppal, K.S. (1991) Field measurements of ammonia volatilization losses of nitrogen from urea applied to wheat, *Fertility News*, 36: 25–28.

Satyanarayana, T., Majumdar, K., Pampolino, M., Johnston, A.M., Jat, M.L., Kuchanur, P., Sreelatha, D., Sekhar, J.C., Kumar, Y. and Maheswaran, R. (2012) Nutrient Expert®: A tool to optimise nutrient use and improve productivity of maize, *Better Crops South Asia*, 6: 18–21.

Schnürer, J., Clarholm, M., and Rosswall, T., (1985) Microbial biomass and activity in an agricultural soil with different organic matter contents, *Soil Biology and Biochemistry*, 17: 611–618.

Shaobing, P., Roland, J.B., Jianliang, H., Xuhua, Z., Yingbin, Z., Jianchang, Y., Guanghuo, W., Yuanying, L., Ruifa, H., Qiyuan, T., Kehui, C., Fusuo, Z. and Achim, D. (2010) Improving nitrogen fertilization in rice by site-specific N management,. *A review. Agronomy for Sustainable Development*, 30: 649–656.

Shapiro, C.A., Francis, D.D., Ferguson, R.B., Hergert, G., Shaver, T. and Wortmann, C. (2013) *Using a chlorophyll meter to improve N management*, Lincoln, NE Univ. Nebraska-Lincoln Extension Bulletin.

Sharma, A.R., Jat, M.L., Saharawat, Y.S., Singh V. P. and Singh, R. (2012) Conservation agriculture for improving productivity and resource-use efficiency: Prospects and research needs in Indian context, *Indian Journal of Agronomy*, 57: 131–140.

Sidhu, H.S., Humphreys, E., Dhillon, S.S., Blackwell, J. and Bector, V. (2007) The Happy Seeder enables direct drilling of wheat into rice stubble, *Animal Production Science*, 47: 844–854.

Singh, V.K., Dwivedi, B.S., Tiwari, K.N., Majumdar, K., Rani, M., Singh, S.K. and Timsina, J. (2014) Optimizing nutrient management strategies for rice–wheat system in the Indo-Gangetic Plains of India and adjacent region for higher productivity, nutrient use efficiency and profits, *Field Crops Research*, 164: 30–44.

Six, J., Conant, R.T., Paul, E.A. and Paustian, K. (2002) Stabilization mechanisms of soil organic matter: Implications for C-saturation of soils, *Plant Soil*, 241: 155–176.

Sommer, R., Thierfelder, C., Tittonell, P., Hove, L., Mureithi, J. and Mkomwa, S. (2014) Fertilizer use should not be a fourth principle to define conservation agriculture: Response to the opinion paper of Vanlauwe et al. (2014) A fourth principle is required to define conservation agriculture in sub-Saharan Africa', *Field Crops Research*, 169: 145–148.

Sprent, J.I. and Sprent, P. (1990) *Nitrogen Fixing Organisms: Pure And Applied Aspects*, Chapman and Hall Ltd., London.

Thomas, G.W., Blevins, R.L., Phillips, R.E. and McMahon, R.L. (1973) Effect of a killed sod mulch on nitrate movement and corn yield, *Agronomy Journal*, 65: 736–739.

Tomich, T.P., Kilby, P. and Johnston, B.F. (1995) *Transforming agrarian economies: Opportunities Seized, Opportunities Missed*, Ithaca, NJ, Cornell University Press..

Tyler, D.D. and Thomas, G.W. (1977) Lysimeter measurements of nitrate and chloride losses from soil under conventional and no-tillage corn, *Journal of Environmental Quality*, 6: 63–66.

Unger, P.W. (1991) Organic matter, nutrient, and pH distribution in no-and conventional-tillage semiarid soils, *Agronomy Journal*, 83: 186–189.

Vanlauwe, B., Wendt, J., Giller, K.E., Corbeels, M., Gerard, B. and Nolte, C. (2014) A fourth principle is required to define Conservation Agriculture in sub-Saharan Africa: The appropriate use of fertilizer to enhance crop productivity, *Field Crops Research*, 155: 10–13.

Verhulst, N., Govaerts, B., Verachtert, E., Mezzalama, M., Wall, P.C., Chocobar, A., Deckers, J. and Sayre, K.D. (2010) Conservation agriculture, improving soil quality for sustainable production systems?, In: Lal, R. *et al.* (eds.), *Advances in Soil Science: Food Security and Soil Quality*, CRC Press, Boca Raton, FL, USA.

Wall, P., Thierfelder, C., Ngwira, A., Govaerts, B., Nyagumbo, I. and Baudron, F. (2013) Conservation agriculture in eastern and southern Africa, In: Ram, A. J. *et al.* (eds.), *Conservation Agriculture: Global Prospects and Challenges,* CABI, Wallingford Oxfordshire.

Yadvinder-Singh, Singh, M., Sidhu, H.S., Humphreys, E., Thind, H.S., Jat, M.L., Blackwell, J. and Singh, V. (2015) Nitrogen management for zero till wheat with surface retention of rice residues in north-west India, *Field Crops Research,* 184: 183–191.

Yao, Y., Miao, Y., Huang, S., Gao, L., Ma, X., Zhao, G., Jiang, R., Chen, X., Zhang, F., Yu, K., Gnyp, M.L., Bareth, G., Liu, C., Zhao, L., Yang, W. and Zhu, H. (2012) Active canopy sensor-based precision N management strategy for rice, *Agronomy for Sustainable Development*, 32: 925–933.

Zagal, E. and Persson, J. (1994) Immobilization and remineralization of nitrate during glucose decomposition at four rates of nitrogen addition, *Soil Biology and Biochemistry,* 26: 1313–1321.

8 Integrated soil management practices

Adaptation and mitigation to climate change

Mehreteab Tesfai, Irene Moed, Inga Greipsland, Thorsten Huber and Niek van Duivenbooden

Introduction

The world's population is estimated to reach 9.2 billion by 2050. Over this period, agricultural production must increase by 70 percent to keep pace with increasing food demand (FAO, 2009; FAO, 2013). More than 95 percent of global food comes from land, so an adequate global food supply depends predominantly on the continued availability of productive soils. However, quality soils are not guaranteed without additional efforts (van Beek *et al.*, 2014). In addition, on-going climate change has increased alterations of weather patterns, affecting soil moisture availability and bringing associated consequences for diseases and pest incidences (e.g. Ben Mohamed *et al.*, 2002). By 2050, climate change is expected to negatively impact at least 22 percent of the cultivated areas of the world's important crops, notably rice and wheat (Campbell *et al.*, 2011), and increase global warming. Global warming is caused by increased atmospheric concentrations of Greenhouse Gases (GHGs), mainly carbon dioxide (CO_2), methane (CH_4) and nitrous oxide (N_2O).

Agriculture is one of the largest contributors to GHG emissions, derived from livestock farming (e.g. enteric fermentation and manure management) and emissions from agricultural soils (i.e. application of excessive N fertilizers and decomposition of organic material). On average, agriculture accounts for about 14 percent of the total global GHG emissions (IPCC, 2007). Contributing factors are poor land management by humans, such as over-cultivation, over-grazing and deforestation (EC-JRC, 2003), draining of peat lands and burning of rainforests. Figure 8.1 shows the cause of climate change and its vicious cycle.

Being part of the problem, agriculture is also part of the solution to climate change impacts. If agricultural soils are properly managed and effective policies are in place, they have the potential to sequester large amounts of carbon from the atmosphere and store it in the soils, thereby mitigating CH_4 and CO_2 emissions (Smith *et al.*, 2007).

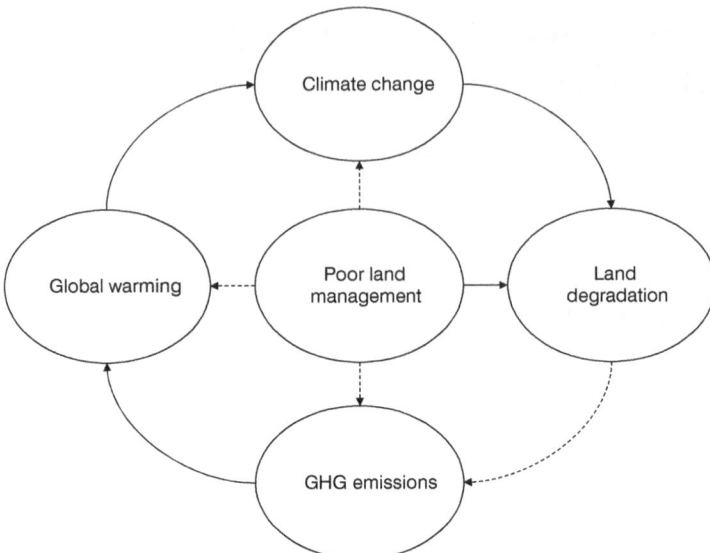

Figure 8.1 The vicious cycle of climate change from human-induced perspectives
Solid arrows indicate direct adverse effects and dotted arrows indicate indirect adverse effects

Source: Authors' own analysis

Soil and climate change

Soils are critical to food security, but are too slowly formed and too quickly lost. Since climatic variables such as rainfall and temperature play an important role in the formation and/or destruction of soils (Brady and Weil, 2007), we need to better understand the impact of climate change on soil processes and properties, and how soil management techniques contribute to climate change (CC) adaptations/resilience, reduction in GHG emissions and increase in agricultural productivity. Soil resilience refers to the magnitude of disturbance (caused by climate change in this case) that can be absorbed or accommodated before the system changes its structure (Seybold *et al.*, 1999). The soil properties and functions that are closely related to soil resilience and mostly affected by CC are soil structure and texture, organic matter content, nutrient dynamics, soil organisms, soil pH and cation exchange capacity (FAO, 2013). For each of these key soil properties, a brief description in relation to CC adaptation/resilience and mitigation potential is provided in Table 8.1.

Soils should be adequately monitored, protected and maintained in order to ensure that the above-mentioned crucial soil properties and functions remain in place. A range of soil management practices, including soil fertility improvements and soil erosion control, have been developed and applied by farmers and researchers in different parts of the world with a goal to achieve sustainable food security. However, a single soil management practice may solve part of the problem

Table 8.1 Key soil properties related to CC adaptation and mitigation potential

Property	CC adaptation and mitigation potential
Soil texture and structure	If soil structure is damaged by tillage operations during wet-dry conditions, the soil texture/structure functions to retain water and nutrients are hampered, and consequently the resilience to CC is reduced (FAO, 2013).
Soil organic matter (SOM)	SOM reduces evaporation, improves infiltration, increases soil aggregate stability; provides nutrients upon mineralization; and reduces surface runoff. These SOM functions increase the required adaptation potential to CC, build up resilience to water stress and help mitigate GHG emissions (Giller *et al.*, 2009; FAO, 2013; Powlson *et al.*, 2011).
Soil nutrients and organisms	Need-based N management reduces losses of N and increases efficiency. Organisms living in soils (like earthworms) make nutrients available for uptake by plants through active involvement in the transformation and decomposition of nutrients. Soil organisms enhance adaptation to CC impacts and increase crop production (Varinderpal-Singh *et al.*, 2012; Ali *et al.*, 2015).
Soil pH	Higher temperatures and evaporation rates increase accumulation of salts in the topsoil and thereby raise the pH levels. Under such conditions, salinity is intensified, and water and nutrient availability is hampered, and soil structure is degraded. On the other hand, substantial increase in rainfall, accompanied by leaching, could lead to increase soil acidity. In consequence, the activity of microorganisms and nutrient release to crop plants is reduced, and does soil resilience to CC (FAO, 2013).

Source: Authors' own compilation

of CC impacts and food security, but not the whole problem. Understanding the status and condition of the soil properties is fundamental to making decisions to adopt or not to adopt soil management practices that contribute to climate smart agriculture (CSA). CSA is based on the simultaneous achievements of three principal objectives: i) adaptation to CC; ii) mitigation of GHGs emissions; and iii) increased agricultural productivity (FAO, 2009). There is a fast growing initiative to integrate the various soil management practices into CSA objectives.

The need for integrated soil management

There is no universal definition adopted for Integrated Soil Management (ISM). It all depends on the particular soil problem in the area. Therefore, there could be ISM for soil fertility improvements, ISM for soil erosion control, and so forth. According to Richardson *et al.* (2014), ISM for soil fertility improvements is:

> a set of soil fertility management practices that entails the use of fertilizer, organic inputs, and improved germplasm combined with the knowledge on how to adapt these practices to local conditions, aiming at maximising

agronomic use efficiency of the applied nutrients and improving crop productivity.

ISM offers the ability to sustainably intensify production and maximize social, economic and environmental benefits. ISM must become the cornerstone of sustainable land management (The Montpellier Panel, 2014).

In this chapter, we define the term ISM under the CSA context as combined practices of soil management that increase adaptation and/or resilience to CC impacts, contribute to reduce GHG emissions and at the same time increase agricultural productivity and incomes in a sustainable way. Not all soil management practices fulfil the three interwoven objectives of CSA. Therefore, a combination of measures of soil management options that complement each other are needed to achieve the said CSA objectives.

For instance, need-based fertilizer applications (based on the congruence between nutrient supply and crop demand) is site specific and it contributes to reduced GHG emissions by avoiding excessive application of chemical fertilizers (Bijay-Singh *et al.*, 2002). Such practice may increase crop production, but does not necessarily build up soil resilience to water stress (e.g. during drought). However, by integrating need-based fertilizer applications with soil and water conservation measures (e.g. stone bunds), we can then build up the required resilience, mitigate GHG emissions and increase food production. Research and pilot projects by the FAO in Ethiopia, Kenya and Tanzania has shown an improved resilience to the changing climate by increasing soil health/fertility, increasing water conservation/use efficiency, and diversifying livelihoods/ farming systems (FAO, 2014). There are, however, only a few published reports describing the contribution of soil management practices to CSA principles. The objectives of this chapter are to review soil management practices that contribute to CSA principles by synthesizing the available information from literature, discussing successful examples from different countries, and suggesting options for wide-spread implementations and adoption of ISM that operationalize CSA principles. In this chapter, the term integrated soil management practices or technologies are used interchangeably.

Climate smart soil management practices

In this section, the role of six 'soil management practices' are described in relation to the CSA principles (i.e. adaptation, mitigation and productivity). Research results from India, Sub-Saharan Africa (SSA) and Burundi on some of these practices are discussed.

The experience from India

Need-based fertilizer applications

Need-based fertilizer application is a system where the amount and type of fertilizer applications to the soils and/or plants is based on soil testing. A leaf colour chart (LCC: Varinderpal-Singh *et al.*, 2012), chlorophyll meter (Ali *et al.*, 2015) are used, or generic yield/nutrient uptake and nutrient applied/nutrient uptake ratio is measured (van Duivenbooden *et al.,* 1996). Need-based fertilizer application is getting more and more attention, as the single blueprint fertilizer doses are not effective (van Beek *et al.*, 2014).

In India, need-based fertilizer management for N in rice via soil testing and LCC have been studied by many researchers (Bijay-Singh *et al.,* 2002; Varinderpal-Singh *et al.,* 2012; Ali *et al.,* 2015). They found that need-based N management through a LCC or chlorophyll meter reduces N requirement of rice from 12.5 to 25 percent, with no loss in yield and higher N use efficiency. This fertilizer N management strategy was also tested by the ClimaAdapt project in farmers' fields in the Ponnaiyar basin, in Tamil Nadu state, South India. Most of the farmers in the project area were following blanket fertilizer application without prior soil testing. Thanks to the project interventions, need-based fertilizer application through soil testing is now being implemented. It has reduced the cost of chemical N fertilizer, while enhancing N use efficiency and crop yield (see Table 8.2). Moreover, LCC guided N management helped in reducing bacterial sheath blight and red stripe diseases (Varinderpal-Singh *et al.,* 2010).

However, the use of chlorophyll meters by smallholder farmers is limited due to the high cost (Ali *et al.*, 2015) despite being a simple, quick and non-destructive method for estimating the N content of rice leaves. Soil testing and LCC require frequent observations and more labour (Varinderpal-Singh *et al.*, 2012).

Table 8.2 Effects of need-based fertilizer applications in relation to CSA principles and corresponding literature

Need-based fertilizer	*CSA addressed*	*References*
• Stored extra soil water of 1392 L/ha	Adaptation	Lakshmanan *et al.* (2015a)
• Saved 20–40 kg N/ha with no yield loss	Adaptation	Bijay-Singh *et al.* (2002)
• Accumulated about 1280 kg C/ha/season	Mitigation	Lakshmanan *et al.* (2015a)
• Reduced fertilizer N related emission and pollutions	Mitigation	Varinderpal-Singh *et al.* (2010)
• Increased yield by 410 to 640 kg/ha in north-western India	Productivity	Bijay-Singh *et al.* (2002)
• Increased yield by 22 to 23%	Productivity	Lakshmanan *et al.* (2015b)
• More extra crop yield (on average) 0.75 ton/ha in south Asia	Productivity	Varinderpal-Singh *et al.* (2010)
• Additional profit US $50 to 140/ ha per season	Productivity	Lakshmanan *et al.* (2015b)

Source: Authors' own compilation

Organic fertilizer applications: green manure, azolla, etc.

There are various forms of organic fertilizers, such as farmyard manure, vermicompost, green manure and *azolla*. These can be applied to soils by farmers, depending on the availability of the resources. One of the major disadvantages of organic fertilizers is their low nutrient content and high variability compared to mineral fertilizers. For example, most organic resources used on farms contain 0.5 to 5 percent N, or 5 to 25 g N/kg, compared with >100 to 460 g N/kg, contained in mineral fertilizer. Organic inputs are therefore required in large quantities if they are to supply significant amounts of nutrients to growing crops (Bationo *et al.,* 2012).

Organic fertilizers improve soil quality by increasing SOM content of soils. This enhances soil carbon sequestration (carbon sinks) and slows the return of stored carbon to the atmosphere. The return of stored carbon is responsible for most of the GHG mitigation potential in agriculture (Thangarajan *et al.,* 2013). However, increased soil organic carbon (SOC) storage (by adding various types of organic materials to soil) does not always imply contribution to CC mitigation. Powlson *et al.* (2011) argue that not all practices that increase SOC content represent a transfer of additional C from atmosphere to land; some are simply a movement of C from one pool in the biosphere to another, with no implications for CC, either positive or negative. They suggest that increases in SOC should normally be termed 'accumulation' and that 'sequestration' be reserved for situations where there is an additional transfer of C from the atmosphere and thus a genuine contribution to CC mitigation.

1   *Green manure (GM):* Table 8.3 gives an example of how changes in SOC (made by applying green manures) modifies the soil's capacity to store water in the Ponnaiyar basin, India. The low organic matter content of the majority of fields (Table 8.3) in the ClimaAdapt project villages triggered the need to apply green manure and green leaf manure. The cultivation of various green manure crops like *Sesbania* has been encouraged through ClimaAdapt since the 2012 *rabi* season[1] in the Ponnaiyar basin, India. The green manure seeds were distributed to the farmers, who ploughed back the green manure crops in their fields before the cultivation of rice. This has resulted in a marginal increase of organic matter content in those fields where green manure was applied (Table 8.3).

The low organic matter content of the majority of fields (Table 8.3) in the ClimaAdapt project villages triggered the need to apply green manure and green leaf manure. After the application, an increase of on average 1392 litres of extra water could be stored per ha in the top 30 cm of soil for every 0.01 percent SOC increase, in addition to about 1.28 t C/ha stock in the soil. Some of the disadvantages of green manure application include extra labour cost and time for growing an extra crop; potential problems with insect pests and diseases due to improper decomposition and lack of seed for GM establishment.

Table 8.3 Measured SOM content before and after applying green manure in the Ponnaiyar basin, India and estimated capacity of the soils to store water and sequester C with a bulk density of 1.2 g/cm^3 (SOC $\cong$ SOM/1.72)

Farmers' field	SOM before (%)	SOM after (%)	Change in SOM (%)	Change in SOC (%)	Extra SOC (g)	Extra soil water (litres/ha)	CO$_2$–C stock (t/ha)
Chinnaanaikaraipatty	0.89	0.91	0.02	0.012	42	1,670.4	1.53
Kalathupatty	1.32	1.34	0.02	0.012	42	1,670.4	1.53
Manpathai	1.02	1.04	0.02	0.012	42	1,670.4	1.53
Periyaanaikaraipatty	0.82	0.83	0.01	0.006	21	835.2	0.77
Ponnaiyar Dam	0.87	0.88	0.01	0.006	21	835.2	0.77
Sakkampatty	1.02	1.04	0.02	0.012	42	1,670.4	1.53
Mean	0.99	1.01	0.02	0.01	350	1,392.0	1.28

Source: Results based on field data from ClimaAdapt Project (http://climaadapt.org/)

The impact of green manure application on rice yield was evaluated in the Kalingarayan basin (KB) and the Ponnaiyar basin (PB) during 2013 *rabi* season. In KB, the rice grain and straw yields from fields with GM application was 6882 and 2255 kg/ha, respectively. Similarly in PB, the grain and straw yields from rice fields with GM application was 5507 and 2019 kg/ha, respectively. This means that the rice grain and straw yields were 23 percent (22) and 9 percent (10) higher than fields without GM in KB and (PB), respectively (Table 8.4).

2   *Azolla* is an aquatic fern that can be used as green manure, fodder for animals and biofertilizer. It grows fast and fixes nitrogen at substantial rates, and decomposes fast so that nutrients are made available rapidly (Raja *et al.*, 2012). *Azolla* supplies plant nutrients like P, K, S, Zn, Fe, Mb and other micronutrients, and it may rendered on average about 1.5 kg/m^2 in a week (Mishra, 2012). Because *azolla* grows in water, it is especially important in rice production.

Bio-fertilizer applications

Biofertilizers are products that contain living cells of different types of microorganisms. When applied to seeds, plant surface or the soil, they colonize the rhizosphere or the interior of the plant and promote growth by increasing the supply or availability of primary nutrients to the plant (Mohammadi and Sohrabi, 2012). Organisms that are most promising to use as biofertilizers are nitrogen fixers like *rhizobium, azospirillum, azobacter,* and blue green algae, and phosphate solubilizers like *psedomonas* and *bacillus* (Mishra *et al.*, 2012). *Rhizobium* fixes nitrogen in association with legumes, while *azospirillium* and *azobacter* are free-living bacteria (Mishra *et al.*, 2012). In the soil, there is normally a huge

Table 8.4 Effects of green manure applications in relation to CSA principles and corresponding literature

Green manure	CSA addressed	References
• Stored extra 1,392 litre/ha soil water	Adaptation	Lakshmanan *et al.* (2015a)
• Less need for chemical fertilizers	Adaptation	Mahajan and Gupta (2009) Cherr *et al.* (2006)
• Increased SOM up to 1% of total soil mass	Adaptation	Mahajan and Gupta (2009) Cherr *et al.* (2006)
• Accumulated 1,280 kg C/ha/ season	Mitigation	Lakshmanan *et al.* (2015a)
• Increased yield by 22–23%	Productivity	Lakshmanan *et al.* (2015b)
• Double the yield of a subsequent cereal crop	Productivity	Giller *et al.* (2009)

Source: Authors' own compilation

amount of unavailable inorganic P. Phosphate-solubilizing bacteria can convert these into soluble forms that can easily be taken up by plants.

Application of biofertilizers like *azospirillum* and cyanobacteria are promoted in the ClimaAdapt project villages. By combining soil test-based fertilizer application with biofertilizer, a saving of 30–35 percent of urea and yield increase between 194 to 271 kg/ha were recorded during *rabi* season 2013. Similar results have been found by Mohammadi and Sohrabi (2012), where plots treated with phosphorus solubilizing bacteria increased chickpea yield by more than 50 percent compared to the control. The plots with biofertilizers applied registered not only higher rice yields, but emitted less methane compared to the conventional (i.e. flooded) cultivated soils (Figure 8.2). The mean methane flux in the control plots (i.e. T7) was 48 mg CH_4-$C/m^2/d$ whilst, in cyanobacteria inoculated plots (i.e. T4 & T5) the methane flux was reduced to 26 mg CH_4-$C/m^2/d$.

The application of phosphobacterium (*bacillus megatherium*) is also promoted in all project villages to increase available P content in the soils. Most of the soils in the project site contain low to medium available phosphorus (i.e. 11 to 20 mg/kg). The available P content was >25 mg/kg in only less than 20 percent of the fields (Lakshmanan *et al.*, 2015a).

The present case study reiterates that biofertilization of rice fields is a potential climate smart agriculture strategy not only due to its promising roles in yield enhancement and nitrogen fixation, which reduces amount and cost of chemical fertilizers, but potentially also in methane flux reduction (Figure 8.2 and Table 8.5). However, there are some limitations in applying biofertilizers. This includes lack of

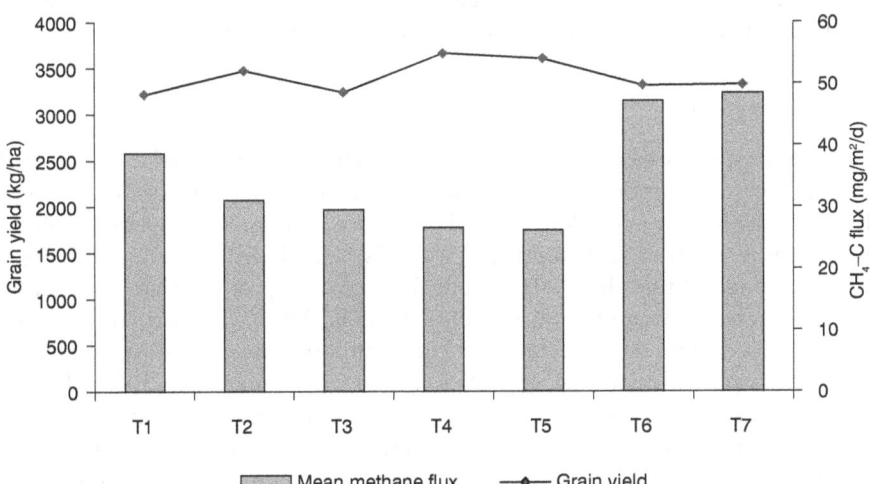

Figure 8.2 Effects of bioinoculants on methane flux and grain yield under conventional rice cultivation in Tamil Nadu, India

Note that T1: *Anabaena* sp. T2: *Nostoc* sp. T3: *Westiellopsis* sp. T4: *Anabaena* sp. + *Nostoc* sp. + *Westiellopsis* sp. T5: *Anabaena* sp. + *Nostoc* sp. + *Westiellopsis* sp. + gypsum T6: without cyanobacteria + with gypsum T7: without cyanobacteria + without gypsum

Source: Adapted from Lakshmanan *et al.* (2015a)

Table 8.5 Effects of biofertilizer applications in relation to CSA principles and corresponding literature

Biofertilizers	CSA addressed	References
• *Pseudomonas* and *azobacter* prevent the growth and establishment of crop pathogens	Adaptation	Mahajan and Gupta (2009) Harman (2006) Mishra *et al.* (2012)
• Improves soil fertility	Adaptation	
• Reduces the need for chemical fertilizers and thus saves money	Adaptation	
• Reduces CH4-C emissions by 50%	Mitigation	Lakshmanan *et al.* (2015a)
• Increases yield by 194 to 271 kg/ha	Productivity	Mahajan and Gupta (2009) Mohammadi and Sohrabi (2012) Lakshmanan *et al.* (2015a)

Source: Authors' own compilation

awareness, trained manpower and low availability of biofertilizers, and high storage and transport cost. In areas with low access to chemical fertilizers, biofertilizers could be a viable option as a fertilizer supplement. More research is needed on crops' response to biofertilizers, storage and transport without destroying the living organisms at a mass scale production (Mahajan and Gupta, 2009).

Applications of biochar

Biochar is a stable, carbon-rich form of charcoal that can be applied to agricultural land as part of agronomic or environmental management. It can be produced by pyrolysis of ligno-cellulosic materials, where biomass such as crop stubble, wood chips, manure and municipal waste is heated with little or no oxygen (Sparkes and Stoutjesdijk, 2011). The high porosity of biochar contributes towards increased water retention (Lehmann *et al.,* 2003). The particulate nature of biochar combined with a specific chemical structure enables it to provide great resistance to microbial degradation in soils (Cheng *et al.,* 2008). Use of biochar may therefore enhance carbon accumulation in soil. Moreover, when applied to sandy soils, some studies suggests that biochar may also contribute to reduced nitrous oxide emissions from agricultural soils as presented in Table 8.6. However, there are also a number of studies showing no changes or even an increase N_2O emissions from soils treated with biochar (Sánchez-García *et al.,* 2014), and this issue needs more research.

Advantages of applications of biochar include increased biological activity; reduced N_2O emissions and enhanced carbon accumulation; and improved water holding capacity of soils (FAO, 2013). The limitations are risks associated

Table 8.6 Effects of biochar applications in relation to CSA principles and corresponding literature

Biochar	CSA addressed	References
• Increased soil water holding capacity by 11%	Adaptation	Cayuela *et al.* (2014)
• Reduced N$_2$O-N emissions by 54%	Mitigation	Bruun *et al.* (2011) Wang *et al.* (2012)
• Increased rice yield by 12%	Productivity	Karhu *et al.* (2011)

Source: Authors' own compilation

with increased soil pH; accumulation of heavy metals in soils; and difficulty to apply in dry soils with lots of dust particles (Wang *et al.,* 2012; Bruun *et al.,* 2011; Sparkes and Stoutjesdijk, 2011). Biochar performance partly depends on type of raw material used to produce the product.

Wang *et al.* (2012) reported that biochar application decreased N$_2$O emissions by 54 percent in rice and 53 percent in wheat cultivated soils. On the contrary, there are a number of studies showing no changes or even an increase in N$_2$O emission from soils treated with biochar. A recent study made by Sánchez-García *et al.* (2014) revealed that addition of biochar increased N$_2$O emissions by 54 percent in Haplic Calcisols, probably because biochar has systematically stimulated nitrification. Predicting which N$_2$O formation pathways predominates in a certain type of soil and source material of biochar are important factors to assure the success of biochar as a N$_2$O mitigation strategy (Sánchez-García *et al.,* 2014).

The experience from SSA

Soil and water conservation measures

A range of soil and water conservation (SWC) measures, both indigenous and/or modern technologies, are practiced by farmers to arrest soil degradation across diverse agroecological zones in different parts of the world. In this section, a brief description on the *zai* planting system in the Sahel and contour stone bunds that are being applied throughout the semi-arid areas of the world is given.

1   *Zai* (also called planting basins) is a soil rehabilitation system applied in the SSA countries (predominantly in Burkina Faso, Niger and Zambia) to concentrate runoff water and organic matter in small pits. The pit typically is 20–30 cm in width, 10–20 cm deep and spaced 60–80 cm apart. The holes are dug manually during the dry season. A handful of organic material (0.3 kg) such as manure, compost, or dry plant biomass is added in the pit hole (Tim *et al.,* 2013).
2   *Contour lines with stone bunds* are built with quarry rock or stones in a series of two or three lines along the natural contour of the land. After 10 to 15 cm of the soil has been removed from the line, they are built to a height of

20 to 30 cm from the ground and spaced 20 to 50 cm apart, depending on the inclination of the terrain. In addition to controlling soil loss on slopes, this practice accumulates fine soil particles (relatively rich in nutrients) from lands upstream, and prevents nutrient runoff further downstream. It is widely practiced in the erosion prone areas in SSA and other dry land areas of the world.

The contour stone bunds and *zai* pits improve soil fertility, soil moisture, increase ground water level, reduce soil erosion, give reliable yields and reduce food insecurity as shown in Table 8.7. On the other hand, the contour stone bunds and *zai* pits require high labour. Digging *zai* takes about 300 man-hours/ ha and construction of contour stone bunds takes between 50 to 100 man-hours/ ha (Tim *et al.*, 2013; Danjuma and Mohammed, 2015). Mechanization is hardly possible in *zai* planting (Roose, 1999) due to the high cost, and there are high recurrent costs for *zai* maintenance, and high costs for production and transport of manure/compost (Reij and Smaling, 2007). Contour bunds are not suitable for uneven or eroded land, and there is a loss of production from areas occupied by stone bunds (Araya and Asafu-Adjaye, 1999).

Although there are some drawbacks when implementing *zai* and/or contour stone bunds, particularly during the early phase of application, the advantages clearly outweigh the disadvantages (Reij *et al.*, 1996). It is often advisable to combine *zai* with contour stone bunds, because the bunds protect the zai against strong runoff, although it adds further labour cost.

Conservation Agriculture

Conservation Agriculture (CA) is a system of agronomic practices that include three main components i) reduced tillage or no-till, ii) permanent organic soil

Table 8.7 Applications of *zai* pit system and contour stone bunds in relation to CSA principles and corresponding literature

Zai pit system	Contour stone bunds	CSA addressed	References
• Improved water use efficiency by a factor of 2 compared to control	• Increases soil water content by 59%	Adaptation	Fatondji *et al.* (2011) Zougmore *et al.* (2014)
• Contributed to GHG mitigation	• n.a.	Mitigation	Zougmore *et al.* (2014)
• Doubled or triples cereals yield	• Increases sorghum yield by 33 to 55%	Productivity	Roose (1999) Tim *et al.* (2013) Zougmore *et al.* (2014)

n.a.: not available

Source: Authors' own compilation

cover by retaining crop residue (mulching), and iii) crop rotations, including growing cover crops (Palm *et al.*, 2014). According to Palm *et al.*, (2014), there is evidence that topsoil organic matter, aggregate stability, soil microbial biomass and the number of macropores increases with CA. Bennett *et al.* (2012) indicate that crops grown in monoculture often suffer from yield decline compared to those grown in longer rotations. An example from Zimbabwe showed an increase of up to 331 percent in water infiltration and a 31 percent greater carbon content in the top 60 cm in CA fields compared to adjacent conventional ploughed fields (Thierfelder *et al.*, 2012). CA gives more stable crop yields and saves water and labour, thus reducing production costs (Table 8.8).

However, even with several reported success stories of CA, Giller *et al.*, (2009) stated that the empirical evidence of the beneficial effects of CA is not clear and consistent. Negative effects of CA do exist, especially in the short term: increased weed competition and residue-borne diseases (Giller *et al.*, 2009). Other limitations of CA include competing multiple use of crop residues for mulching, livestock feed and building materials; shortage of organic residues; lack of capital to apply herbicides; increased labour demand for weeding and residue-borne diseases; increased herbicide use; soil compaction from heavy wet soils; waterlogging; soil acidity; and aluminium toxicity. In dry parts of SSA, the adaptation of CA (for e.g. zero tillage) has been limited due to a long lasting dry season resulting in little opportunity for cover crops to survive and crust-prone soils to adapt (Rockström *et al.*, 2009). Mulching is also difficult to achieve in savannah agroecosystems due to free grazing conditions (Rockström *et al.*, 2009). To achieve full benefits of CA in SSA, more emphasis on water conservation may be needed (Rockström *et al.*, 2009).

Reduced tillage or no-till reduces erosion by minimizing the time the soil is exposed to wind, rainfall and runoff. In addition, it increases water infiltration into the soil and decreases runoff. Several studies on reduced or no-till have shown a decrease in losses of soil particles, nitrogen and phosphorus to nearby streams (Delaune and Sij, 2012). In a no-till system, the soil is never tilled and therefore the soil always has a plant cover protecting the soil. No-till farming is widely adopted in South America, where the area under no-till covers 47 percent, and about 38 percent of North America is under no-till (Derpsch *et al.*, 2010). For example in Brazil (Paraná), no-till plots are reported to give three

Table 8.8 Effects of applications of CA in relation to CSA principles and corresponding literature

Conservation agriculture	CSA addressed	References
• Improved water infiltration by 70–238%	Adaptation	Bayer *et al.* (2006) Thierfelder *et al.* (2012)
• Accumulated 350–600 kg C /ha/yr	Mitigation	Bayer *et al.* (2006) Thierfelder *et al.* (2012)
• Increased maize yields by 20 to 120%	Productivity	Rockström *et al.*, 2009 Chauhan *et al.* (2012)

Source: Authors' own compilation

times more wheat and soybean yield than conventionally ploughed plots, and they reduce erosion by up to 90 percent (Altieri *et al.*, 2011). However, physical weed control becomes more difficult when the soil is untilled, often leading to increased herbicide use (Chauhan *et al.*, 2012). In a reduced-till system, tillage is kept at a minimum and/or limited to certain period of the year.

The experience from Burundi: how farmers build resilience to climate change

Burundi is a small landlocked country in Africa that ranks amongst the poorest countries in the world. The country faces a variety of problems from climate change to population pressure, with associated increased pressure on natural resource base. Poor land management practices on steep hills have led to erosion and soil degradation, with concomitant effects of food insecurity under a fragile socioeconomic environment that is prone to internal conflicts and unrest (MINAGRIE *et al.*, 2014). In Burundi, the main impacts of CC include change in precipitation patterns, loss in soil fertility and failure of crops. The most vulnerable are the agricultural ecosystems, farming communities and the physical infrastructure (roads, houses etc.).

Climate change effects

A concept like climate change is not an easy issue to discuss with local farmers, since it is difficult to find out whether people mean a changing climate or just bad climatic conditions (climate risks they face). In general, about 80 percent of the interviewed farmers are aware of the changing climate and 75 percent mention effects of climate change. Usually older people say that the climate is unpredictable now. In Figure 8.3, the effects of climate changed noticed by farmers in percent are presented. The dominant effects of climate change mentioned include hail, violent winds, prolonged drought and excessive rainfall.

In Burundi, within the framework of the project Fanning the Spark[2], soil and water management is integrated in the PIP approach (Plan Intégré du Paysan) which is synonymous with an integrated farm plan (Kessler *et al.*, 2015). This approach is an innovative way of transforming small-scale subsistence farm households into more productive and sustainable farms. It is based on a visionary integrated farm plan, which is developed and drawn on a map by all family members, as well as a concrete action plan of how to realize that vision. Changing farmers' mind-sets and making them aware that they can transform their reality by conscious collective action is at the core of the PIP approach. To make these communities more resilient towards climate change, it is crucial to focus on increased water and nutrient use efficiencies. Often, PIP requires collaboration with fellow farmers. A crucial issue within the PIP approach is scaling-up, which is realized through Farmer to Farmer (F2F) training. Preliminary research carried out among 21 farmers in Gitega province demonstrates encouraging first results of the effect of F2F training on the adoption of agricultural practices and associated SWC measures, which could be seen as a first step to build resilience

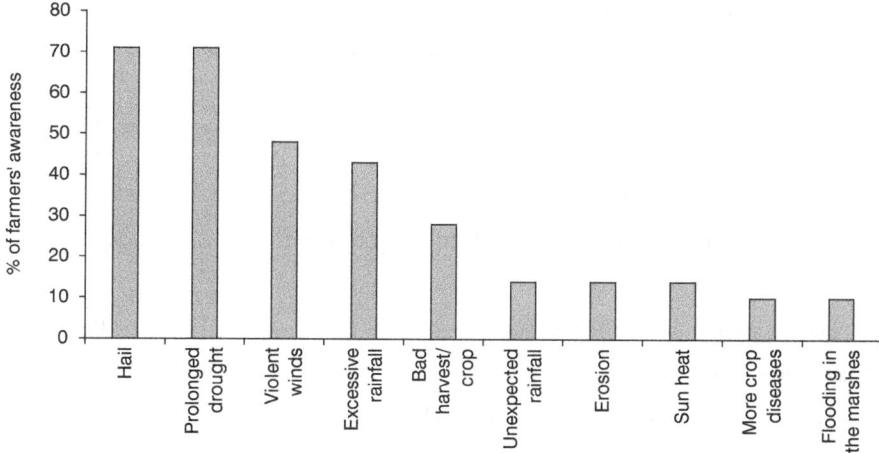

Figure 8.3 Percentage of farmers (*n* = 21) who are aware of climate change effects in Gitega province, Burundi

Source: Based on results from the project Fanning the Spark (Wageningen UR, 2015)

to climate change. The most common implemented adaptation measures to CC are the cultivation of drought resistant crops, implementation of erosion control measures and planting agro forestry trees (Table 8.9).

Knowledge related to SWC measures and its implementation

There is a lack of knowledge on modern agricultural techniques in general, and in the various SWC measures in particular (IMF, 2006). One notable example is the contour lines constructed by the Burundian government in 2009 in Gitega province, which have been destroyed by farmers since then. This is because farmers did not understand why the structures were constructed. Farmers knew that erosion leads to a loss of fertility, and mentioned that it washes the top soil, fertilizer and crops away and does not permit crops to grow. Around 50 percent of the farmers seemed to be aware of the seriousness of erosion. Other farmers expressed that only their neighbours were affected by erosion. Thus, farmers can recognize erosion, but are not always aware of the seriousness. Some of the barriers causing farmers not to implement erosion measures include lack of money, materials and time. Other barriers mentioned were free grazing goats that destroy grass strips, rocky soils and lack of help from agricultural extension agencies.

Thanks to the project 'Fanning the Spark', farmers have now gained knowledge on modern agricultural techniques including SWC measures. Almost all participating farmers (*n* = 21) in the project have erosion control measures in their mind as one of their first focal points to implement in their fields. Farmer's awareness and the rate of adoption of some recommended components of SWC measures after having received F2F training are shown in Table 8.10.

Table 8.9 Percentage of farmers who apply adaptation measures to CC in Gitega province, Burundi

Adaptation measures to CC	% farmers (n = 21)
Cultivate climate resistant crops such as cassava and sweet potato	43
Erosion control measures	38
Plant trees on mountain tops – diminish effects of drought	24
Stock food in case of drought	19
Collect rain water for kitchen gardening	10
Staggered sowing	5
Spread the risk by growing crops on several locations in the village	5
Crop rotation	5

Note: Percentages are computed independently hence do not add up to 100

Source: Results from the project Fanning the Spark (Wageningen UR, 2015)

Knowledge related to soil fertility measures and its implementation

Farmers' knowledge on soil fertility measures was ascertained from F2F training and extensions conducted by the project Fanning the Spark (2015). Farmers were asked which soil fertility measures they currently use. About 67 percent of the farmers apply compost to their fields. Agroforestry (as soil fertility measure) is mentioned by almost 76 percent of the farmers but currently only 5 percent of the farmers apply the measure (Table 8.11).

More than 75 percent of the farmers have good knowledge of agroforestry. However, the adoption rate by farmers is low (i.e. 5 percent), probably due to factors such as shortage of farmland, lack of awareness, limited knowledge dissemination, germplasm availability, time-lag between costs and benefits and limited incentives. To a smaller extent, farmers have animals in their stables, and apply mulch and mineral fertilizers. Farmers' knowhow on the adoption of mulching is very low due to lack of training and weak extension services.

Adoptions and upscaling integrated soil management technologies

The previous section has shown how ISM practices increase farmers' adaptation to CC impacts (e.g. by increasing the buffering capacity of soils to water and nutrients), contribute to reduce GHG emissions (by increasing biomass production and, consequently, carbon accumulation in the soils and increase nutrient use efficiency), and enhance food security (by increasing crop yields and farmers income). This section will assess the ISM technologies uptake and adoption from the feasibility, accessibility, scalability and sustainability perspectives.

Table 8.10 Percentage of farmers who are aware and/or are practicing some soil and water conservation measures in Gitega province, Burundi

SWC measures	Percentage of farmers (n = 21)	
	Aware	Practicing
Contour lines with grass strips	100	90
Agroforestry	76	14
Using compost	67	67
Ditches along contour lines	29	29
Cultivation banana in ditches	10	24
Cultivation crops on ridges	10	19
Mulching	5	19

Note: the percentages are computed independently hence do not add up to 100

Source: Results from the project Fanning the Spark (Wageningen UR, 2015)

Table 8.11 Percentage of farmers who are aware and/or are practicing some soil fertility measures in Gitega province, Burundi

Soil fertility measures	Percentage of farmers (n = 21)	
	Aware	Practicing
Agroforestry	76	5
Production of compost (piles)	67	67
Use of organic and mineral fertilizer	58	14*
Mulching	5	19

Note: The percentages are computed independently hence do not add up to 100. * includes only mineral fertilizers

Source: Results based on project Fanning the Spark (Wageningen UR, 2015)

'Feasibility': do SM technologies work?

Once a soil management technology has been proven successful (under a research environment) the technology should be further tested under real farm conditions, with active participation of relevant stakeholders, before transferring the technology to other similar areas. In the process, background information is needed on the farming system of the area and farm households via a situation analysis study. Farming systems are highly variable at different scales, and local adaptation to crop and site specific soil conditions are important. Adjustments may be required on the application of the soil management technology to adapt to local conditions. However, a SM technology can be very effective from a technical point of view, but less feasible from a socioeconomic point of view because the cost of the technology (capital, labour) is high and unaffordable by resource poor farmers or the technology is socially or culturally not accepted (e.g. use of human manure, vermicomposting).

The performance levels of SM technologies depend on multiple factors, among others the knowledge and capacity of the adopter/implementer to appropriately apply the technology. Lack of knowledge is one of the main barriers to adoption of SM technologies, as mentioned in the case of Burundi (Tables 8.10 and 8.11). Awareness creation and/or raising of the benefits of a technology is a commonly acknowledged prerequisite for farmers to decide whether to adopt a technology (Odendo *et al.*, 2009). In addition, the approach used to implement the technology determines success rate and the adoption level. It is often not the technology *per se* to blame for not being adopted by farmers, but rather the approach, or the way the technology was applied in the area. For instance, SWC measures are conservation effective, but if these measures are not linked to the entire value chain (i.e. to the production-processing-marketing-consumption continuum), their adoption rate remains low and slow, because farmers are often keen to obtain more yield and more income, not only just controlling erosion and runoff. Therefore, the answer to the above-mentioned question is that it depends on how the technologies were applied in the area.

'Accessibility': can SM technologies reach the intended beneficiaries?

One of the main bottlenecks for adoption of SM technologies (like CA) by smallholder farmers in SSA is the inaccessibility and unavailability of organic residues, particularly crop residues. Crop residues are mainly used as fodder, but also for fuel and construction purposes. On top of this, shortage of crop residues and lack of capital makes dry land farmers (with mixed crop-livestock farming) often reluctant to add the required amount of organic residues into the soils (Palm *et al.*, 2014), despite farmers knowing the benefits of incorporating the residues into the soils. In effect, adoption rates for SM are limited unless alternative sources of crop residues are made available, and the right incentives are given to farmers to retain the residues. Incentives, such as offering (micro) credits to those farmers who manage their soils well, could motivate them to adopt the technique.

Low cost SM technologies that are environmentally friendly, economically feasible and socially acceptable are easily adapted by local farmers and are often sustainable. For example, combining indigenous SWC measures, such as contour bunds and *zai* with manure applications, are more easily implemented by farmers than biofertilizers or biochar applications, which are relatively expensive and require trained personnel.

'Scalability': can the SM technology be adjusted over a wide range of conditions?

There are several success stories available from implementing various SM technologies on a local scale, as are shown in the previous sections. For instance, a survey of 89 projects aimed at improving food production in developing countries showed an average per project increase in per hectare food production of 93 percent (Pretty *et al.,* 2003). The most common mechanisms were yield improvements with regenerative technologies such as legumes and integrated pest management, and new and locally appropriate crop varieties and animal breeds. Increased water use efficiency, improvements to soil health and fertility, and pest control with minimal or zero-pesticide use also played a substantial role.

A certain soil management technology may be suitable to a specific agroecological zone but not to diverse ecological conditions. For example, biochar is more effective in poor soils or infertile sandy soils in improving soil fertility, water-holding capacity, reducing GHG emissions particularly N_2O, and also increasing yield but it is less responsive to clayey soils that are relatively fertile (Cayuela *et al.,* 2014).

The limitation of SM technologies should be identified during the testing and validation phase. Different SM technologies can offer substantial benefits in certain locations. Many farmers easily adopt soil management technologies that can be applied and/or adjusted to a wide range of environmental conditions over a wider area. For example, physical SWC measures such as contour bunds can be implemented at plot, field, farm, catchment and even at river basin scales as opposed to CA, which would only fit in a limited set of socio-ecological niches.

'Sustainability': do SM technologies continue to operate without external support?

A soil management technology can be feasible, accessible and scalable but may be unsustainable due to some external factors imposed on the land user. One notable driving factor for sustainable adoption and upscaling SM technologies is the security of land tenure issue. In most African countries, the state owns the land and takes responsibility to administer it. Under such conditions, farmers do not have an incentive to invest in long-term structures on their plots of land. Long-term lease regulations and protection of tenancy rights are essential to ensure tenure security and sustainability of SM technologies.

In addition, most agricultural projects including SM have been initiated and run by project funds obtained from external donors. The project may come up with a viable SM technology that is feasible, accessible and adjustable over a wide range of conditions (i.e. scalable) but not sustainable. In many projects, there is no continuity of the work once the project is over. To avoid or least to limit this effect, key relevant stakeholders (from farmers and importantly local extension workers to policy makers) need to be involved from the onset of a project and a detailed exit strategy needs to be prepared with the aim to ensure project activities will continue and, if successful, even scaled up by the local/national government.

Mainstreaming SM technologies into policies and programs

Under the CSA context, mainstreaming refers to the integration of CC adaptation and mitigation measures (related to soil management) into established or ongoing development programs and policies. One good example is the ClimaRice project in India. Recommendations on growing green manure crops, biofertilizer applications and conservation agriculture practices have been included in the final draft of State Climate Change Action Plan for Tamil Nadu state of India (http://www.climarice.org/).

Conclusions and research needs

The discussed six SM practices fulfil one or more of the CSA principles of adaptation, mitigation and productivity (Tables 8.1 to 8.7), and can therefore be labelled as 'climate smart soil management technologies'. Any of the ISM practices that increase SOC content is likely to have beneficial effects on soil properties and functioning, although some could have little benefit when it comes to CC mitigation. Combined applications of SM technologies are likely to generate more CSA co-benefits than applying a single SM technology.

Climate change is a reality in Burundi that aggravates anthropogenic factors leading to severe soil degradation, declining soil fertility and agricultural productivity. For most Burundian farmers engaged in the 'Fanning the Spark' project CC is a rather complex concept and they do not have the knowledge yet for increasing their resilience. In order to turn back this trend, urgent measures are required using for example the PIP approach. The negative impacts of CC need to be identified and concerted actions are needed. The national agriculture research institute of Burundi needs to play an active role in supporting farmers e.g. in the identification of alternative crops or improved varieties that are better adapted to future CC. In order for SM technologies to be sustainable and applicable on a wider scale, the technologies must be mainstreamed into national policy making processes. For this to happen, the role of SM technologies in adapting and mitigating CC impacts has to be better understood by stakeholders at all levels. Capacity building and information dissemination on CC adaptation and identification of limitations to adoption of CSA practices are thus required. The main knowledge gaps and/or future research needs of ISM practices to adapt and mitigate CC are as follows:

* Which different SM technologies work best in adapting and mitigating GHG emissions while increasing crop production, and why?;
* Feasibility, accessibility, scalability and sustainability of different SM technologies in promoting the core principles of CSA; and
* Integrated effects of the different SM technologies on CC adaptation, mitigation and food security developments.

Notes

1 The *rabi* cropping season (in India) is from October to March (winter).
2 The project Fanning the Spark (SCAD in French; *projet de Solidarité Communautaire pour l'Auto Développement*) is a collaborative FDOV funded project set up by Achmea, Alterra

and HealthNetTPO, which has employed an integrated approach to improve the food, financial and social security of the local population since July 2013. Alterra provides the expertise in agriculture and soil and water management, Achmea in banking (micro-credits, savings) and crop and health insurances, and HealthNetTPO in healthcare.

References

Ali, A.M., Thind, H.S., Sharma, S. and Singh, Y. (2015) Site-specific nitrogen management in dry direct-seeded rice using chlorophyll meter and leaf colour chart, *Pedosphere*, 25 (1): 72–81.

Altieri, M.A., Marcos, L.A., Bittencourt, H.V., Kieling, A.S., Comin, J.J. and Lobsyo, P.E. (2011) Enhancing crop productivity via weed suppression in organic no-till cropping systems in Santa Catarina, Brazil, *Journal of Sustainable Agriculture*, 35: 1–15.

Araya, B. and Asafu-Adjaye, J. (1999) Returns to farm level soil conservation on tropical steep slopes: the case of the Eritrean highlands, *Journal of Agricultural Economics*, 50 (3): 589–605.

Bationo, A., Fairhurst, T., Giller, K., Kelly, V., Lunduka, R., Mando, A., Mapfumo P., Oduor G., Romney, D., Vanlauwe, B., Wairegi, L. and Zingore, S. (2012) *Handbook on soil fertility management*, Africa soil Health Consortium, CAB International, India.

Bayer, C., Martin-Neto, L., Mielniczuk, J., Pavinato, A. and Dieckow, J. (2006) Carbon sequestration in two Brazilian Cerrado soils under no-till, *Soil Tillage Research*, 86: 237–245.

Bennett, A. J., Bending, G. D., Chandler, D. and Hilton, S. (2012) Meeting the demand for crop production: the challenge of yield decline in crops grown in short rotations, *Biological Reviews*, 87: 52–71.

Ben Mohamed, A., van Duivenbooden, N. and Abdoussallam, S. (2002) Impact of climate on agricultural production in the Sahel – Part 1: Methodological approach and case study for millet in Niger, *Climatic Change*, 54: 327–348.

Bijay-Singh., Yadvinder-Singh, Ladha, J.K., Bronson, K.F., Balasubramanian V., Singh, J. and Khind C.S. (2002) Chlorophyll meter- and leaf color chart-based nitrogen management for rice and wheat in north-western India, *Agronomy Journal*, 94: 821–829.

Brady, N.C., and Weil, R.R. (2007) *The Nature and Properties of Soils*, 14th Ed., Prentice Hall, Upper Saddle River, NJ, USA.

Bruun, E.W., Müller-Stöver, D., Ambus, P. and Hauggaard-Nielsenet, H. (2011) Application of biochar to soil and N_2O emissions: potential effects of blending fast-pyrolysis biochar with anaerobically digested slurry, *European Journal of Soil Science*, 62: 581–589.

Campbell, B., Mann, W., Meléndez-Ortiz, R., Streck, C. and Tennigkeit, T. (2011) *Agriculture and Climate Change: A Scoping Report*, Meridian Institute, Washington, DC.

Cayuela, M.L., van Zwieten, L., Singh, B.P., Jeffery, S., Roig, A. and Sánchez-Monedero, M.A. (2014) Biochar's role in mitigating soil nitrous oxide emissions: A review and meta-analysis, Agriculture, *Ecosystems and Environment*, 191: 5–16.

Chauhan, B. S., Singh, R. G. and Mahajan, G. (2012) Ecology and management of weeds under conservation agriculture: A review, *Crop Protection*, 38: 57–65.

Cheng, C.H., Lehmann, J., Engelhard, M. (2008) Natural oxidation of black carbon in soils: changes in molecular form and surface charge along a climosequence, *Geochimica et Cosmochimica Acta*, 72: 1598–1610.

Cherr, C.M., Scholberg, J.M.S. and McSorley, R. (2006) Green Manure approaches to crop production: a synthesis, *Agronomy Journal*, 98: 302–319.

Danjuma, M.N. and Mohammed, S. (2015) Zai pits system: a catalyst for restoration in the dry lands, *Journal of Agriculture and Veterinary Science*, 8(2): 2319–2380, DOI: 10.9790/2380-08210104.

Delaune, P.B. and Sij, J.W. (2012) Impact of tillage on long-term no-till wheat system, *Soil and Tillage Research*, 124: 32–35.

Derpsch, R., Friedrich, T., Kassam, A. and Hongwen, L. (2010) Current status of adoption of no-till farming in the world and some of its main benefits, *International Journal of Agricultural and Biological Engineering*, 3 (1): 1–26.

European Commission, Joint Research Centre (EC-JRC) (2003) In: Jones, R.J.A. and Montanarella, L. (eds.) *Land Degradation in Central and Eastern Europe*: European Soil Bureau Research Report No.10, Office for Official Publications of the European Communities, Luxembourg.

Fatondji, D., Martius, C., Vlek, P.L.G., Bielders, C.L. and Bationo, A. (2011) Effect of zai soil and water conservation technique on water balance and the fate of nitrate from organic amendments applied: a case of degraded crusted soils in Niger. In: *Innovations as Key to the Green Revolution in Africa*, Springer, Dordrecht.

Food and Agricultural Organization (FAO) (2009) *Food Security and Agricultural Mitigation in Developing Countries: Options for Capturing Synergies*, Rome, Italy, online, http://www.fao.org/docrep/012/i1318e/i1318e00.pdf (accessed 5 August 2015).

Food and Agricultural Organization (FAO) (2013) *Module 4: Soils and their management for climate smart agriculture, Climate-smart Agriculture Source Book*, Rome, Italy, online, http://www.fao.org/3/a-i3325e.pdf (accessed 5 August 2015).

Food and Agricultural Organization (FAO) (2014) Adapting to climate change through land and water management in Eastern Africa, results of pilot project in Ethiopia, Kenya and Tanzania. FAO, Rome, online, http://www.fao.org/3/a-i3781e.pdf (accessed 5 August 2015).

Giller, K.E., Witter, E., Corbeels, M. and Tittonell, P. (2009) Conservation agriculture and smallholder farming in Africa: The heretics view, *Field Crops Research*, 114: 34–34.

Harman, G.E. (2006) Overview of mechanisms and uses of Trichoderma spp., *Phytopathology*, 96: 190–194.

Intergovernmental Panel on Climate Change (IPCC) (2007) *Climate Change 2007: Synthesis Report. International Panel on Climate Change*, Cambridge University Press, Cambridge.

International Monetary Fund (IMF) (2006) Burundi: Selected issues and statistical appendix, Washington D.C., USA, online, https://www.imf.org/external/pubs/ft/scr/2006/cr06307.pdf (accessed 30 July 2015).

Karhu, K., Mattila, T., Bergström, I. and Regina, K. (2011) Biochar addition to agricultural soil increased CH_4 uptake and water holding capacity- Results from a short-term pilot field study, *Agriculture, Ecosystems and Environment*, 140: 309–313.

Kessler, C.A., van Duivenbooden N., Nsabimana, F. and van Beek, C.L. (2015) Bringing ISFM to scale through an integrated farm planning approach: a case study from Burundi, *Nutrient Cycling Agroecosystems*, DOI 10.1007/s10705-015-9708-3.

Lakshmanan, A., Geethalakshmi, V. and Nagothu, U.S. (2015a) Impacts of microbial inoculants on productivity, methane flux and carbon sequestration in rice ecosystem, Technical brief No. 14, Tamil Nadu Agricultural University, Coimbatore.

Lakshmanan, A., Geethalakshmi, V. and Nagothu, U.S. (2015b) Comparison of different systems of rice cultivation for water productivity and yield, Technical brief No.15, Tamil Nadu Agricultural University, Coimbatore.

Lehmann, J., da Silva Jr., J.P., Steiner, C., Nehls, T., Zech, W. and Glaser, B. (2003), Nutrient availability and leaching in an archaeological Anthrosol and a Ferralsol of the Central Amazon basin: fertilizer, manure and charcoal amendments, *Plant and Soil,* 249: 343–357.

Mahajan, A. and Gupta, R.D. (eds.) (2009) *Integrated Nutrient Management (INM) in a Sustainable Rice-Wheat Cropping System,* Springer Publishing, Palampur, India.

MINAGRIE, CAPAD, FOPABU (2014). *Rapport synthèses du forum national sur l'agriculture familiale,* Ministère de l'Agriculture et de l'Elevage (MINAGRIE), Bujumbura, Burundi.

Mishra, D.J., Rajvir, S., Mishra, U.K. and Kumar, S. (2012) Role of bio-fertilizer in organic agriculture: a review, *Research Journal of Recent Sciences,* 2: 39–41.

Mohammadi, K. and Sohrabi, Y. (2012) Bacterial biofertilizers for sustainable crop production: A review, *Journal of Agricultural and Biological Science,* 7 (5): 307–316.

Odendo M., Obare G. and Salasya B. (2009) Factors responsible for differences in uptake of integrated soil fertility management practices amongst smallholders in western Kenya, *African Journal of Agricultural Research,* 4 (11): 1303–1311.

Palm, C., Blanco-Canqui, H., DeClerck, F., Gatere, L. and Grace, P. (2014) Conservation agriculture and ecosystem services: An Overview, *Agriculture, Ecosystems and Environment,* 187: 87–105.

Powlson, D.S., Whitmore, A.P. and Goulding, K.W.T. (2011) Soil carbon sequestration to mitigate climate change: a critical re-examination to identify the true and the false, *European Journal of Soil Science,* 62: 42–55.

Pretty, J.N., Morrison, J.I.L. and Hine, R.E. (2003) Reducing food poverty by increasing agricultural sustainability in developing countries, *Agriculture, Ecosystems and Environment,* 95: 217–234.

Raja, W., Rathaur, P., John, S.A. and Ramteke, P. (2012) Azolla: An aquatic pteridophyte with great potential, a review article, *International Journal of Research in Biological Sciences,* 2 (2): 68–72.

Reij, C. and Smaling, E.M.A. (2007) Analyzing successes in agriculture and land management in Sub-Saharan Africa: is macro-level gloom obscuring positive micro-level change? *Land Use Policy,* 25: 410–420.

Reij, C., Scoones, I. and Toulin, C. (1996) *Sustaining The Soil: Indigenous Soil and Water Conservation in Africa,* EarthScan, London.

Richardson, A.E., Kirby, C.A., Banerjee, S. and Kirkegaard, J.A. (2014) The inorganic nutrient cost of building soil carbon, *Carbon Management,* DOI:10.1080/17583004.2 014.923226.

Rockström, J., Kaumbutho, P., Mwalley, J., Nzabi, A.W., Temesgen, M., Mawenya, L., Barron, J., Muta, L. and Damsgaard-Larsen, S. (2009) Conservation farming strategies in East and Southern Africa: Yields and rain water productivity from on-farm action research, *Soil and Tillage Research,* 103: 23–32.

Roose, E. (1999) Zai practice: A West African traditional rehabilitation system for semiarid degraded lands, a case study in Burkina Faso, *Arid Soil Research and Rehabilitation,* 13: 343–355.

Sánchez-García, M., Roig, A., Sánchez-Monedero, M.A. and Cayuela, M.L. (2014) Biochar increases soil N_2O emissions produced by nitrification-mediated pathways, *Frontiers in Environmental Science,* DOI:10.3389/fenvs.2014.00025.

Seybold, C.A., Herrick, J.E. and Brejda, J.J. (1999) Soil resilience: a fundamental component of soil quality, *Soil Science,* 164 (4): 224–234.

Smith, P., Martino, D., Cai, Z., Gwary, D., Janzen, H., Kumar, P., McCarl, B., Ogle, S., O'Mara, F., Rice, C., Scholes, B. and Sirotenko, O. (2007) Agriculture. In: Metz, M. *et al.* (eds.) *Climate Change 2007: Mitigation. Contribution of Working Group III to the Fourth Assessment Report of the IPCC,* Cambridge University Press, Cambridge.

Sparkes, J. and Stoutjesdijk, P. (2011) Biochar: implications for agricultural productivity. *ABARES Technical Report,* 11 (6). Australian Bureau.

Thangarajan R., Bolan N.S., Tian G.L., Naidu R., Kunhikrishnan A. (2013), Role of organic amendment application on greenhouse gas emission from soil. *Science of the Total Environment,* 465, 72–96. http://www.sciencedirect.com/science/article/pii/S0048969713000417

The Montpellier Panel (2014) No ordinary matter: conserving, restoring and enhancing Africa's soils, Agriculture for impact, online, http://ag4impact.org/news/no-ordinary-matter-conserving-restoring-and-enhancing-africas-soils-2014/ (accessed 5 December 2014).

Thierfelder, C., Cheesman, S. and Rusinamhodzi, L. (2012) A comparative analysis of conservation agriculture systems: Benefits and challenges of rotations and intercropping in Zimbabwe, *Field Crops Research,* 137: 237–250.

Tim, M., D'Aiuto, C. and Lingbeek, B. (2013) Zai pit system, technical note #78, ECHO, online, https://c.ymcdn.com/sites/members.echocommunity.org/resource/collection/27A14B94-EFE8-4D8A-BB83-36A61F414E3B/TN_78_Zai_Pit_System.pdf (accessed 5 November 2014).

van Beek, C.L., van Duivenbooden, N., Noij, G.J. and Heesmans, H., (2014) More food from fertile grounds: integrating approaches in order to improve soil fertility. Alterra-Wageningen UR, Wageningen. http://knowledge4food.net/wp-content/uploads/2014/12/FGI-Booklet_Nov14.pdf (accessed 10 December 2014).

van Duivenbooden, N., de Wit, C.T. and van Keulen, H. (1996) Nitrogen, phosphorus and potassium relations in five major cereals reviewed in respect to fertilizer recommendations using simulation modelling, *Fertilizer Research,* 44: 37–49.

Singh, V., Singh, B., Singh, Y., Thind, H.S. and Gupta, R.K. (2010) Need based nitrogen management using the chlorophyll meter and leaf colour chart in rice and wheat in South Asia: a review, *Nutrient Cycling Agroecosystems,* 88: 361–380.

Singh, V. Singh, B., Singh, Y., Thind, H.S., Singh, G. Kaur, S., Kumara, A. and Vashistha, M. (2012) Fixed-time adjustable dose site-specific fertilizer nitrogen management in transplanted irrigated rice (*Oryza sativa* L.) in South Asia, *Field Crops Research,* 130: 109–119.

Wageningen UR (2015) *Fanning the spark: Increasing food production through integrated farm planning in Burundi,* online, https://www.wageningenur.nl/en/project/Fanning-the-spark-Increasing-Food-Production-through-Integrated-Farm-Planning-in-Burundi.htm (accessed 4 April 2015).

Wang, J., Pan, X., Liu, Y., Zhang, X., and Xiong, Z. (2012) Effects of biochar amendment in two soils on greenhouse gas emissions and crop production, *Plant Soil,* DOI: 10.1007/s11104-012-1250-3.

Zougmore, R., Jalloh, A., and Tioro, A., (2014) Climate smart soil water and nutrient management options in semiarid West Africa: a review of evidence and analysis of stone bunds and *Zaï* techniques, *Agriculture and Food Security,* 3 (16): 1–8.

9 Water productivity under different rice growing practices

Results from farmer-led field demonstrations in India

Johannes Deelstra, Krishna Reddy K.,
Suresh Reddy K., Udaya Sekhar Nagothu,
Geethalakshmi V., Lakshmanan A. and
Arasu M. S.

Introduction

Food production in India is largely dependent on irrigated agriculture. The increase in population and urbanization, in addition to climate change will have a significant impact on the available fresh water resources. This includes both surface and groundwater used for irrigation, domestic use and industrial purposes. Competition between water uses has increased in recent years, aggravating the ongoing water use conflicts in many regions across the world, and has become highly politicized, more so in India (Kreamer, 2013; Nagothu *et.al*, 2012). These conflicts will become intensive with climate change-induced water scarcity. With the federal structure of governance in India, it will become even more difficult to resolve the conflicts if common platforms for negotiations and policy dialogue are not established. There is an urgent need to improve the current water productivity (WP) levels in the country and manage water resources sustainably. A number of measures, including the large-scale drip irrigation projects were initiated by the Government of India in 2008, with the primary objective of improving WP. However, there are several data and technology gaps, and in the case of some crops, the basic data related to WP does not exist. Our study attempts to address some of data gaps in WP using data from field measurements in three southern states in India. The study also attempts to demonstrate the use of different types of WP measurements and their constraints. Water productivity plays a crucial role in modern agriculture that aims to increase yield production per unit of water used, both under rainfed and irrigated conditions. As part of the initiatives to introduce climate resilient water or crop management systems, measuring WP provides useful data to managers and farmers.

The per-capita water availability in India has declined from 1,816 to 1,544 m³ between 2001 and 2011. It is projected to further decrease to 1,401 and 1,191 m³ by the years 2025 and 2050 respectively, merely based on population projections, and with the assumption that the availability of water resources shall not change in the future. However, climate change with an increase in temperature and a more unpredictable monsoon can lead to additional challenges in agriculture. It is expected that the total water demand will increase 13–19 percent by 2025 compared to 2010 (CWC, 2013). The projected decrease in water availability and increasing demands from other sectors will severely impact irrigated agriculture and food production in the country. By 2025, north and central India (about 8.4 million ha) will experience 'physical water scarcity', while in addition another 2 million ha of the dry-season irrigated rice in central India will suffer physical scarcity. A recent article stated India will become a water scarce country within 10 years from now (The Economic Times, 2015). A study by Singh *et al.* (2010) states that an improvement of the water productivity in agriculture is necessary to compensate for additional water withdrawals over the next 25 years in order to meet the additional food demand.

Groundwater and local and inter-state rivers constitute the available fresh water resources of Andhra Pradesh and Tamil Nadu states in India where the present study was conducted. In Andhra Pradesh (including the new Telangana state), these sources represent a total of about 108 billion m³ of water, but the usable amount is not more than 78 billion m³ (Water Vision Team, 2003). Further withdrawals are constrained by high costs and serious environmental concerns. By the year 2025, Andhra Pradesh's water resource demands will have used and even exceeded the available supplies. This can seriously affect agricultural production. According to Tuong and Bouman (2003) water scarcity already is present in major rice growing areas, with constant over exploitation of groundwater having caused serious problems. The groundwater table in Tamil Nadu and hard rock southern India has dropped by about 1 metre per year.

The anticipated change in water use between the different sectors requires water to be used more efficiently in agriculture. This is also reflected in the objectives of the National Mission for Sustainable Agriculture (NMSA) and the National Water Mission (NWM), the latter clearly stating the need to improve the water use efficiency by 20 percent (Government of India, 2008). The NMSA is to devise strategies making agriculture more resilient to climate change, to be made possible through, among others, developing new crop varieties, crop rotations, alternative cropping and irrigation systems, integrating traditional knowledge and capacity building. Rice is the largest consumer of water, having the lowest WP among the dominating food grain crops in India (Yadav *et al.*, 2000). A study by Singh *et al.*, (2010) concludes that an improvement of WP is necessary to compensate for the need for additional water withdrawals over the next 25 years in order to meet the additional food demand. New rice growing techniques such as a system of rice intensification (SRI) (Stoop *et al.*, 2002; Uphoff, 2003; Uphoff *et al.*, 2011) and alternating wetting drying (AWD) (Bouman *et al.*, 2007) use less water compared to traditional paddy rice and have a higher WP. The use of these systems is gaining ground in India and other Asian countries where rice is the major crop and staple diet of the people. As rice is the major consumer of water in irrigated agriculture, this study focused on water productivity in rice cropping systems. There are four alternative methods of rice cultivation that are practiced by farmers in India and other regions. A brief description of the practices and water application in the four systems are described in the following section.

Traditional paddy rice cultivation/normal irrigation (NI)

A majority of the farmers growing rice in India use this method. This is also one of the reasons why it uses most of the water meant for irrigation. Under paddy rice cultivation, the rice field is permanently flooded with water varying from 5 to 10 cm depth during the period after transplanting until 2 weeks before harvest. In this system, rice fields are puddled before the transplanting of rice seedlings. The main objective of puddling is to reduce the percolation losses and control weed growth. The recommended plant density is 33 hills/m² during *kharif* (rainy season) and 44 hills/m² during *rabi* (winter) season in which a hill represents 2 or 3 seedlings (Cheralu, 2011).

Direct seeded rice (DSR)

DSR is another promising method that farmers practice in many regions. In the DSR method, raising seedlings in the nursery and subsequent transplanting is not practiced. The land is prepared under dry conditions and hence puddling is not required. The rice seeds are directly sown into the dry soil, manually or by machine-operated seed drills at a depth of 2–3 cm below soil surface, before or immediately after the pre-monsoon rain showers. The spacing between rows is normally set to 20 cm. Irrigation water is not applied until 45–60 days after sowing, the length of this period depending on both the monsoon onset and the timing of irrigation water release in the canal system. A study by Gurava R.K. *et al.* (2013a) showed that, compared to NI, water use under DSR was reduced by 23 percent. The crop matures 10 days earlier than the paddy rice (Bhushan *et al.*, 2006). Compared to NI, weed growth is higher in DSR, requiring the use of more herbicides (Chauhan and Opeña, 2012). However, the total cost of cultivation is less compared to traditional rice cultivation. Thus, farmers prefer DSR in areas where the monsoon onset is delayed and water is not adequate for transplanting. In Guntur district, where the current study was carried out, DSR has started to replace normal paddy rice cultivation during the last 10 years to adapt to the erratic monsoon patterns.

Modified system of rice intensification (MSRI)

In MSRI, the seedlings are grown separately on a mat nursery. This method is more or less similar to the system of rice intensification, SRI (Stoop *et al.*, 2009), with a few modifications. In MSRI, the seedlings are transplanted using a machine transplanter and irrigation water is applied as practiced under AWD (see next section). A spacing of 25 x 16 cm is used in MSRI and (Gurava *et al.*, 2013b) water application could be reduced by almost 16 percent as compared to normal paddy rice. Land levelling is essential in MSRI to maintain the thin layer of water in the initial stages of the transplantation. MSRI can save 50 percent of the seedling mortality, improves tillering, increases panicle number and reduces incidence of pest and diseases. However, a major challenge in the adaptation of MSRI is the increased need for manual labour and constraints in land levelling. Small farmers find it difficult to get support from agro-service centres if they decide to use machinery for transplantation.

Alternating wetting drying (AWD)

In AWD, the transplanting of rice seedlings is similar to that of traditional paddy rice. However, the main difference is the absence of a standing water layer on the field. Water is provided for the first 30 days after transplantation to overcome the weed problem. The fields are then intermittently dried every 10– 15 days depending on the water level that is observed in the perforated plastic observation tubes (Bouman *et al.*, 2007; Rejesus *et al.*, 2011). The plant density followed in AWD is 33 hills.m^{-2} during the *kharif* and 44 hills.m^{-2} during *rabi* season. A normal practice is to apply irrigation water when the water level has dropped 5–15 cm below the soil surface.

One of the objectives of the ClimaAdapt[1] project was to estimate water productivity under these different rice-growing systems. This chapter presents results from field measurements on water productivity of the above-mentioned four rice-growing practices.

Materials and methods

Location of the study areas

The study was carried out at different locations in the Andhra Pradesh (AP), Telangana and Tamil Nadu states of India for the period from *rabi* (winter season) 2012–13 to *kharif* (monsoon season) 2014 (Figure 9.1). In AP and Telangana, the field measurements were carried out in the command area (irrigated area downstream from the reservoir) of the Nagarjuna Sagar Project (NSP). The Project, located in the lower Krishna River Basin, was commissioned in the 1960s, and it is one of the largest multipurpose irrigation projects in the country. It distributes water through the Nagarjuna Sagar Project Left Canal (NSLC) and the Nagarjuna Sagar Project Right Canal (NSRC) to the Telangana and Andhra Pradesh states respectively, covering a total area of 890,000 hectares. The NSP has significantly changed the socio-economic profile of the communities in the region. With the availability of water for irrigation, crops such as rice, tobacco, cotton and chillies were introduced. The area experiences water scarcity, and is ideal for measuring WP. Measurements were carried out in selected clusters

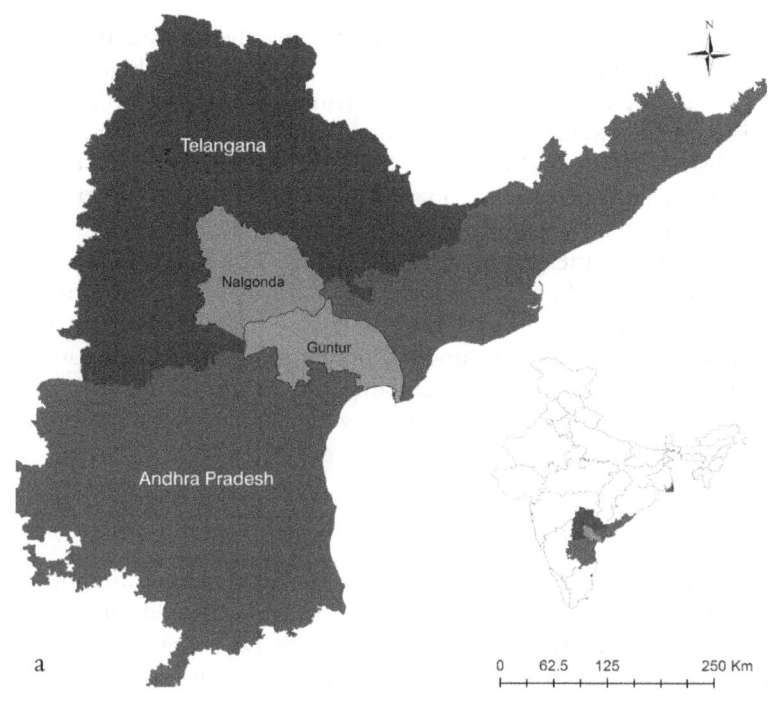

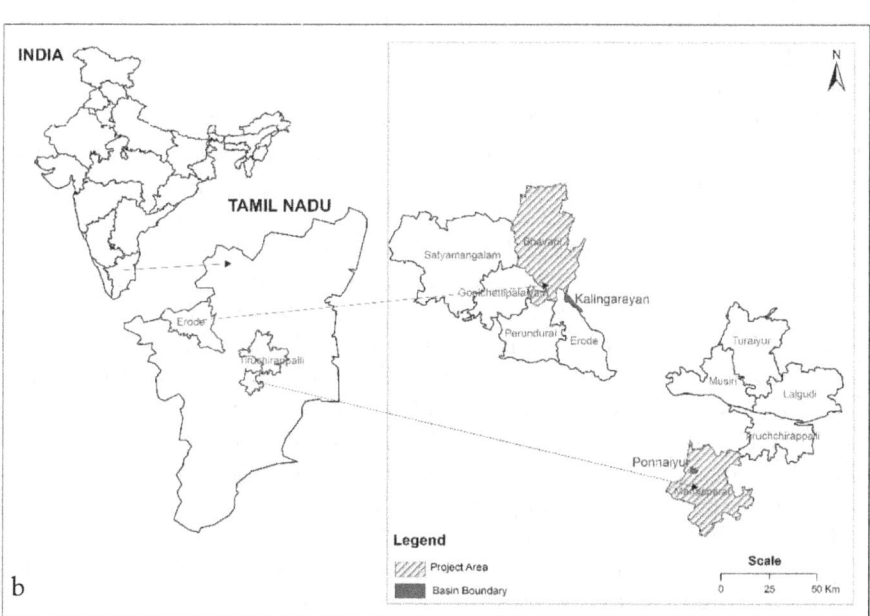

Figure 9.1 Maps showing ClimaAdapt project sites in: a) Andhra Pradesh and Telangana states in India; and b) Tamil Nadu, India

Source: Developed by the authors

under the sub-canal systems, located approximately 140 km from the main reservoir. In the NSRC, distributary canal 21 (DC 21, Chilakaluripet sub-district) was selected, having a total command area of approximately 9652 ha, administered by 9 water user associations (WUAs) and covering 29 villages in Guntur district. In the NSLC, DC 4 (Wazirabad distributary) was selected from Nalgonda district. The system under DC 4 has a total command area of about 8497 ha, administered by 6 WUAs and covering 19 villages.

In Tamil Nadu, measurements were carried out at a location under the command area of the Ponnaniyar Dam, approximately 70 km south-west from the town of Tiruchirappalli. Two small plots were selected for the measurement of water use efficiency (WUE) under different rice growing alternatives. Since the reservoir was not open for irrigation due to shortage of water during the study period, groundwater was used for irrigating the farmer plots and measurements were done using flow meters.

Crop evapotranspiration

Weather data, needed to calculate the reference crop evapotranspiration (ET_0), was obtained from two weather stations, one located at the Agricultural Research Station (KVK), Kampasagar, closer to DC 4, and the other at the Regional Agricultural Research Station (RARS), at Lam, located closer to DC 21. Since the weather data did not cover the complete measurement period, data collected at the International Crop Research for Semi-Arid Tropics (ICRISAT) at Hyderabad was supplemented to calculate the evapotranspiration for the missing data periods.

The ET_0 calculator (Raes, 2009) has been used to calculate the daily reference crop evapotranspiration, being the sum of the evaporation and transpiration from a reference crop of grass covering the soil completely and grown under optimal conditions. The weather data for Ponnaniyar basin in Tamil Nadu were obtained from the Grand Anicut climatological station. The weather data included air temperature, relative humidity, wind speed and sunshine hours.

There is a similar temporal variation in the monthly ET_0 for the three locations calculated when using available data (Figure 9.2). To obtain daily values for the ET_0 for RARS covering the period with missing data, a linear relation was established between the daily ET_0 for Hyderabad and RARS-Lam, being $ET_{0,RARS} = 0.83 \times ET_{0,Hyd} + 0.4417$ ($r^2 = 0.61$), which was used to obtain the ET_0 for the periods with missing weather data. A similar procedure was used to obtain the daily ET_0 for Kampasagar. In this case, the relation was $ET_{0,Kampasagar} = 0.5002 \times ET_{0,Hyd} + 1.3431$ ($r^2 = 0.51$).

The crop evapotranspiration was calculated (Allen *et al.*, 1998) as the product of the crop coefficient (k_c) representing a growth stage of the rice crop and the calculated ET_0 using the formula:

$$ET_{crop} = k_c \times ET_0 \tag{9.1}$$

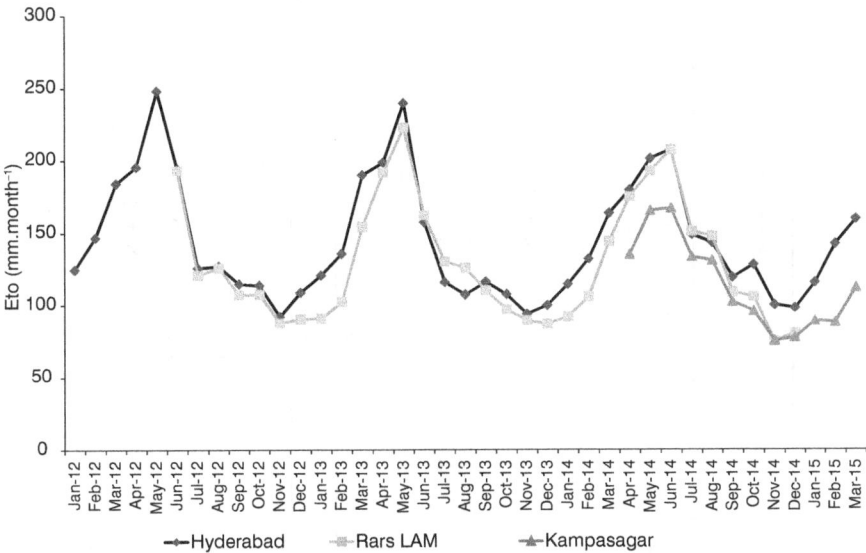

Figure 9.2 Reference crop evapotranspiration values

Source: Authors analysis from ClimaAdapt project data (www.climaadapt.org)

Table 9.1 Length of growth stages and crop coefficients for a rice crop

	Length of different growth stages	*Crop coefficients* k_c
Initial	30	1.05
Development	30	1.125
Mid-season	60	1.2
Late season	30	0.75
Total length growing season	150	–

Source: ClimaAdapt project (www.climaadapt.org)

The length of the different growth stages and their representative crop coefficients are available for a standard paddy rice crop (Table 9.1). The values for the crop coefficient representing the development and late season stage were in this case assumed to have a constant value, but in reality developed as a function of the growth stage. Alberto *et al.* (2011) when comparing the aerobic and flooded rice obtained a slightly lower crop coefficient for aerobic rice compared to flooded rice, showing 0.95 and 1.04 respectively. Water surfaces have a higher evaporation rate than soil surfaces. Practices such as AWD reduce the duration of the time when the field is flooded, and might therefore decrease the amount of evaporation. However, quantitative data of the impact on the evaporation

Table 9.2 Results of water application to alternative rice growing practices

Canal		Cluster	Kg/ha	WP (kg/m³)	Water applied (mm)	ET_crop (mm)	Water balance (mm)	Main soil type	Source
DC 21	DSR	Muppalla	6,580	0.448	1,468	686	781	Silty clay loam	C1
Kharif 2013	AWD	Kunduri Palem	6,570	0.537	1,224	312	911	Silty clay loam	C
	MSRI	Rangareddy Palem	5,026	0.491	1,024	389	635	Silty clay loam	B2
	AWD	Kanaparru	5,439	0.571	953	361	591	Silty clay loam	B2
	NI	Kavuru	3,626	0.219	1,652	406	1247	Silty clay loam	B
Kharif 2014	DSR	Muppalla	6,348	0.588	1,079	448	631	Silty clay loam	C
	AWD	Rangareddy Palem	7,781	0.772	1,008	311	697	Silty clay loam	C
	AWD	Kanaparru	7,608	0.765	994	391	603	Silty clay loam	B
DC 4	AWD	Kondrapolu	8,645	0.533	1,621	456	1,165	Sandy loam	B
Rabi 2012	AWD	Narsapur	8,151	0.458	1,778	493	1,285	Sandy loam	B
	AWD	Balajinagar thanda	7,978	0.395	2,021	443	1,578	Sandy loam	B
	AWD	KJR colony	8,139	0.424	1,919	495	1,425	Silty loam	B
	NI	Nimya thanda & Dilawarpur	7,175	0.237	3,023	478	2,545	Silty loam	B
Kharif 2013	AWD	Kondrapolu	5,706	0.319	1,790	375	1,416	Sandy loam	C
	AWD	Balajinagar - I	4,323	0.240	1,802	389	1,413	Sandy loam	C
	AWD	Balajinagar - II	5,026	0.257	1,955	400	1,554	Sandy loam	C

Season	Method	Location						Soil	Cat.
	AWD	Balajinagar - III	5,439	0.256	2,128	424	1,704	Sandy loam	C
	AWD	KJR colony	3,626	0.288	1,257	381	876	Silty loam	C+B
	NI	Rallawagu	3,112	0.124	2,512	415	2,097	Sandy loam	C+B
Rabi 2013	AWD	Kondrapolu	5,706	0.394	1,447	413	1,034	Sandy loam	C
	AWD	Balajinagar - I	5,706	0.279	2,045	422	1,622	Sandy loam	C
	AWD	Balajinagar - II	5,026	0.282	1,784	409	1,375	Sandy loam	C
	AWD	Balajinagar - III	5,439	0.615	885	428	457	Sandy loam	C
	AWD	KJR colony	5,212	0.515	1,011	409	603	Silty loam	C+B
	MSRI	Rallawagu	6,402	0.668	958	446	512	Sandy loam	B
Kharif 2014	AWD	Kondrapolu	6,279	0.458	1,370	379	991	Sandy loam	C
	AWD	Balajinagar - I	5,706	0.397	1,437	403	1,034	Sandy loam	C
	AWD	Balajinagar - II	6,195	0.458	1,354	390	964	Sandy loam	C
	AWD	Balajinagar - III	5,565	0.442	1,260	379	881	Sandy loam	C
	AWD	KJR colony	5,212	0.398	1,311	388	923	Silty loam	C
	NI	Rallawagu	4,928	0.276	1,784	420	1,363	Sandy loam	C+B
Kharif 2014	AWD	Ponnaniar	6,867	0.863	796	413	382	Sandy clay	B
Tamil Nadu.	NI	--	5,083	0.559	909	413	496	--	B

Source: Based on results from ClimaAdapt project (www.climaadapt.org)

losses are not adequately available (Tuong *et al.*, 2005). In the calculation of the ET_{crop} under alternative rice growing techniques, a value for $k_c = 1.05$ for the initial phase is assumed. In our case, the length of the different growth stages as indicated in Table 9.1 were changed relative to the real and observed length of the growing season during the measurements.

Crop yields in the measured plots

At the end of the growing season, information on crop yield was obtained from farmers in all the clusters where measurements were made and the data was used to estimate the water productivity or WUE. During crop development at 25, 45, 65 days after transplantation, the number of hills/m² and number tillers per hills were recorded. The panicle length, number of grains per panicle and 1000 grain weight was also collected. The average yield for rice in the various cluster under different rice growing practices is presented in Table 9.2.

Water productivity

Confusion exists concerning the terminology related to the effective use of water in agricultural production, more specifically water use efficiency (WUE) versus water productivity. WUE actually represents the actual water use by the crop as a fraction/percentage of the water applied. WP represents the amount of crop produced per unit of water applied (Molden *et al.*, 2010; Ragab, 2014) and in this study calculated based on the water applied to the field and the observed yield as:

$$WP = \frac{\text{crop yield (kg)}}{\text{irrigation and precipitation (m}^3)} \quad \text{or} \quad \frac{\text{crop yield (gg)}}{\text{irrigation and precipitation (litres)}}$$

Water balance

Deep percolation and seepage losses can significantly influence the water use of the rice crop (Tuong *et al.*, 1994), particularly the depth of the standing water affecting the percolation losses. Since alternative rice growing techniques (such as AWD, SRI and MSRI) result in a reduction of the standing water on the field, it can affect the percolation losses. The water balance was estimated as follows:

Water balance = irrigation water application – crop evapotranspiration

The water balance includes, among others, the percolation losses. A difference in the water balance provides information about the percolation losses under different rice growing alternatives. The irrigation water application is the sum of water supplied through irrigation and precipitation.

Results and discussion

Water productivity

Compared to other grain crops, water losses under irrigated rice can be considerable leading to a rather low water productivity (Singh *et al.*, 2010). Alternative rice growing practices such as SRI (Stoop *et al.*, 2009) and AWD (Bouman *et al.*, 2007, Lampayan *et al.*, 2015) are introduced with the objective to save water while increasing the yield or at least keep it at the same level. However, measurements indicate that alternative practices of rice growth might lead to an increase in the WP (more crop per drop) but at the same time give a decrease in the yield per unit area. Bouman and Tuong (2000) carried out an analysis of experimental data, mainly from the central northern India and the Philippines, on water saving irrigation practices in rice made by reducing ponded water depths to soil saturation and AWD. In India, the WP varied from 0.2 – 0.4 kg/m³ or g/L under traditional practices. This increased up to 1.9 g/L under water saving irrigation practices. However, with a decrease in yield it did not produce more rice with less water on the same field. Also Tuong and Bouman (2003) report similar results. A study carried out in West Bengal, India by Sinha and Talati (2007) showed that SRI significantly increased the yield of rice compared to traditional paddy cultivation. However, in this case no information was available on the water use and hence WP. Swarup *et al.* (2008) also report an increase in WP under different water management treatments in rice, but showed a decrease in the yield per unit area. Radha *et al.* (2009), when studying the effects of different alternative rice growing practices on water use in Andhra Pradesh observed that water savings were largest under SRI, obtaining a water productivity of 0.9 g/L while farmers practicing traditional NI obtained a water productivity of 0.6 g/L. The yield per unit area increased under alternative rice growing practices with the highest yield obtained under SRI. Results by Zhao *et al.* (2010) also showed that the yield increased with a reduction in water use, resulting in a considerable increase in WP.

The measurements carried out as part of the ClimaAdapt project showed large variations in both yield and water application between different rice growing practices but also within a single practice (Table 9.2, Figure 9.3). The minimum and maximum yield of rice under AWD varied from 3.6 to 8.6 ton/ha, with an average of 6.1 ton/ha. For normal paddy rice, the yields varied from 3.1 to 7.2 ton/ha with an average of 4.8 ton/ha. For DSR and MSRI the average yields were 6.5 ton and 5.7 ton/ha respectively. Although there is an overlap in yield between the different rice growing practices, yields under alternative improved rice practices are higher than those obtained under normal practice (NI). This is similar to the results obtained by Sinha and Talati (2007), Radha *et al.* (2009) and Zhao *et al.* (2010).

Water application

The irrigation water application was in general lower in alternative rice growing practices than water used in traditional paddy rice cultivation. Water application

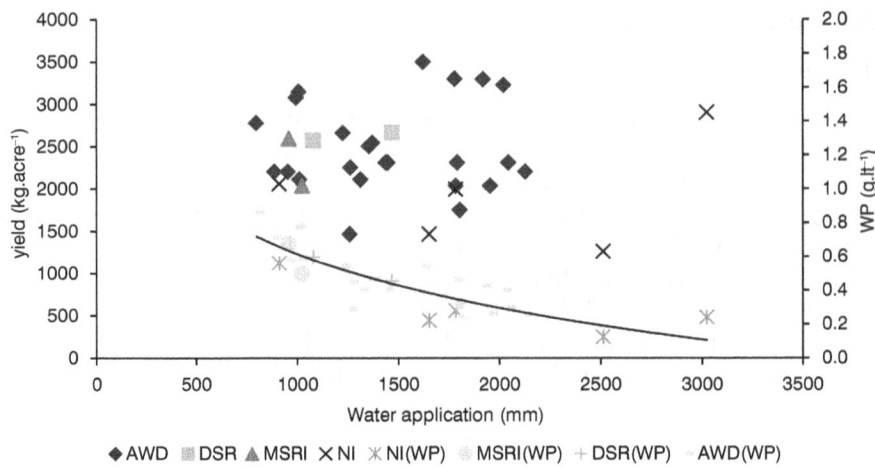

Figure 9.3 Water application, yield and WP under different rice growing practices

Source: Authors analysis from ClimaAdapt project data (www.climaadapt.org)

under AWD varied from 796 to 2128 mm with an average of 1464 mm. For traditional paddy rice cultivation, the water application varied from 909 to 3023 mm with an average of 1976 mm while the average applications to DSR and MSRI were 1273 and 991 mm, respectively. Compared to other measurements, the lowest water applications to both AWD and NI were measured in the case of Tamil Nadu (TN). This may be attributed to sandy clay soil type dominating in TN that has a lower saturated hydraulic conductivity than the soil types in AP and Telangana. However, this has not been confirmed through soil sampling and laboratory measurements. As for the yield, there is also an overlap in water use when comparing the different alternatives. The alternative rice growing practices have the lower water use as compared to the normal rice growing practice. This is in agreement with findings by Radha *et al.* (2009) and Zhao *et al.* (2010).

Based on the water delivery measurements and yield observations, the calculated WP under AWD varied from 0.240 to 0.863 g/L with an average of 0.455 g/L. For traditional paddy rice cultivation, the WP varied from 0.124 to 0.559 g/L with an average of 0.283 g/L. For DSR and MSRI the WP was 0.518 and 0.580 g/L (Table 9.2, Figure 9.3). A study by Belder *et al.* (2004) reported that under conditions with soil water potentials not less than -10kPa and the shallow groundwater table stays within 0–30 cm below soil surface, AWD can reduce water use by as much as 15 percent without affecting the yield.

The results from our study showed that the largest water application led to the lowest WP (Figure 9.3). A general trend in our results was that under reduced water application, mainly under alternative rice growing practices, the yield increased when water application was reduced, leading to a higher WP. This is similar to the results obtained by Radha *et al.* (2009) and Zhao *et al.* (2010).

Alternative options to improve the WP can be attained through improvement of the harvest index (HI), being the ratio of marketable grain yield to the total crop biomass. Considerable improvements were made in HI for many crops during the last years. For wheat and maize for example, the HI was raised from 0.35 before the 1960s to 0.5 in the 1980s. However, there seems to be little potential left for further improvements (Molden *et al.*, 2010).

One strategy to improve the WP can be through genotype improvement. Research carried out by Su *et al.* (2015) showed that by adding a transcription factor gene from barley to rice led to a shift in the carbon flux into the above ground biomass instead of the root system. This resulted in increased biomass production, while at the same time reducing the potential for methane production. Approaches to increase rice productivity and reduce methane emissions are also beneficial in a future climate, with rising temperatures resulting in increased methane emissions from paddies. Bodelier (2015), in a comment to the findings by Su *et al.* (2015), considered the results very beneficial for both food production and to reduce methane production from rice cultivation. But Bodelier (2015) also commented that additional research is needed as it might affect processes involving soil carbon, nutrients and microbial activity, with knock-on effects for the sustainability of rice cultivation.

In a study carried out by Kumar *et al.* (2012) a set of advanced breeding lines were tested under a diverse set of conditions, ranging from favorable irrigated conditions to conditions with moderate to severe reproductive stage drought in the drought-prone eastern Indian region. A set of improved genotypes were identified, having the ability to obtain stable high yields across variable environments for rainfed rice ecosystems.

Water balance

Research has shown that under paddy rice cultivation, a significant amount of water can get lost through seepage and percolation (Tuong *et al.*, 1994). Wopereis *et al.* (1994) reported losses in the order of 3,500 mm during one crop cycle in a well-puddled rice field. Seepage in this case is the loss through bunds, while percolation is the vertical loss through the soil profile. Percolation losses depend on the soil and hydrological characteristics to a large extent. Significant amounts of percolation can occur in clay soils during the land preparation phase, through cracks or in permeable sandy soils having high permeability (Tuong, 1999). Puddling reduces the percolation losses, but still a positive correlation exists between the percolation losses and the depth of the standing water in the rice field (Tabbal *et al.*, 1992). On more coarse soils, the effects of puddling are limited. A change to alternative rice growing techniques might reduce these losses due to the absence of standing water on the field. Belder *et al.* (2004) and Singh *et al.* (2010) indicated that percolation losses can be reduced by reducing the depth of the standing water similar to the hydrostatic water pressure.

However, different water balance studies show varied results. Lu *et al.* (2000) report that under intermittent irrigation regimes, enhanced percolation

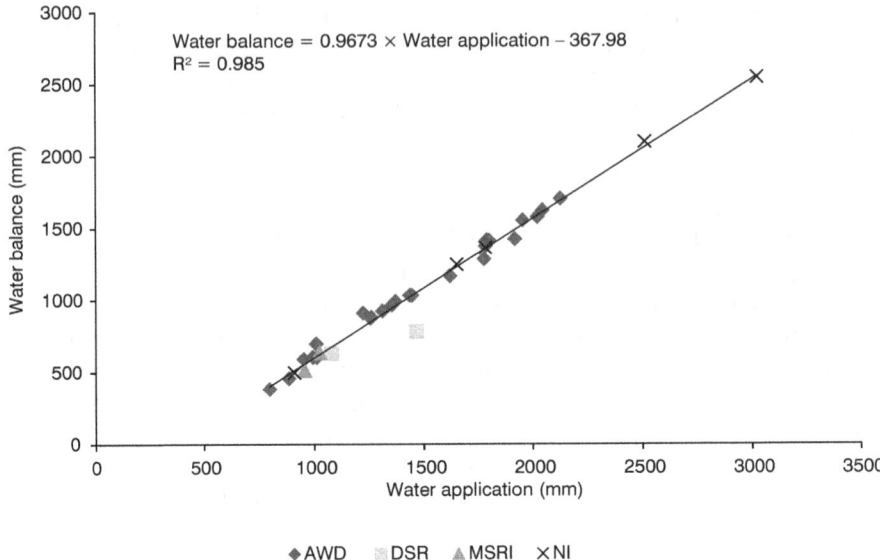

Figure 9.4 Water application and water balance results for measurements in AP and TN

Source: Authors analysis from ClimaAdapt project data (www.climaadapt.org)

occurred due to the formation of cracks in the soil. Similar results were obtained by Tournebize *et al.* (2006) from their observations in traditional mid-season drainage in paddy rice cultivation. Savings in water up to 70 percent were reported due to the introduction of intermittent irrigation by Belder *et al.* (2007) and Feng *et al.* (2007).

There is a large variation in the water balance between different rice growing practices but also within the individual practice (Table 9.2). The evapotranspiration varied from 311 to 686 mm for the growing season, with an average of 417 mm. Our measurements showed that the average water application under alternative rice growing practices showed a reduction in water application. The effect of a reduced water application led to considerable reduction in the water balance (Figure 9.4), most likely due to a reduction in percolation losses. The relation between the water balance and water application is:

Water Balance = 0.9673 × Water Application − 367.98

The relation indicates that with a water supply of 380 mm, being approximately similar to the average crop evapotranspiration, the water balance or water loss would be zero. Such an application is not possible in practice as there will be water losses with each irrigation. However, it indicates that water losses can be decreased significantly using alternative rice practices without a loss in yield. This is promising for growing rice under water scarcity conditions. Adaptation strategies should consider promoting alternate rice growing technologies and provide needed resources and training to farmers to improve their capacity to adopt the new systems.

Effects of alternative rice growing practices

Groundwater

Alternative rice growing practices have impacts on groundwater recharge in a particular area. The objective of the alternative practices is to use less water by reducing the losses. However, there might be many beneficiaries of this lost water under normal irrigation practice, such as downstream users where the water flows back to the river or canals system. Also the water losses can be used to maintain a so-called minimum flow requirement in rivers to sustain the ecosystem and provide livelihood for, among others, fishery (Perry *et al.*, 2009; Ward and Pulido-Velazquez, 2008). Groundwater is also used extensively for irrigation purposes in India for e.g. in the ClimaAdapt project sites (Table 9.2). When introducing alternative water saving rice growing practices, the recharge of groundwater might decrease, which in the long run can affect water availability for bore wells users.

Water quality

Percolation loss from rice fields might also have a negative effect on water quality, for example leading to an increase in salinity, nitrogen and fluorine content. Groundwater is also used for other purposes than agriculture, for example as drinking water. Excessive concentration of nitrogen in ground water for drinking is leading to health problems. Recommendations are that the concentration should not exceed 11 mg/L (WHO, 2011). In September 2014 a number of water samples were taken in the Nalgonda district from bore wells and open wells. The nitrogen concentrations varied from 0.91 to 7.11 mg/L with an average of 3.59 mg/L, well below the maximum allowable concentration (Figure 9.5). However, it might be worth collecting and testing water samples at regular intervals to see its development over time. One sample showed a rather high salinity concentration (2,000 μS/cm). Despite high percolation losses, there seems to be no immediate problems for agricultural production. Fluoride concentrations in DC 4 study area varied from 0.88 to 2.36 mg/L with an average of 1.51 mg/L, and with seven samples exceeding the maximum allowable concentration of 1.5 mg/L suggested by WHO (1996). This has the potential to cause serious health problems if used for drinking water purposes.

Uncertainties in the measurements

The water delivery was based on discharge measurements using RBC (Replogle, Bos, Clemmens) flumes when water supply was through the irrigation canal system, and flow meters when water supply was from bore wells. Flow meters can effectively measure the water delivery through daily readings of the metering system. However, when using RBC flumes, errors might occur as the total volume of water delivered to the farmer fields is calculated as the product of the length of time of water application and the manually measured

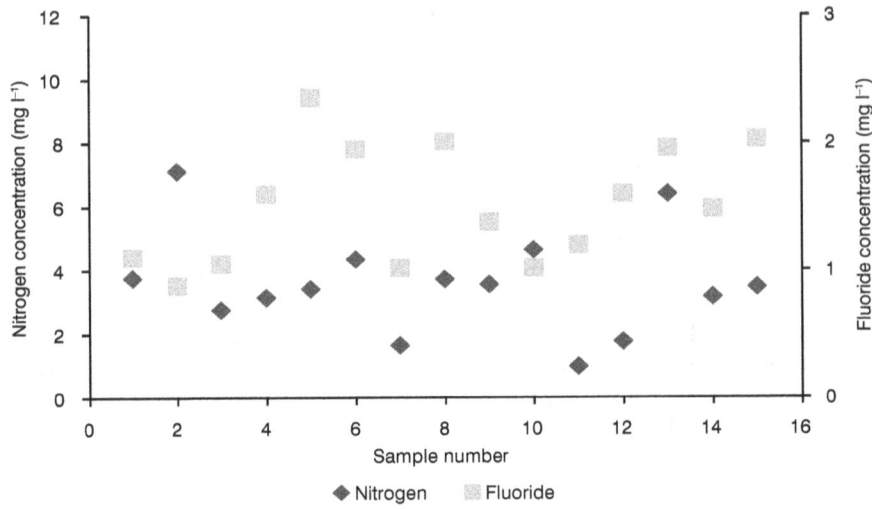

Figure 9.5 Nitrogen and fluoride concentration in water samples collected in Nalgonda district

Source: Authors analysis from ClimaAdapt project data (www.climaadapt.org)

discharge, carried out once a day. The calculation of the crop evapotranspiration was based on the crop coefficients for paddy rice (Allen *et al.*, 1998). However, knowledge about crop coefficients for alternative rice growing practices possibly could influence the calculated evapotranspiration. Improved results about crop evapotranspiration also could have been obtained using the Aquacrop model (Steduto *et al.*, 2012), but this requires additional information such as crop canopy development and soil physical parameters, which were lacking.

Summary and Conclusions

A slight increase in yield was obtained under alternative rice growing practices when compared to conventional paddy rice cultivation, while at the same time a considerable reduction in water delivery was observed. This meant an increase in the WP, hence more crop per drop. This also means that so-called 'new' water can become available for other agricultural crops and/or domestic or industrial uses. This is a very positive result, taking into consideration the need for increased food production and the increased need for water in other sectors of the economy.

The obtained results from this study are very useful for water scarce areas, and provide important information to policy makers, farmers, Agriculture Departments and water management boards devising climate smart adaptation and mitigation strategies. However, care should be taken while using the results in other areas, as the agro-climatic conditions could vary from place to place. The reduction in water delivery under alternative rice growing practices could have been impacted by the reduction in percolation losses. Water lost through percolation recharges the groundwater, and some of it also contributes to the river systems thus helping in maintenance of the ecosystems services. It is recommended to study the effects of the reduced percolation under alternative rice growing practices and the implications of losses to groundwater and river flows.

The study recommends carrying out further detailed measurements to address some of the uncertainties that were beyond the scope of this study. Additional studies could provide information that could support the recommendations regarding the improvement of the WP as observed in the present study. Our measurements were embedded with some errors, for example concerning the consistency in water delivery through the canal systems. Our experience shows that it is important to use systems that can provide reliable data concerning the amount of water delivered to the farmer's field. To improve the higher level of accuracy of the water delivery measurements, water sensors could be used together with discharge measurement structures such as the RBC flume. Measurement systems are indispensable in obtaining information to provide inputs to policy interventions and can play an important motivation role for water saving. As it is often said, *'what we cannot measure, we cannot manage.'*

When introducing the new rice growing alternatives, it is recommended to have the possibility to record water levels in the field to obtain improved water delivery to the farmer's field. Knowledge about the water level in the rice field is decisive as to when the farmer needs the next irrigation water application. However, this also needs a more flexible approach in water delivery. The present system in the project areas is mainly a supply driven system, in which farmers receive water at fixed intervals, irrespective of their use. When using rice growing practices like AWD or SRI, this is not practical. The irrigation water supply has to be switched to a demand-based water supply, both in terms of the time when the water is needed and also the quantity. This requires a shift in the current water distribution systems. Close coordination between the concerned departments and farmers will be necessary while upscaling alterative rice growing practices.

Water demand is local, depending on the evaporative demand, soil type, sowing dates and growth stage of crop. As part of the ClimaAdapt project, sensors were used on an experimental basis to measure the water level in the Bouman well in AWD rice fields for the first time in the project area. The results were positive, and continued efforts on improving such measurements with high-levels of accuracy and linking with the Irrigation Departments could be useful. This then could form the basis to move towards a demand driven irrigation water distribution.

Modernization of the irrigation canal systems might be needed to shift to a demand driven system. This means that the irrigation system has to be flexible both with respect to new rice irrigation systems but also with respect to alternative crops if WP has to be increased. When new systems are introduced, farmers, agricultural and irrigation engineers all have to be trained, and awareness has to be raised about agricultural water use and the need to increase the water productivity.

Note

1 www.climaadapt.org

References

Alberto, M.C.R., Wassmann, R., Hirano, T., Miyata, A., Hatano, R., Kumar, A., Padre, A. and Amante, M. (2011) Comparisons of energy balance and evapotranspiration between flooded and aerobic rice fields in the Philippines, *Agricultural Water Management,* 98: 1417–1430.

Allen, R., Pereira, L.S., Raes, D. and Smith, M. (1998) Crop evapotranspiration: Guidelines for computing crop requirements, Irrigation and Drainage Paper No. 56, FAO.

Belder, P., Bouman, B.A.M., Cabangon, R., Guoan, L., Quilang, E.J.P., Yuanhua, L., Spiertz, J.H.J. and Tuong, T.P. (2004) Effect of water-saving irrigation on rice yield and water use in typical lowland conditions in Asia, *Agricultural Water Management,* 65: 193–210.

Belder, P., Bouman, B.A.M. and Spiertz, J.H.J. (2007) Exploring options for water savings in lowland rice using a modelling approach, *Agricultural Systems,* 92: 91–114.

Bhushan, L., Ladha., J.K., Gupta., R.K., Singh., S., Tirol-Padre., A., Saharawat., Y.S., Gathala, M. and Pathak, H. (2006) Saving of water and labor in a rice–wheat system with no-tillage and direct seeding technologies, *Agronomy Journal,* 99: 1288–1296.

Bodelier, P.L.E. (2015) Bypassing the methane cycle: movies of a growth, *Nature,* 523: 2.

Bos, M.G., Repogle, J. and Clemmens, A.J. (1984) *Flow Measuring Flumes For Open Channel Systems,* John Wiley and Sons, New York.

Bouman, B.A.M., Lampayan, R.M. and Tuong, T.P. (2007) Water management in irrigated rice : coping with water scarcity. IRRI Los Baños, Philippines. online. http://books.irri.org/9789712202193_content.pdf (accessed 23 June 2015).

Bouman, B.A.M. and Tuong, T. (2000) Field water mangement to save water and increase its productivity in irrigated lowland rice, *Agricultural Water Managagement,* 1615: 1–20.

Chauhan, B.S. and Opeña, J. (2012) Effect of tillage systems and herbicides on weed emergence, weed growth, and grain yield in dry-seeded rice systems, *Field Crops Research,* 137: 56–69.

Cheralu, C. (2011) Status paper on rice in Andhra Pradesh, Rice Knowledge Management Portal. http://www.rkmp.co.in/sites/default/files/ris/rice-state-wise/Rice State Wise Andhra Pradesh_0.pdf (accessed 27 October 2015).

CWC (Water Resources Information System Directorate) (2013) Water and related statistics. http://www.cwc.nic.in/main/downloads/Water and Related Statistics-2013.pdf (accessed 28 October 2015).

Feng, L., Bouman, B.A.M., Tuong, T.P., Cabangon, R.J., Li, Y., Lu, G. and Feng, Y. (2007) Exploring options to grow rice using less water in northern China using a modelling approach. I. Field experiments and model evaluation, *Agricultural Water Management,* 88: 1–13.

Government of India (2008) National Action Plan on Climate Change, online, http://www.moef.nic.in/sites/default/files/Pg01-52_2.pdf (accessed 30 October 2015).

Gurava, R.K., Kakumanu, K.R., Palanisami, K. and Nagothu, U.S. (2013a) Can mechanization reduce labour and water demand in agriculture? a case of rice transplanters in Andhra Pradesh. International Conference. Extension Educational Strategies for Sustainable Agricultural Development – A Global Perspective. University of Agricultural Sciences, Bangalore.

Gurava R.K., Kakumanu, K.R., Tirapamma, S., Palanisami, K. and Nagothu, U.S. (2013b) Partnership as key for enhancing technology transfer: A case of direct sowing rice in Guntur district, Andhra Pradesh, India. Compendium: National Seminar on Futuristic Agricultural Extension for Livelihood Improvement and Sustainable Development. January 19–21. Rajendranagar, Hyderabad, India.

Kreamer, D.K. (2013) The past, present, and future of water conflict and international security, *Journal of Contemporary Water Research & Education*, 149 (1): 87–95.

Kumar, S., Surinaidu, L., Pavelic, P. and Davidson, B. (2012) Integrating cost and benefit considerations with supply and demand-based strategies for basin scale groundwater management in South-West India, *Water International*, 37: 460–477, doi:10.1080/0250 8060.2012.708601, 2012.

Lampayan, R.M., Rejesus, R.M., Singleton, G.R. and Bouman, B.A.M. (2015) Adoption and economics of alternate wetting and drying water management for irrigated lowland rice, *Field Crops Research,* 170: 95–108.

Lu, J., Ookawa, T. and Hirasawa, T. (2000) The effects of irrigation regimes on the water use, dry matter production and physiological responses of paddy rice, *Plant Soil*, 223: 207–216.

Molden, D., Oweis, T., Steduto, P., Bindraban, P., Hanjra, M.A. and Kijne, J. (2010) Improving agricultural water productivity: Between optimism and caution, *Agricultural Water Management*, 97: 528–535.

Nagothu, U.S., Gosain, A.K., Barton D.N., Palanisami, K., Tirupathaiah, K., Reddy, K.K., Stålnacke, P., Deelstra, J. and Gupta, S. (2012) *Climate Change and Impacts on Water Resources: Guidelines for Adaptation in India*, BioforskAs, Norway.

Perry, C., Steduto, P., Allen, R.G. and Burt, C.M. (2009) Increasing productivity in irrigated agriculture: Agronomic constraints and hydrological realities, *Agricultural Water Management*, 96: 1517–1524.

Radha, Y., Reddy, K.Y., Rao, G.S., Chandra, S.R. and Babu, G.K. (2009) Water-saving rice production technologies in krishna western delta command of Andhra Pradesh – An economic analysis 1, *Agricultural Economics Research* review, 22: 397–400.

Raes, D. (2009) *ET_o Calculator.* Food and Agriculture Organization of the United Nations Land and Water Division FAO, Rome.

Ragab R. (2014) A note on water use efficiency and water productivity. http://www. water4crops.org/water-use-efficiency-water-productivity-terminology/ (accessed 29 October 2015).

Rejesus, R.M., Palis, F.G., Rodriguez, D.G.P., Lampayan, R.M. and Bouman, B.A.M. (2011) Impact of the alternate wetting and drying (AWD) water-saving irrigation technique: Evidence from rice producers in the Philippines, *Food Policy*, 36: 280–288.

Singh, R., Kundu, D.K. and Bandyopadhyay, K.K. (2010) Enhancing agricultural productivity through enhanced water use efficiency, *Journal of Agricultural Physics*, 10: 1–15.

Sinha, S.K. and Talati, J. (2007) Productivity impacts of the system of rice intensification (SRI): A case study in West Bengal, India, *Agricultural Water Management*, 87: 55–60.

Steduto, P., Hsiao, T.C., Fereres, E. and Raes, D. (2012) *Crop Yield Response To Water*, FAO Irrigation and Drainage Paper No.66.

Stoop, W.A., Adam, A. and Kassam, A. (2009) Comparing rice production systems: A challenge for agronomic research and for the dissemination of knowledge-intensive farming practices, *Agricultural Water Management*, 96: 1491–1501.

Stoop, W.A., Uphoff, N. and Kassam, A. (2002) A review of agricultural research issues raised by the system of rice intensification (SRI) from Madagascar: Opportunities for improving farming systems for resource-poor farmers, *Agricultural Systems,* 71: 249–274.

Su, J., Hu, C., Yan, X., Jin, Y., Chen, Z., Guan, Q., Wang, Y., Zhong, D., Jansson, C., Wang, F., Schnürer, A. and Sun, C. (2015) Expression of barley SUSIBA2 transcription factor yields high-starch low-methane rice, *Nature,* 523: 19.

Swarup, A., Panda, D., Mishra, B. and Kundu, D.K. (2008) Water and nutrient management for sustainable rice production. In: Singh, D.P. (Ed.) *Rice Research Priorities and Strategies for Second Green Revolution,* Central Rice Research Institute, Cuttack, India.

Tabbal, D.F., Lampayan, R.M. and Bhuiyan, S.I. (1992) Water efficient irrigation technique for rice. In: Murty, V.N. and Koga, K. (Eds.) Soil and water engineering for paddy field management. *Proceedings of the International Workshop on Soil and Water Engineering for Paddy Field Management,* Asian Institute of Technology, Bangkok.

The Economic Times (2015) India becoming water scarce http://articles.economictimes. indiatimes.com/2015-07-31/news/65074360_1_water-sector-average-annual-water-availability-utilisable-water (accessed 28 October 2015).

Tournebize, J., Watanabe, H., Takagi, K. and Nishimura, T. (2006) The development of a coupled model (PCPF-SWMS) to simulate water flow and pollutant transport in Japanese paddy fields, *Paddy Water Environment* 4: 39–51.

Tuong, T.P. (1999) Productive water use in rice production, *Journal of Crop Production,* 2: 241–264.

Tuong, T. and Bouman, B. (2003) Rice production in water-scarce environments. In: Kijne, J.W., Barker, R. and Molden, D. (Eds). *Water Productivity in Agriculture: Limits and Opportunities for Improvement. The Comprehensive Assessment of Water Management in Agriculture Series.* Volume 1, CABI Publishing, Wallingford, UK.

Tuong, T.P., Bouman, B.A.M. and Mortimer, M. (2005) More Rice, less water-integrated approaches for increasing water productivity in irrigated rice-based systems in Asia, *Plant Production Science,* 8: 231–241.

Tuong, T.P., Marquez, J.A., Kropff, M.J. and Wopereis, M.C.S. (1994) Mechanisms and control of percolation losses in irrigated puddled rice fields, *Soil Science Society of America Journal,* 58: 1794–1803.

Uphoff, N. (2003) Higher yields with fewer external inputs? The system of rice intensification and potential contribution to agricultural sustainability, *International Journal of Agricultural Sustainability,* 1: 38–50.

Uphoff, N., Amir, K. and Richard H. (2011) SRI as a methodology for raising crop and water productivity: productive adaptations in rice agronomy and irrigation water management, *Paddy and Water Environment,* 9: 3–11.

Ward, F.A. and Pulido-Velazquez, M. (2008) Water conservation in irrigation can increase water use, *Proceedings of the National Academy of Sciences,* 105: 18215–18220.

Water Vision Team (2003) Andhra Pradesh Water Vision, Andhra Pradesh Water Vision Document. Mission Support Unit, Water Conservation Mission, Government of Andhra Pradesh. https://www.academia.edu/10786480/Andhra_Pradesh_Water_Vision_-_a_shared_water_vision._Volume_1 (accessed 29 October 2015).

WHO (1996) Fluoride in drinking water, online, http://www.who.int/water_sanitation_health/dwq/chemicals/fluoride.pdf (accessed 04 November 2015).

WHO (2011) *Nitrate and Nitrite in Drinking-water: WHO Guidelines for Drinking-water Quality,* WHO Press, Geneva, Switzerland. http://www.who.int/water_sanitation_health/dwq/chemicals/nitratenitrite2ndadd.pdf

Wopereis, M.C.S., Bouman, B.A.M., Kropff, M.J., ten Berge, H.F.M. and Maligaya, A.R. (1994) Water use efficiency of flooded rice fields I. Validation of the soil-water balance model SAWAH, *Agricultural Water Management,* 88 (1–3): 1–13.

Yadav, R.L., Singh, S.R., Prasad, K., Dwivedi, B.S., Batta, R.K., Singh, A.K., Patil, N.G. and Chaudhari, S.K. (2000) Management of irrigated agro-ecosystem. In: Yadav, J.S.P. and Singh, G.B. (Eds.) *Natural Resource Management for Agricultural Production in India,* Indian Society of Soil Science, New Delhi.

Zhao, L., Wu, L., Li, Y., Animesh, S., Zhu, D. and Uphoff, N. (2010) Comparisons of yield, water use efficiency, and soil microbial biomass as affected by the system of rice intensification, *Communications in Soil Science and Plant Analysis,* 41: 1–12.

10 Climate smart rice production systems

Studying the potential of alternate wetting and drying irrigation

Krishna Reddy K., Mehreteab Tesfai, Andrew Borrell, Udaya Sekhar Nagothu, Suresh Reddy K. and Gurava Reddy K.

Introduction

Asia is the biggest producer and consumer of rice in the world. Almost 90 percent of the total world's rice is produced using irrigation (IRRI, 2003; IRRI, 2009). By 2025, about 15 to 20 million hectares of irrigated rice fields will suffer some degree of water scarcity as a result of climate change and competing water uses (Richards and Sander, 2014), which requires rethinking of the current management practices of rice. Rice grown in a traditional paddy is water intensive and requires between 700–1,500 mm of water per cropping season, depending on soil type and texture (Bhuiyan, 1992). The traditional paddy practice often leads to losses of water via surface run-off, seepage and percolation that accounts for 50–80 percent of total water input (Chapagain and Riseman, 2011). Downstream rice farmers may use part of the lost water through percolation. On top of this, the growing competition for water use for industrial and domestic purposes exacerbates the water scarcity for agricultural use.

In India, rice covers about 44 million hectares of cultivated land with an average total production of 103 million tons (MoA, 2012). To meet the growing food demand, India may need to produce at least 130 million tons of milled rice by 2030 (Gujja and Thiyagarajan, 2009). There is a need for the development of efficient rice cultivation methods that require relatively less water than normal irrigation (NI), while at the same time sustaining rice production to meet the ever-increasing national food demand. In this chapter, the term normal irrigation (NI) refers to the traditional paddy system where rice is grown under continuous flooding.

There are a number of improved rice cultivation methods, such as Alternate Wetting and Drying (AWD) irrigation developed by the International Research Rice Institute (IRRI) in cooperation with national research institutions. AWD irrigation is a water management system where rice fields are not kept continuously submerged, but are allowed to dry intermittently during the rice growing stage, and water is applied after disappearance of flooded or ponded water (Borrell *et al.*, 1991; Borrell *et al.*, 1997; IRRI, 2009; Yamaji, 2011). In this practice, irrigation is provided when the water level drops to 5–15 cm below the soil surface. The number of days when the field is allowed to be non-flooded before the next irrigation is applied can vary from 1 day to more than 10 days, depending on the soil type and crop growth stage (Rejesus, *et al.*, 2011). AWD promotes better root growth, facilitates nutrient uptake and increases land and water productivity (Sarkar, 2001; Uphoff, 2006). AWD is considered as a Climate Smart Agriculture (CSA) practice because it generates multiple benefits in terms of increased productivity by promoting more effective tillering and stronger root growth of rice plants (Dong *et al.*, 2008; Bouman *et al.*, 2007; Liu *et al.*, 2013; Richards and Sander, 2014), mitigating Green House Gases (GHGs) emissions, particularly methane (Jain *et al.*, 2013), and enhancing adaptation to water scarcity by consuming less irrigation water (Salim and Shehzad, 2008; Siopongco *et al.*, 2013). Liu *et al.* (2013) showed that AWD significantly increased grain yield (at p = 0.05 level) and other yield components such as number of panicle per m^2, filled grains (percent) and 1,000 grain weight are greater compared to NI, the continuously flooded rice fields. However, other studies reported a yield penalty associated with AWD (Borrell *et al.*, 1997; Tabbal *et al.*, 2002; Towprayoon *et al.*, 2005; Dong *et al.*, 2012).

Saturated soil culture (SSC) is another water saving technology. Initially developed for soybean production in the semi-arid tropics, it was adapted to rice production in northern Australia (Borrell *et al.*, 1997) and Indonesia (Mahrup *et al.*, 2005; Ma'shum *et al.*, 2009). SSC comprises growing rice on raised beds of height 20 cm and width 120 cm, with water maintained in the furrows (30 cm wide) some 10 cm below the bed surface (Borrell *et al.*, 1997). The Australian studies found that it was not necessary to flood rice to obtain high grain yield and quality. There was no significant difference in grain yield and quality between SSC and traditional flooded production (NI), although SSC used about 32 percent less water in both the wet and dry seasons (Borrell *et al.*, 1997). Similarly, the studies in Indonesia found that SSC reduced water requirement by 50 percent (Wakan, Lombok) and 44 percent (Kawo, Lombok), compared with NI (Mahrup *et al.*, 2005). Furthermore, water use efficiency (WUE) for grain production in SSC was increased by 90 percent at Wakan and 56 percent at Kawo, compared with NI. The extra water saved with SSC was then used to irrigate high value secondary crops, highlighting the value of this water saving technology to the whole cropping system (Mahrup *et al.*, 2005).

Recent studies suggest that rice cultivation is an important anthropogenic source of not only atmospheric CH_4, but also of N_2O. Rice soils that are flooded for long periods of the year tend to accumulate soil organic carbon, even with complete removal of above-ground plant biomass (Wassmann *et al.*, 2009). Changing water management from continuously flooding to AWD irrigation appears to be the most promising mitigation option, and is particularly suited to reducing emissions from irrigated rice paddies. AWD can reduce methane emissions by up to 50 percent, owing to the periodic aeration of soil that inhibits methane-producing bacteria (IPCC, 2006; FAO, 2007; FAO, 2010; Uprety *et al.,* 2012; Jain *et al.,* 2013). While attempting to reduce CH_4 emissions, there is a trade-off of increasing emission of N_2O from AWD fields when they are heavily fertilized, but the net effect on the Global Warming Potential is reduced under AWD (Jain *et al.,* 2013).

AWD technology can reduce the number of irrigations needed during the crop growing stages compared with NI and can lower irrigation water consumption by at least 25 percent (Borrell *et al.*, 1997; Veeraputhiran *et al.*, 2012; Ndiiri *et al.*, 2013; Thiyagarajan & Biksham, 2013; Siopongco *et al.*, 2013). Water savings in AWD compared with NI could be as high as 45 percent, thus resulting in reallocation of the saved water to downstream fields for irrigation or other purposes (Siopongco *et al.,* 2013). Furthermore, AWD can save farmers money spent on pumping costs by using less irrigation water and reducing labour cost through improved field conditions thereby increased net return for farmers (Richards and Sander, 2014). AWD generates multiple benefits related to GHG emissions reduction (mitigation), reducing water use (adaptation where water is scarce), increasing productivity and efficiency of water use for grain production (Borrell *et al.,* 1997; Mahrup *et al.*, 2005), contributing to food security (Bouman *et al.,* 2007), and controlling water-borne disease-causing organisms (Van der Hoek *et al.,* 2001). Nevertheless, there is a lack of adequate information that addresses the multiple benefits of AWD in coping with climate change, reducing GHG emissions, saving farmers' money, and saving water for other high-value enterprises within the farming system while maintaining rice production in a holistic way.

The objective of this chapter is to demonstrate the potential of AWD to adapt to changing water availability and its related co-benefits of CSA (i.e. enhanced water use efficiency and mitigated GHG emissions) by reviewing and presenting case studies from rice growing countries in Asia that apply AWD practice. The description of AWD performance is based on an extensive literature review and analysis of research results in some countries of Asia. The description of AWD in India is primarily based on the analysis of the recent research trials carried out in the ongoing ClimaAdapt project in Andhra Pradesh and Telangana States in Southern India.

Review of literature

In this section, the AWD performance with respect to the three objectives of CSA (i.e. adaptation to water scarcity, mitigation to methane and nitrous oxide emissions, and increased productivity and food security) are assessed in comparison with the conventional rice cultivation method (here after referred to as NI) in some selected rice growing countries of Asia and Australia where data is available.

AWD as adaptation measure to water scarcity

Table 10.1 provides some examples of AWD's role in saving water from irrigated rice fields and increasing water productivity in comparison to the conventional

Table 10.1 AWD performance in water consumption during the rice growing period

Country (site)	NI (mm)	AWD (mm)	Water saving (mm)	Water saving (percent)	References
Philippines (Guimba)	2,352	905	1,447	62	Tabbal *et al.* (2002)
India (Andhra Pradesh)	2,248	1,245	1,003	45	This study
India (New Delhi)	1,750	980	770	44	Jain *et al.* (2013)
Bangladesh (Mymensingh[a])	1,172	897	275	23	Oliver *et al.* (2008)
Philippines (Canarem)	750	590	160	21	Toung (2007)
China (Zhangbang)	876	780	96	11	Yao *et al.* (2012)
Australia (dry season)	1,351	764	587	43	Borrell *et al.* (1997)
Australia (wet season)	1,286	873	413	32	Borrell *et al.* (1997)

[a] A farm located at Bangladesh Agricultural University

Source: Authors' own compilation

rice cultivation methods in some selected countries of Asia. The rice water consumption by AWD and NI varied among the countries (i.e. 750 to 2,352 mm in NI fields and from 590 to 1245 mm under AWD treatment, Table 10.1). For all countries in Table 10.1, AWD fields consumed less water than NI. For example, field research carried out by Borrell *et al.* (1997) in Australia during the dry season, Tabbal *et al.* (2002) in the Philippines, and Jain *et al.* (2013) in India has shown that from 43–62 percent of irrigation water can be saved by applying the AWD method relative to NI.

The water productivity of AWD fields (i.e. grain yield (kg) per m^3 of total water used) varied between the sites (Table 10.2). For example in Japan, water productivity was higher by 31 percent at Chiba and by 52 percent at Fukuoka over the NI method. Field research by Tabbal *et al.* (2002) and the current study has shown that AWD can increase water productivity by 84–88 percent. However, AWD had little impact on water productivity in Australian experiments.

AWD as mitigation measure to GHGs emissions

CH$_4$ emissions

Table 10.3 presents examples of the role of AWD in reducing CH$_4$ emissions in some rice growing countries of Asia. In all cases where information is available,

Table 10.2 AWD performance in water productivity during the rice growing period

Country (site)	NI (kg/m³)	AWD (kg/m³)	References
Japan (Chiba)	1.30	1.70 (+31 %)	Yamaji (2011)
Philippines (Muñoz)	1.20	1.48 (+23 %)	Belder *et al.* (2004)
China (Tuanlin)	0.99	1.07 (+8 %)	Belder *et al.* (2004)
Australia (Queensland[a])	0.65	0.66 (+2 %)	Borrell *et al.* (1997)
Japan (Fukuoka)	0.50	0.76 (+52 %)	Sujono *et al.* (2011)
Australia (Queensland[b])	0.48	0.41 (-15 %)	Borrell *et al.* (1997)
Philippines (Guimba)	0.25	0.46 (+84 %)	Tabbal *et al.* (2002)
India (Andhra Pradesh)	0.24	0.45 (+88%)	This study

[a] Dry season; [b] Wet season. Percentages in brackets indicate positive/negative change by AWD over NI field

Source: Authors' own compilation

CH_4 emissions from NI fields were greater than AWD fields. The CH_4 emissions from AWD rice fields varied among the countries and between the sites. By applying the AWD method, CH_4 emissions from rice fields can be reduced by 13 percent to 56 percent (Table 10.3).

Table 10.3 Some examples of measured CH_4–C (g C/m²) average emissions from AWD and NI rice fields

Country (site)	NI	AWD	Reduction in CH_4 emission (%)	References
India (Tamil Nadu)	3.9	3.4	13	Lakshmanan *et al.* (2015)
India (Punjab)	5.0	2.2	56	Sidhu and Benbi (2011)
Indonesia (Central Java)	1.6	1.1	31	Hidayah *et al.* (2008)
Philippines (Laguna)	n.a.	3.3	–	Alberto *et al.* (2014)
Bangladesh (Mymensingh)	12.4	9.0	27	Ali *et al.* (2013)
Thailand (Samutsakorn)	24.4	15.7	36	Towprayoon *et al.* (2005)
Indonesia (West Java)	40.9	21.3	48	Husin *et al.* (1995)

Total emissions per season for 120 growing days. n.a.: not available

Source: Authors' own compilation

Table 10.4 Some examples of measured N_2O-N emissions (mg N/ m^2) from AWD and NI rice fields. Total emissions per season for 120 growing days

Country (site)	NI	AWD	References
India (Tamil Nadu)	28.8	27.6	Lakshmanan *et al.* (2015)
India (New Delhi)	9.6	14.6	Jain *et al.* (2013)
Thailand (Samutsakorn)	37.2	38.5	Towprayoon *et al.* (2005)
Bangladesh (Mymensingh)	55.0	98.0	Ali *et al.* (2013)
Japan (Ibaraki)	6.5	6.5	Yagi *et al.* (1996)

Source: Authors' own compilation

N_2O emissions

There are conflicting results reported regarding nitrous oxide emissions from AWD fields, (Cai *et al.*, 1997; Zou *et al.*, 2005; Sharma, 2008). For instance, Sharma (2008) reported that considerable emissions of nitrous oxide from AWD rice fields in northern India using a simulation modelling approach. In Tamil Nadu, India, there was very little difference in N_2O emissions between AWD and NI plots, where experimental plots planted with the rice variety CO-48 emitted, on average, 2.35 g N_2O-N/ha/d from AWD fields and 2.40 g N_2O-N/ha/d from NI fields (Lakshmanan *et al.*, 2015). On the other hand, Jain *et al.* (2013) reported nitrous oxide emissions from AWD rice fields in New Delhi measured 14.6 mg N/m^2 N_2O-N, which were greater than NI fields (9.6 mg N/m^2 N_2O-N Table 10.4). While attempting to reduce CH_4 emissions, however, there may be a trade-off resulting in increased emissions of N_2O from AWD fields due to the changing redox potential in the soil pore system (e.g. Jain *et al.*, 2013; Ali *et al.*, 2013). However, the net effect on the Global Warming Potential is reduced under AWD (Jain *et al.*, 2013; Richards and Sanders, 2014).

AWD role in increasing agricultural productivity and food security

Table 10.5 shows examples of the role of AWD in increasing crop yields in some rice growing countries. For example in China, AWD fields sown with a rice variety Yangjing 4038 (Japonica) gave higher grain yield and also rendered a greater percentage of productive tillers, filled grains and 1,000-grain weight when compared to NI fields. While rice yields from AWD fields were higher than NI fields in China, Madagascar and India, they were lower relative to NI in Vietnam, Thailand, Philippines and Australia (Table 10.5).

Case study from India

Site descriptions

The study was carried out in the Andhra Pradesh and Telangana States of the Krishna river basin, India, during the *rabi* seasons of 2012–13 and 2013–14

Table 10.5 Impact of AWD on crop yield in some Asian and African countries

Country (site)	NI (ton/ha)	AWD (ton/ha)	References
China (Yangzhou University)	9.6	10.2	Liu *et al.* (2013)
Madagascar (Ambatondrazaka)	6.4	7.4	McHugh *et al.* (2002)
Vietnam (Mekong river delta)	6.0	5.7	Dong *et al.* (2012)
India (Andhra Pradesh)	5.7	6.4	This study
Australia (Queensland[a])	8.8	5.1	Borrell *et al.* (1997)
Australian (Queensland[b])	6.1	3.6	Borrell *et al.* (1997)
Philippines (Guimba)	5.3	4.0	Tabbal *et al.* (2002)
Thailand (Samutsakorn)	4.3	3.9	Towprayoon *et al.* (2005)

[a] Dry season; [b] Wet season

Source: Authors' own compilation

(December/January to March/April) and the *kharif* seasons of 2013 and 2014 (June/July to December). In the basin, the Nagarjuna Sagar Project was selected to demonstrate AWD co-benefits of CSA, covering two districts, Guntur (15°44'N, 80°55'E) from Andhra Pradesh State and Nalgonda (16°25'N, 80°05'E) from Telangana State. The Guntur and Nalgonda districts are located on the right and left canal of the Nagarjuna Sagar Project, respectively.

The long-term mean annual rainfall at Guntur is 815 mm and at Nalgonda is 753 mm, of which 70 percent falls between June and September from the southwest monsoon, and the remaining 30 percent from the northeast monsoon (October to December). Long-term mean annual air temperature is high in April and May (36 to 43°C), and milder in August to October (17 to 35°C). Weather data for the study areas were acquired from nearby meteorological stations located at the Regional Agricultural Research Station, Lam, in Guntur and the Agricultural Research Station, Kampsagar, in Nalgonda districts. The rainfall, mean, minimum and maximum temperatures during the crop growing season (July–October) recorded at Guntur site is given in Figure 10.1. At this site, during the *kharif* season, the cumulative rainfall was 641 mm. The maximum temperature (34.4°C) recorded was in July and the minimum temperature (17.7°C) was in October. The mean temperature varied between 24 to 27°C over the growing season.

Four sub-districts (locally called *mandals*) from Guntur and three *mandals* from Nalgonda district were selected on the basis of the Water Users' Associations (WUAs) under Andhra Pradesh Famers Management of Irrigation Systems Act, 1997. Interested farmers from those *mandals* were pooled to form clusters. The clustered farmers used water from the same irrigation canal outlets. In each

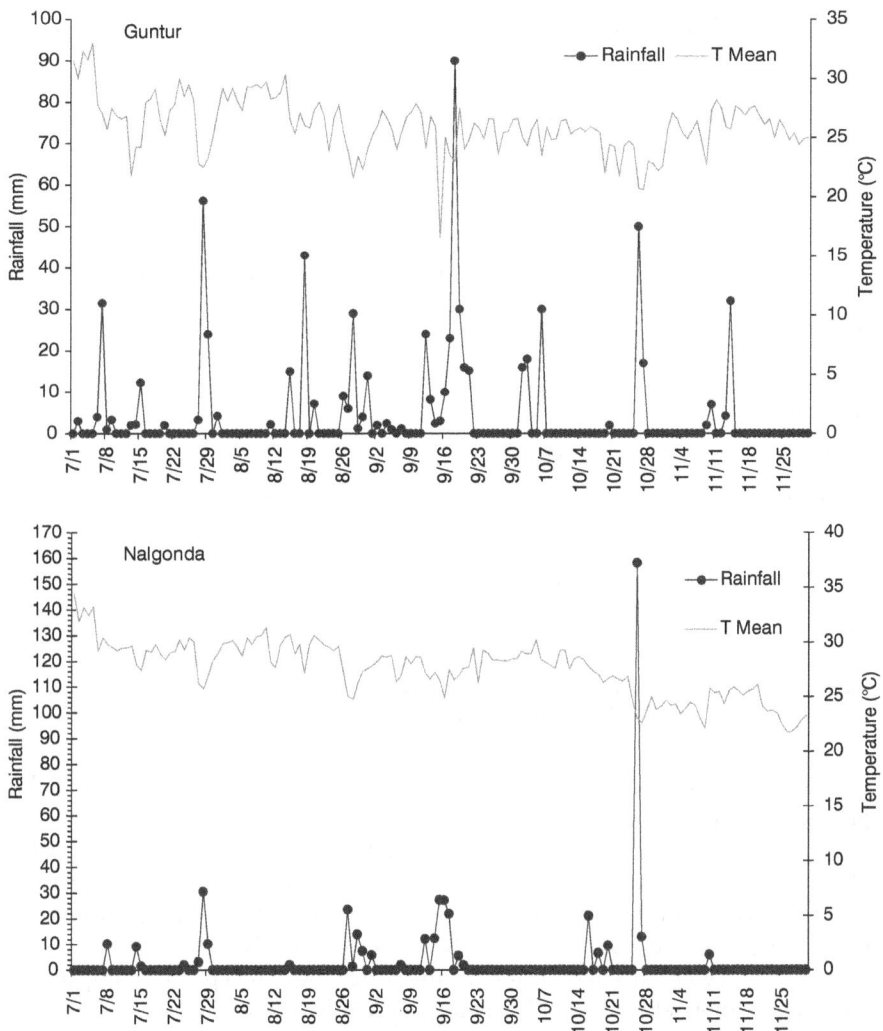

Figure 10.1 Precipitation and mean air temperatures of the study sites at Guntur and Nalgonda during growing season *kharif* 2014

Source: Results based on field data from ClimaAdapt project, (www.climaadapt.org)

cluster, land preparation (i.e. ploughing, puddling and levelling) was undertaken in the NI and AWD fields. In total, 207 and 126 farmers' fields were monitored during the *kharif* and *rabi* seasons respectively, on selected agronomic and water utilization parameters. The sampling schemes comprised of 34 percent from Guntur (n = 114) and 66 percent (n = 219) from Nalgonda district. These two sets of farmers have more or less similar land holding sizes and educational levels within the groups. The average land holding size of the majority of farmers in Guntur and Nalgonda varies from 1 to 2 ha. In both districts, more

Table 10.6 Some soil characteristics for Guntur and Nalgonda sites

Characteristics	Site	
	Guntur (n = 67)	Nalgonda (n = 42)
Sand (%)	20-40	50-70
Silt (%)	30-60	10-20
Clay (%)	30-70	20-30
Org. C (%)	0.55 (0.14-0.98)	0.49 (0.07-0.78)
pH (H$_2$O)	8.1-9.3	6.8-8.6
N (kg/ha)	95.8	63.2
P (kg/ha)	13.9	14.9
K (kg/ha)	34.2	20.2
Zinc (mg/kg)	5.01	1.42
Boron (mg/kg)	0.24	0.37
SO$_4^{2-}$ (%)	0.008	0.002
Mg (%)	0.013	0.021
Fe (%)	0.0003	0.004

Source: Based on data from ClimaAdapt project (www.climaadapt.org)

than 90 percent of the households are literate and only <10 percent have below primary education.

Soil samples were collected from the topsoil layer in the farmers' fields before planting rice seedlings in Guntur (n = 67) and in Nalgonda (n = 42) depending on the variability of the soils in the study areas. The sampled soils were analysed for major plant nutrients i.e. NO_3^-, NH_4^+, PO_4^{3-} and K^+ and some important micronutrient elements like Zn at KKB micro-testing labs (a private company and accredited national laboratory) following the standard procedures.

The soils of the Guntur area are dominated by deep black cotton soils with a silty clay loam texture, and contain, on average, low organic carbon (0.55 percent), and pH range 8.1–9.3 (Table 10.6). In contrast, shallow red soils with sandy loam texture cover a large area in Nalgonda district containing, on average, low organic carbon (0.49 percent), and pH range 6.8–8.6. The fields in Guntur district are located at the tail ends of the right canal with *kharif* (June to October) rice cultivation only. While, fields in Nalgonda district are located on the upper reaches of the canal in the Nagarjuna Sagar Project command area, with *kharif* as well as *rabi* (November to March) rice cultivation.

The major water resources in the study area are canal, bore well and lift irrigation. There was a shortage of water in the canals during the 2013 *rabi* season due to drought conditions. During this period, most of the farmers irrigated their fields from bore well water. The common rice variety grown during the *rabi* season is MTU-1010, a short duration variety (120 to 125 days). In the 2013 and 2014 *kharif* seasons, the long duration rice variety BPT 5204 (150 to 155 days) and short duration variety NLR 34449 (120–125 days) were grown depending on the onset of the monsoon and canal water availability. The main reason for selecting BPT 5204 during *kharif* is because of its quality and premium price. The biophysical observations on the rice plant, water utilization and market price varied between *kharif* and *rabi* seasons and years, and between the study sites. Results from this study compared the NI method (as a control field) and AWD during the *kharif* and *rabi* seasons, and the following measurements and estimations were carried out.

Agronomic measurements

The rice seedlings (30 days old) were transplanted at a spacing of 20 cm x 15 cm during *kharif* and 15 cm x 10 cm in the *rabi* season. The higher crop density in *rabi* is to compensate the yield difference which may arise from the short duration variety planted. Fertilizers were applied at the rate of 150 to 200 kg N/ha in the form of urea, 75 kg P/ha in the form of diammonium phosphate and 75 kg K/ha in the form of muriate of potash. All the phosphotic and half of the potassium fertilizers were applied during the last land preparation. The remaining half potassium dose was applied at the panicle initiation stage. The nitrogen (in the form of urea) was split into two doses and applied during tillering and panicle initiation stages. In some instances, phosphorus was applied in split doses, along with nitrogen in the form of complex fertilizers. In addition, some farmers added farm yard manure to their fields at a rate of 5 tons/ha.

Agronomic parameters such as number of hills/m^2, days after transplanting (DAT), number of tillers after transplanting till maximum tillering (at 20 days interval), number of panicles per hill, panicle length (cm), number of grains per panicle, 1,000 grains weight (g) and grain yield (kg/ha) were measured in the *kharif* and *rabi* seasons. In addition, cost of seeds, fertilizers, herbicides, pesticides and labour were also recorded. Cost benefit analysis and cost benefit ratio was calculated for AWD and NI treatments.

Water flow measurements

RBC flumes (Replogle, Bos, Clemmens)[1], were installed at both inlets and outlets of the canals in order to measure the irrigation water added to the fields. Flow meters of 5 to 15 cm size were used to monitor the volume of water pumped from the boreholes and lift irrigations. The irrigation water used in each successive growing stage of the rice plant was registered using a flow meter and RBC flumes. To derive a first approximation of the impact of the AWD method on irrigation water use, the number of pumping hours used for growing the rice crop was selected as a proxy indicator to understand the differences in water use between AWD and NI cultivation (Adusumilli and Laxmi, 2011). Farmers without flow meters were advised to record the number of irrigations and duration of irrigation after measuring the discharge rates from bore wells to estimate the irrigation water used. The irrigation water used was estimated using the following equation:

Water used = Discharge × Irrigation time × Irrigation frequency

The discharge rate is expressed in L/s, irrigation time (hours) and irrigation frequency in numbers. In the 2013 and 2014 *kharif* seasons, effective rainfall (i.e. 60 percent of the total rainfall) as suggested by Kung (1971) was also considered in the calculation of the total water application.

In the command area (i.e. cluster), each farmer's field was equipped with AWD perforated water tubes[2] to monitor the water level in the field and regulate the irrigations and drainage water during the *kharif* and *rabi* seasons. Farmers were advised to keep the AWD fields submerged with water for the first 15–30 days after transplanting to avoid weed growth until the rice seedlings were established. Careful monitoring of crop growth is required during AWD to avoid weed problems. As a rule of thumb, the farmers were asked to irrigate only after finding hairline cracks in the soil (soil moisture suction up to −10 kPa) or when the water level in the AWD tube was 5–10 cm deep below ground level. The depth of water in the tube for the next irrigation depended on the stage of the crop. To overcome uncertainty in the release of water into canals and subsequent yield loss due to stress at critical periods, it was suggested to keep continuous irrigation during flowering (Sadeghi and Danesh, 2011). Observations of water flows through RBC flumes were taken daily, and flow meter readings were taken according to the crop growth stages. Water levels from both inlets and outlets of the selected command areas were monitored daily through RBC flumes and the difference between the two discharges (i.e. inlet-outlet) was taken as net water used from the command area.

The Water Productivity (WP) of AWD and NI fields were calculated as follows:

WP = Crop yield (kg)/Water used (m3).

Water used refers to the total water applied for irrigation through canals, water pumped from bore wells and effective rainfall during the growing period.

Estimating CH_4 emissions

In India, the annual CH_4 emissions from paddy fields has been estimated using Inter-governmental Panel on Climate Change (IPCC) 1996 guidelines, as the product of the harvested area under different paddy water regimes as classified by IPCC with the corresponding seasonally integrated CH_4 flux under a paddy-cropping season in g/m² (Houghton *et al.*, 1997). Indian paddy cultivation areas, which consist of mainly low to medium levels of soil organic carbon (<0.7%) without any organic amendments, CH_4 emissions were estimated at 17.48 ± 4 g CH_4-C/m² for continuously flooded fields and 2.01 ± 1.49 g CH_4-C /m² for intermittently flooded fields with multiple aerations (Gupta *et al.*, 2008).

In this study, gas samples were not collected from AWD and NI fields, and gas measurements were not carried out due to lack of facilities. The IPCC default values for emission factors (Tier 1 approach) were used to estimate CH_4 emissions from NI and AWD fields. Accordingly, the total seasonal CH_4 emissions of 11 g CH_4-C/m^2 (Mitra, 1992 cited in IPCC, 2006) from rice NI fields of Madras (Tamil Nadu) was assumed to represent the study areas. To estimate the methane emissions from AWD fields with multiple aerations, a 0.2 scaling factor for NI fields was used (IPCC, 1996). However, these emission factors should be used with caution since methane fluxes from rice fields vary substantially from day to day and even during day and night time, and spatially as well (Neue and Sass, 1994 cited in IPCC, 1996).

Data analysis

The data on agronomic performance, cumulative irrigation water used during the whole growing period, cultivation costs and crop returns are presented as mean data for all AWD and NI fields monitored in Guntur and Nalgonda study sites during the 2013 and 2014 *kharif* and *rabi* seasons, despite soil type differences between the two sites. We propose to give mean values because users can easily understand the benefits from AWD. All the original data in acres were converted into ha to use the Standard International units of measurements.

Results and discussion

Rice water consumption

Figure 10.2 shows the volume of water used at different growth stages of the rice plant under NI and AWD regimes in *rabi* 2012–13 and *kharif* 2013 seasons. Water consumption of rice under AWD was less than NI at all stages of crop growth, except for the land preparation stage in the *rabi* season. The volume of water consumed during *rabi* 2012–13 by NI (2,901 m^3) was greater than AWD (1,802 m^3) at the panicle initiation to flowering stage. It was also higher during this stage in *kharif* 2013, with NI (2251 m^3) and AWD (1569 m^3). On the other hand, during the nursery period, a smaller volume of water was used by NI (*rabi*: 778 m^3 *kharif*: 661 m^3) and AWD (*rabi*: 453 m^3, *kharif*: 383 m^3). In general, rice water consumption under AWD was lower than NI throughout the growing stages of the rice plant. Rice under NI consumed almost twice as much water compared with AWD at later stages of crop growth, particularly from tillering to panicle initiation and from flowering to maturity. The higher volume of water used for land preparation of AWD fields compared with NI fields during the *rabi* season (Figure 10.2) is probably due to additional water needed for ploughing, puddling and refilling soil water lost via seepage and percolation.

In total, the NI rice consumed 25,281 m^3 of water during *rabi* 2013 and 2014, whereas about 17,324 m^3 of water was consumed by AWD (i.e. lower by 32 percent). This means that about 7,957 m^3 of water was saved by AWD during the two years. Similarly in *kharif* 2013 and 2014, NI rice consumed 22,477 m^3 of water and AWD reduced the water use to 12,453 m^3, saving about 10,024 m^3

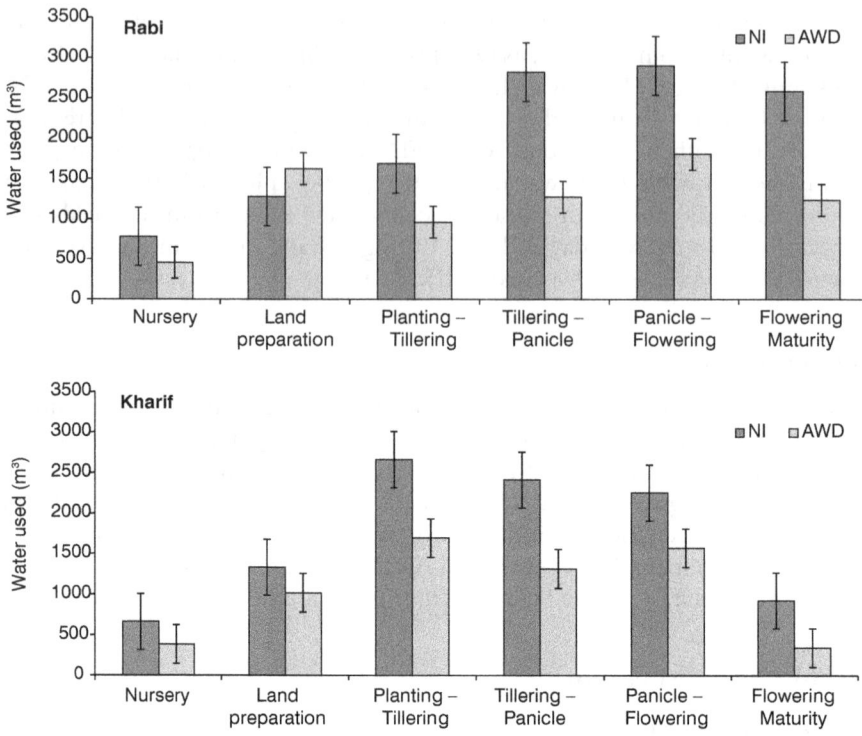

Figure 10.2 Water consumption at different growth stages of rice under AWD and NI during: a) *rabi* 2013; and b) *kharif* 2013

Source: Results based on field data from ClimaAdapt project, (www.climaadapt.org)

of water. The water saved by AWD could be used to grow other supplementary food crops with short duration by reallocating the water to downstream fields for irrigation, enabling the irrigation service area to be expanded for other purposes. In the current study, the total water consumption in AWD and NI was higher than recorded in other rice growing regions as shown in Table 10.1.

Agronomic performance of AWD and NI

Table 10.7 presents selected agronomic characteristics of BPT 5204 (during *kharif*) and MTU 1010 (during *rabi*) grown under NI versus AWD fields in 2013 and 2014. The average number of tillers at maximum tillering stage in the *rabi* season were 24 and 28 in NI and AWD, respectively, and in the *kharif* season were 21 and 23 in NI and AWD, respectively. Similarly the number of productive tillers i.e. panicles per hill was more by 11 percent and 12 percent in AWD than in NI during *rabi* and *kharif* seasons respectively. The average yield of BPT 5204 variety (4.6 to 5.5 kg/ha) during *kharif* was lower when compared to MTU 1010 (6.5 to 7.4 kg/ha) in *rabi* seasons. The average rice yield under AWD

Table 10.7 Average values of growth and yield parameters of rice grown under NI and AWD fields during *rabi* and *kharif* season

Parameters	Rabi (MTU 1010)		Kharif (BPT 5204)	
	NI (n = 126)	AWD (n = 126)	NI (n = 207)	AWD (n = 207)
Number of hills/m²	30	30	24	22
Number of tillers	24	28	21	23
Number of panicles/hill	15	17	11	13
Panicle length (cm)	27	27	22	22
Number of grains/panicle	150	146	156	167
1000 grains weight (g)	25	25	19	20
Grain yield (ton/ha)	6.5	7.4	4.6	5.5

Number of tillers recorded 65 DAT

Source: Based on data from ClimaAdapt (www.climaadapt.org)

was 7.4 tons/ha during *rabi* and 5.59 tons/ha during *kharif* which was higher than NI (6.5 and 4.6 tons/ha during *rabi* and *kharif*, respectively). Except for number of grains per panicle, all the other yield components and grain yield were higher in *rabi* season than *kharif* in AWD as well as NI for both years, probably due to more sunshine hours in *rabi* and varietal differences.

AWD fields exhibited higher agronomic performance with respect to number of tillers, particularly at 65 DAT, the number of panicles per hill, and grain yield during *rabi* as well as *kharif* seasons, compared to rice plants grown under the NI treatment (Table 10.7). Higher yields were also measured from AWD studies made by Liu *et al.* (2013) in China and McHugh *et al.* (2002) in Madagascar. The possible explanation for the higher agronomic performance of AWD in these studies could be:

- AWD improves the root system environment by allowing better access for oxygen and water absorption by the rice plant during the growing period;
- AWD improves the micro environment at the plant base and root zone which, in turn, reduces the conditions for pest and diseases incidence; and
- AWD enhances water and fertilizer use efficiency by reducing the water and nutrient losses via deep percolation and seepage. However, nitrogen losses may increase in AWD due to the multiple cycles of nitrification and denitrification, particularly if the application of fertilizer is not properly managed.

The effective utilization of the available water, nutrients and oxygen in the soil by the rice plant would result in more productive tiller formation, higher grain yields and biomass production. On the other hand, some studies have reported lower grain yield in AWD compared with NI (Borrell *et al.,* 1997; Tabbal *et al.,* 2002; Towprayoon *et al.,* 2005; Dong *et al.,* 2012).

Water productivity for rice production

AWD increased rice yields, on average, by almost 6 percent and 26 percent compared to the NI method during *rabi* and *kharif,* respectively (Figure 10.3), supporting results from other studies including China (Liu *et al.,* 2013) and Madagascar (McHugh *et al.,* 2002) (see Table 10.5). The increased yields in AWD were mainly attributed to increased tiller numbers, fewer pests and diseases during the crop period, as some studies indicate (Nasiruddin and Roy, 2012). Similar trends were observed during *kharif* 2013, but yields are comparatively lower than *rabi* due to the varietal traits, outbreak of brown plant hopper (BPH) infestation, and uneven monsoon rains during the harvest period in *kharif* 2013. Rice variety MTU 1010 grown during *rabi* gave more number of tillers and higher yields than BPT 5204 in *kharif* under AWD as well as NI fields (see Table 10.7).

To produce one kilogram of rice grain, the water consumed under NI (3.86 m^3/kg in *rabi* and 4.8 m^3/kg in *kharif*) was significantly greater than AWD (2.5 m^3/kg in *rabi* and 2.1 m^3/kg in *kharif*) (data not shown). Averaged across seasons, this is almost double the amount of water consumed under NI than AWD. In other words, the WUE of AWD ranged from 0.40 to 0.48 kg/m^3/ha, which was almost 50 percent greater than NI (0.21 to 0.26 kg/m^3/ha). Similar results have been found in other rice growing countries where the rice water productivity

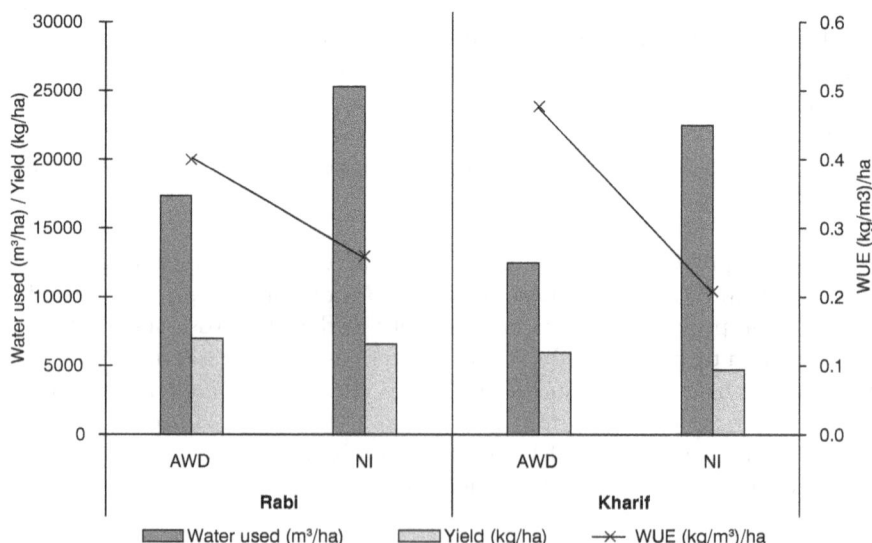

Figure 10.3 Average rice yields and water use in AWD and NI irrigation methods during *rabi* and *kharif* 2013 and 2014

Source: Results based on field data from ClimaAdapt project (www.climaadapt.org)

is greater under AWD than NI, despite considerable variation among the sites (see Table 10.2).

Cost-benefit analysis of AWD and NI

The costs and benefits of AWD and NI during *rabi* and *kharif* 2013 and 2014 are summarized in Table 10.8. The total costs between the AWD and NI methods were almost the same in *rabi,* but in the *kharif* season the costs were slightly lower in AWD. The profits were higher by 35 percent for AWD during the *rabi* season and 74 percent for the *kharif* season compared to NI (Figure 10.4). The rice yields were also increased by 6 to 26 percent in AWD. AWD cultivation during the *rabi* season seems more profitable than the *kharif* season due to the high yielding character of the varieties selected, less incidence of the major pests and diseases, and also less water usage in general. The benefits derived in the *rabi* and *kharif* seasons were primarily due to the increased yield and reduced cost. The total cost during *kharif* seasons was high due to higher planting cost, manures and fertilizer, weed management, and pest management activities (Figure 10.4).

The benefit cost ratio is the financial term that measures farming profitability. The benefit cost ratio of AWD (1.55 and 1.32) was superior to rice produced under the NI treatment (1.36 and 1.08) in both *rabi* and *kharif* seasons, respectively. AWD reduced the number of irrigations with cluster level control

Table 10.8 Production costs of AWD and NI for rice in *rabi* 2012–13, 2013–14 and *Kharif* 2013, 2014 seasons (Indian Rupees/ha)[a].

	Rabi		Kharif	
	AWD	NI	AWD	NI
Seed and planting	5,779	5,759	7,371	7,509
Land preparation	12,681	12,053	10,892	11,393
Manures and fertilizer	8,070	9,301	11,984	11,890
Weed management	2,651	2,049	3,681	4,535
Pest management	4,925	4,879	6,014	7,242
Irrigation	2,250	2,250	1,812	2,250
Harvesting	9,608	9,279	11,081	9,967
Fixed cost	15,875	15,875	16,896	16,542
Total cost	61,839	61,445	69,731	71,328
Income	95,770	83,638	92,205	77,081
Benefit cost ratio	1.55	1.36	1.32	1.08

[a] Currently, 1 Indian rupee (INR) is equivalent to 0.0157 US $ (or 1 US $ to 63.7 INRs)

Source: Based on data from ClimaAdapt (www.climaadapt.org)

by the irrigation department. Otherwise, the cost for irrigation is almost the same for AWD and NI due to fixed water rates adopted in the region and country. The slight variation in irrigation cost seen in AWD during *kharif* was due to the labour utilization in controlling the water at farm level. Farmers paying a fixed rate for irrigation based on land area rather than quantity of water would not immediately benefit from using less irrigation water by AWD (Price *et al.,* 2013), although there might be long-term or broader-scale benefit via reduced use of the water resource in general.

AWD consumes less water than NI (Figure 10.2) and the productivity per unit area is higher (Figure 10.3). The cost of cultivation should be less in the case of AWD compared to NI due to less investment in pest management and nutrient management. However, in the study area there was no difference found in the cost of cultivation in *rabi*, with a slight difference recorded in *kharif*. This was due to less pest incidence in *rabi* and controlled irrigation from the reservoir itself. The difference in *kharif* was mainly due to less incidence of brown plant hopper (BPH) in AWD fields compared to NI. As studies in Indonesia found, the extra water saved by growing rice on permanent raised beds under saturated (but not flooded conditions) could be used to irrigate secondary crops (Mahrup *et al.,* 2005). Water savings from AWD may also be reallocated for other domestic and/or industrial uses. This could help to safeguard the tail-end farmers growing commercial crops like cotton and chillies. In summary, the increased productivity of AWD compared with NI should contribute to improved food security and increased farmers' adaptation to water scarcity induced by climate change.

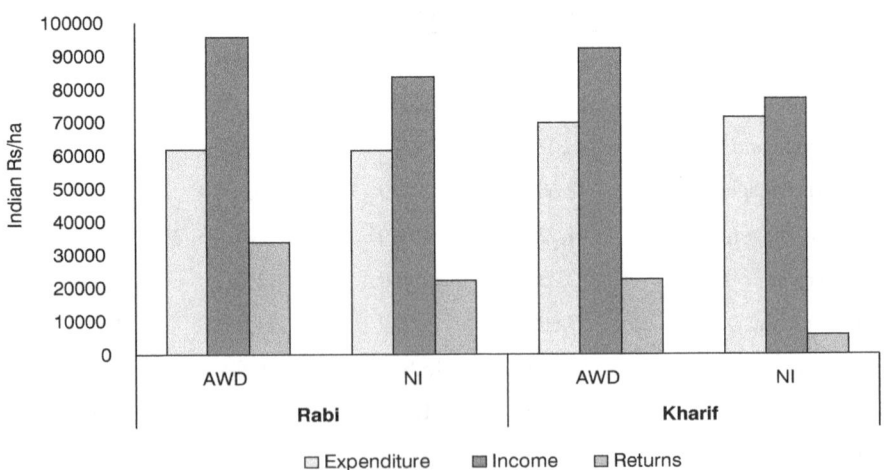

Figure 10.4 Total expenditure, income and returns of rice grown under AWD and NI during *rabi* and *kharif* 2013 and 2014

Source: Results based on field data from ClimaAdapt project (www.climaadapt.org)

Methane emission estimates

As shown in Table 10.9, the total methane emissions from NI fields were estimated to be 220 kg CH_4-C during *rabi* from a small farmer land-holding size of 2 ha field. In the *kharif* season, emissions from a big farmer with a land-holding size of 6 ha field were estimated to be 60 kg CH_4-C, assuming the default value of 11 g CH_4-C /m^2 (Mitra, 1992). In contrast for AWD, the total estimated methane emissions were 44 kg CH_4-C in *rabi* and 132 kg CH_4-C in *kharif*, representing 20 percent of the emissions incurred under NI. However, these estimates appear too low for AWD when compared with measured data by Rajkishore *et al.* (2013) who found CH_4-C emission averages 37.7 kg/ha under System of Rice intensification (SRI) in Tamil Nadu, India. The total CH_4 emissions and total CO_2 equivalents from AWD fields estimated in the study area were much lower than the NI fields in *kharif* and *rabi* seasons (Table 10.9). Hence AWD has potential to reduce the Global Warming Potential (GWP) by at least 1/3 of the NI rice field (IRRI, 2009). GWP is a measure of the total energy that a gas absorbs over a particular period of time (usually 100 years), compared to CO_2 equivalents.

There is considerable literature showing that methane emissions from AWD rice fields are comparatively lower than continuously flooded paddy fields (Yagi *et al.*, 1996; Sidhu and Benbi, 2011; Jain *et al.*, 2013), supporting the lower estimated methane emissions from AWD compared with NI in the current study. The practice of AWD helps reduce methane emissions from rice soils due to the periodic removal of anaerobic conditions, thereby halting the production of CH_4. However, the alternating cycles of aerobic and anaerobic conditions may trigger substantial N_2O emissions as a result of bacterial conversion of other nitrogen compounds to N_2O, and its subsequent release from the soil. AWD causes more N_2O emissions than the continuously flooded NI fields by creating nearly saturated soil conditions with increased soil redox potential, promoting N_2O production (e.g. Jain *et al.*, 2013; Ali *et al.*, 2013). Australian studies showed that while the soil of unflooded rice systems like AWD remained

Table 10.9 Estimated total CH_4-C emissions and CO_2 equivalent from NI and AWD rice fields during *rabi* and *kharif* seasons

	Rabi 2013 (n = 2 ha from small farmer)		Kharif 2013 (n = 6 ha from big farmer)	
	kg CH_4-C	kg CO_2 equiv.	kg CH_4-C	kg CO_2 equiv.
NI	220	5,500	60	16,500
AWD	44	1,100	132	3,300

Calculated using the global warming potential of CH_4 = 25 times of CO_2 in 100 year time horizon (IPCC, 2007). 0.2 scaling factor used relative to emission factors of continuously flooded fields

Source: Based on data from ClimaAdapt (www.climaadapt.org)

aerobic throughout the experiment as evidenced by Eh values in excess of 400 mV, the flooded treatments successively became weakly reducing (<300 mV) upon flooding. Proper nitrogen management should, to some extent, curtail N_2O losses (Bhatia *et al.,* 2009).

Adoption and upscaling AWD practice: opportunities and constraints

Opportunities

There are a number of opportunities that could be used to disseminate and upscale AWD in rice growing areas of India. Some of these are discussed below:

1   *Increased water productivity and WUE:* Water scarcity is one of the major problems rice farmers are now facing and will continue to face in the future due to climate change. The continuous flooding of paddy fields consumes about twice as much water as AWD irrigation (Figure 10.2), with lower productivity per unit of water consumed. In contrast, AWD fields showed higher water productivity and water use efficiency (Figure 10.3). Therefore, AWD is more likely to be adopted by rice farmers who pay higher prices for pump irrigation, indicating that this technology could be applicable in a much wider area.

2   *Costs of production:* In the current study, there was no difference in the total costs of rice production, including the costs of labour, fuel for pumping and pesticides in AWD compared with NI (Figure 10.4). It might be expected that, due to the lower amount of water required by the rice crop in AWD (Figure 10.2), the total productions costs would be less. However, farmers paying a fixed rate for irrigation based on land area rather than quantity of water would not immediately benefit from using less irrigation water by AWD (Price *et al.,* 2013). There may, however, be long-term or broader-scale benefits from reduced water use. Also, it is likely that some rice varieties such as BPT 5204 will be preferred by farmers during the *kharif* season using AWD due to the high quality of the grain, ultimately commanding a higher price. The benefit cost ratio of AWD was superior to rice produced under NI in both the *rabi* and *kharif* seasons. Hence, AWD has the potential to increase farmers' income and contribute to achieving food security. These benefits accrued by farmers should increase the likelihood that AWD will be readily accepted, adopted and diffused to many farmers and to a larger area at the same time.

3   *Reduce greenhouse gases, particularly CH_4:* GHG mitigation potential by AWD has an indirect positive effect to farmers because AWD primarily reduces methane emissions by increasing carbon storage in soils and, to some extent, reduces nitrous oxide if more organic fertilizers are applied. In doing so, AWD could increase fertilizer use efficiency of crops and improve soil pH, nutrient and water retention capacity, increase soil redox potential,

and reduce micronutrient toxicity (Salim and Shehzad, 2008). Above all, reducing the emissions of CH_4 and N_2O contributes to the mitigation of climate change effects such as GWP derived from CH_4 and N_2O concentrations in the atmosphere, improving resilience to extreme weather events like flooding and drought. Mitigating GHGs through the AWD method can also help farmers to earn carbon credits if policy mechanisms are developed. The total benefit of AWD gives an opportunity for policy makers and farmers to promote the up-scaling of AWD.

4 *Expansion of farm mechanization:* AWD creates a management system conducive to mechanization in the rice fields, for example, for transplanting seedlings, for soil cultivation in inter-rows, for mechanical harvesting and other farm operations.

Constraints

Despite the above mentioned opportunities for up-scaling AWD technology, the adoption rate is low (i.e. <10 %) among farmers and it has not been applied by rice farmers at a wider scale, for example in the Andhra Pradesh and Telangana part of the Krishna river basin in India. The major constraints to adoption of AWD technology are as follows:

1 *Water management/scheduling:* Under AWD, farmers need to apply only the required amount of water throughout the crop growing stages by monitoring the soil moisture and water depth using the AWD field water tubes, recording the discharge water meters and regular field observations. This requires a higher level of water management than NI. In this regard, farmers lack the experience and knowledge on field moisture monitoring, and need-based water applications. Furthermore, weed control may be a challenge for rice farmers due to lack of knowledge in controlling weeds at the initial stages of crop growth. But then, once the rice plant becomes established, AWD plants grow vigorously and fully cover the soil surface (IRRI, 2009).

2 *Supply of irrigation water and power:* There is uncertainty with regards to supply of irrigation water, particularly with farmers who irrigate their crops using boreholes and tube wells, as it requires well-tuned irrigation intervals and measures. In addition, the power cuts often seen in the irrigation schemes deter farmers from applying AWD, particularly during the stages of critical crop water requirement such as flowering.

3 *Weed and/or pest infestation:* This often exerts a more severe threat to crop establishment under AWD relative to transplanted rice crops. Normally, weed growth will be enhanced under the aerobic conditions of AWD fields compared with flooded fields since the drained soil environment is conducive to weed growth. For example in Australian studies, weed dry mass was higher (P<0.05) in AWD (160 g/m²) than in the flooded rice control (17 g/m²) during the wet season (Borrell *et al.,* 1997). Maintaining

flooded conditions until around 2 weeks after transplanting discourages the growth of weeds while the rice plant becomes established (Richards and Sander, 2014). Since more labour is required to control weeds for small-holder farmers with limited resources, such farmers may refrain from applying AWD due to lack of capital to access farm implements for weeding (e.g. cono-weeders), marking and levelling. Increased population of rats has also been observed in AWD fields due to decreased flooding (Cabangon *et al.*, 2014). The rats may create holes in the basin bunds and increase water losses through seepage.

4   *Trade-off between CH_4 mitigation and N_2O emission enhancement:* There is a large uncertainty with respect to the performance of AWD in mitigating N_2O emissions from rice soils. In fact, some studies show that AWD triggers more N_2O emissions than the continuously flooded NI fields by creating nearly saturated soil conditions and increased soil redox potential (Borrell, 1993), which promote N_2O production (e.g. Jain *et al.*, 2013; Ali *et al.*, 2013).

Conclusions and recommendations

This study has demonstrated that AWD fields have, on average, 26 percent lower irrigation water requirements compared with NI, while achieving higher yields and profits during both the *rabi* and *kharif* seasons. Therefore, it is possible to save irrigation water and improve water productivity in rice grown under the AWD irrigation regime. The water productivity, number of tillers and number of panicles were high in AWD compared to NI cultivation. The case studies in Andhra Pradesh and Telangana States have proved that AWD irrigation is a climate smart agricultural practice with multiple benefits. Nonetheless, farmers will only adopt water saving technologies such as AWD on a large scale if they actually benefit from the use of the technology in terms of reduced cost of production and increased income (Figure 10.4). The following measures could help promote adoption and up-scaling of AWD:

- initiate training and demonstrations on field moisture monitoring, weed and pest control, need-based water and fertilizer applications;
- increase awareness of the principles and practices of AWD and its benefits regarding water, energy and money savings to change the mind-set of farmers and other stakeholders;
- empower extension staff by equipping them with the necessary resources to conduct farmer training on AWD;
- provide an adequate and reliable supply of energy to manage irrigation more efficiently in the AWD fields; and
- increase AWD trials (including GHGs measurements) at multi-locations and system level water control to develop management strategies and analyse the trade-offs between CH_4 mitigation and N_2O emissions in AWD in order to come up with 'win-win' scenarios.

Notes

1 RBC flume of 50 lps capacity is designed to use in small field channels for more accurate flow measurements. A calibration chart was developed to correlate the depth of water flow above the crust (placed at the bottom of flume with a height of 10cm for free flow condition) with flow of water in litre per second.
2 AWD tube was made of PVC pipe or bamboo with dimensions of 30 cm length and 15 cm diameter. Holes of 2 cm x 2 cm were made and up to 50 percent of the tube was pushed into the soil, with 10 cm of the tube protruding above soil level. The water levels below the soil were observed through the tube to indicate when the next AWD irrigation should occur.

References

Adusumilli, R. and Laxmi, S.B. (2011) Potential of the system of rice intensification for systemic improvement in rice production and water use: the case of Andhra Pradesh, India, *Paddy Water Environment* DOI 10.1007/s10333-010-0230-6.

Alberto, M.C.R., Wassmann, R., Buresh R.J., Quilty, J.R., Correa Jr, T.Q., Sandro J.M. and Centeno, C.A.R. (2014) Measuring methane flux from irrigated rice fields by eddy covariance method using open-path gas analyse, *Field Crops Research,* 160: 12–21.

Ali, M.A., Hoque, M. A. and Kim, P.J. (2013) Mitigating global warming potentials of methane and nitrous oxide gases from rice paddies under different irrigation regimes, *AMBIO,* 42: 357–368.

Belder, P. Bouman, B.A.M., Cabangon, R., Guoan, L., Quilang, E.J.P., Yuanhua, L., Spiertz, J.H.J. and Tuong, T.P. (2004) Effect of water saving irrigation on rice yield and water use in typical lowland conditions in Asia, *Agricultural Water Management,* 65: 193–210.

Bhatia, A., Pathak, H., Aggarwal, P.K. and Jain, N. (2009) Trade-off between productivity enhancement and global warming potential of rice and wheat in India, *Nutrient Cycling in Agroecosystems,* DOI 10.1007/s10705-009-9304-5.

Bhuiyan, S.I. (1992) Water management in relation to crop production: Case study on rice, *Outlook on Agriculture,* 21: 293–299.

Borrell, A.K. (1993) Improving efficiencies of nitrogen and water use for rice in semi-arid tropical Australia, PhD Thesis. The University of Queensland, Brisbane, Australia.

Borrell, A.K., Fukai, S. and Garside, A.L. (1991) Alternative methods of irrigation for rice production in tropical Australia, *International Rice Research Newsletter,* 16 (3): 28.

Borrell, A.K., Garside, A.L. and Fukai, S. (1997) Improving efficiency of water use for irrigated rice in a semi-arid tropical environment, *Field Crops Research,* 52: 231–248.

Bouman, B.A.M., Lampayan, R.M. and Tuong, T.P. (2007) *Water Management in Irrigated Rice: Coping with Water Scarcity,* International Rice Research Institute, Los Baňs, Philippines.

Cabangon, R., Lampayan, R., Bouman, B. and Tuong, T.P. (2014) Water saving technologies for rice production in the Asian region. Food & Fertilizer Technology Center http://www.fftc.agnet.org/library.php?func=view&style=type&id=20140303145242 (accessed 30 March 2015).

Cai, Z., Xing, G., Yan, X., Xu, H., Tsuruta, H., Yagi, K. and Minami, K. (1997) Methane and nitrous oxide emissions from rice paddy fields as affected by nitrogen fertilizers and water management, *Plant and Soil,* 196: 7–14.

Chapagain, T., Riseman, A. and Yamaji, E. (2011) Achieving more with less water: Alternate wet and dry irrigation (AWD) as an alternative to the conventional water management practices in rice farming, *Journal of Agricultural Science,* 3 (3): 3–13, DOI 10.5539/jas.v3n3p3.

Dong, B., Loeve, R., Li, Y.H., Chen, C.D., Deng, L. and Molden, D. (2008) Water productivity in Zhanghe irrigation system at different scales. In: Salim, M. and Shehzad, F.D. *Alternate Wetting and Drying (AWD) of NI fields: A Water Saving Technology*, Science, Technology and Development, 27(3 & 4), Islamabad, Pakistan.

Dong, N.M., Brandt, K.K., Sørensen, J., Hung, N.N., Hach, C.V., Tan, P.S. and Dalsgaard, T. (2012) Effects of alternating wetting and drying versus continuous flooding on fertilizer nitrogen fate in rice fields in the Mekong Delta, Vietnam, *Soil Biology & Biochemistry*, 47: 166–174.

Food and Agriculture Organization (FAO) (2007) *Adaptation to Climate Change in Agriculture, Forestry And Fisheries: Prospective Framework and Priorities*, Interdepartmental Working Group on Climate Change, Rome.

Food and Agriculture Organization (FAO) (2010) *"Climate-Smart" Agriculture: Policies, Practices and Financing for Food Security*, Adaptation and Mitigation, Rome.

Gujja, B. and Thiyagarajan, T.M. (2009) *New Hope for Indian Food Security? The System of Rice Intensification*, Gate keeper 143, International Institute for Environment and Development, London.

Gupta, P.K., Gupta, V., Sharma, C., Das, S.N., Purkait, N, Adhya, T.K., Pathak ,H., Ramesh, R., Baruah, K.K., Venkatratnam, L., Singh, G. and Iyer, C.S.P. (2008) Development of methane emission factors for Indian paddy fields and estimation of national methane budget, *Chemosphere*, DOI 10.1016/j.chemosphere.2008.09.042

Hidayah, S., Agustina, D.A. and Joubert, M.D. (2008) Intermittent irrigation in system of rice intensification potential as an adaptation and mitigation option of negative impacts of rice cultivation in irrigated paddy field, Indonesia. http://www.rid.go.th/thaicid/_6_activity/Technical-Session/SubTheme2/2.10-Susi_H-Dewi_AA-Marasi_DJ-Soekrasno.pdf (accessed 30 March 2015)

Houghton, J.T., Meira Filho, L.G., Callander, B.A., Harris, N., Kattenberg, A. and Maskell, K. (1997) *Climate Change 1996: The Science of Climate Change. IPCC.* Cambridge University Press, Cambridge.

Husin, Y.A., Murdiyarso, D., Khalil, M.A.K., Rasmussen, R.A., Shearer, M.J., Sabhiham, S., Sunar, A. and Adijuwana, H. (1995) Methane flux from Indonesian wetland rice: The effect of water management and rice variety, *Chemosphere*, 31 (4): 3153-3180.

Intergovernmental Panel on Climate Change (IPCC) (1996) *Revised 1996 IPCC Guidelines for National Greenhouse Gas Inventories Reference manual for Agriculture*, 3 (4): 1–20.

Intergovernmental Panel on Climate Change (IPCC) (2006) *Guidelines for National Greenhouse Gas Inventories Volume 4 Agriculture, Forestry and Other Land Use*, Cambridge University Press, Cambridge.

Intergovernmental Panel on Climate Change (IPCC) (2007) *Climate Change 2007: The Physical Science Basis Contribution of Working Group I to the Fourth Assessment Report of the IPCC*, Cambridge University Press, New York.

International Rice research Institute (IRRI) (2003) Annual Report, *International Rice Research Institute*, Los Banos, Laguna Philippines.

International Rice research Institute (IRRI) (2009) Saving water: Alternate wetting and drying (AWD), Rice fact sheets, online, http://www.knowledgebank.irri.org/factsheetsPDFs/watermanagement_FSAWD3.pdf (accessed 25 October 2013).

Jain, N., Dubey, R., Dubey, D.S., Singh, J., Khanna, M., Pathak, H. and Bhatia, A. (2013) Mitigation of greenhouse gas emission with system of rice intensification in the Indo-Gangetic Plains, *Paddy Water Environment*, DOI 10.1007/s10333-013-0390-2.

Kung, P. (1971) *Irrigation Agronomy in Monsoon Asia*, AGPC MISC/2, FAO, Rome.

Lakshmanan, A., Geethalakshmi, V., Nagothu, U.S., (2015) Impacts of microbial inoculants on productivity, methane flux and carbon sequestration in rice ecosystem, Technical brief No. 14, Tamil Nadu Agricultural University, Coimbatore.

Liu, L., Chen, T., Wang, Z., Zhang, H., Yang, J. and Zhang, J. (2013) Combination of site-specific nitrogen management and alternate wetting and drying irrigation increases grain yield and nitrogen and water use efficiency in super rice, *Field Crops Research*, 154: 226–235.

Mahrup, M., Borrell, A., Mansur, M., Kusnata, I., Sukartono, M., Tisdall, J. and Gill, J. (2005) Soil management systems improve water use efficiency of rainfed rice in the semi-arid tropics of southern Lombock, Eastern Indonesia, *Plant Production Science*, 8 (3): 342–344.

Ma'shum, M., Tisdall, J.M., Borrell, A.K., McKenzie, B.M., Gill, J.S., Kusnarta, I.G.M., Mahrup, M., Sukartono, M. and Van Cooten, D.E. (2009) Rice responses to soil management in a rice-based cropping system in the semi-arid tropics of southern Lombok, Eastern Indonesia, *Field Crops Research*, 110: 197–206.

McHugh, O. V., Stenhuis, T.S., Barosin, J., Fernandes, E.C.M. and Uphoff, N.T. (2002) Farmer implementation of alternate wet-dry and non-flooded irrigation practices in the System of Rice Intensification (SRI). In: Bouman, B.A.M. *et al.* (eds.), *Water-wise Rice Production,* Los Baños (Philippines), International Rice Research Institute, pp. 89–102.

Ministry of Agriculture (MoA) (2012) Agricultural statistics at a glance, Directorate of Economics and Statistics, Department of Agriculture and Cooperation (DAC), Ministry of Agriculture, Government of India on line, http://dacnet.nic.in (accessed 7 January 2013).

Mitra, A.P. (1992) *Greenhouse Gas Emission in India: 1991 Methane Campaign.* Science Report No.2 Council of Scientific and Industrial Research and Ministry of Environment and Forest, New Delhi, India.

Nasiruddin, M. and Roy, R.C. (2012) Rice field insect pests during the rice growing seasons in two areas of Hathazari Chittagang, *Bangladesh Journal of Zoology*, 40 (1): 89–100.

Ndiiri, J.A., Mati, B.M., Home, P.G., Odongo, B. and Uphoff, N. (2013) Adoption, constraints and economic returns of paddy rice intensification in Mwea, Kenya, *Agricultural Water Management*, 129: 44–55.

Neue, H.U. and Sass, R. (1994) Trace gas emissions from rice fields. In: Prinn R.G. (ed.), *Global Atmospheric-Biospheric Chemistry, Environmental Science Res.,* 48, Plenum Press, New York.

Oliver, M.M.H., Talukder, M.S.U. and Ahmed, M. (2008) Alternate wetting and drying irrigation for rice cultivation, *Journal of Bangladesh Agricultural University*, 6 (2): 409–414

Price, A.H., Norton, G.J., Salt, D.E., Ebenhoeh, O., Meharg, A.A, Meharg, C., Islam, M.R., Sarma, R.N., Dasgupta, T., Ismail, A.M., McNally, K.L, Zhang, H., Dodd, I.C. and Davies, W.J. (2013) Alternate wetting and drying irrigation for rice in Bangladesh: Is it sustainable and has plant breeding something to offer? *Food Energy Security*, 2: 120–129.

Rajkishore, S.K., Doraisamy, P., Subramanian, K.S. and Maheswari, M. (2013) Methane emission patterns and their associated soil microflora with SRI and conventional systems of rice cultivation in Tamil Nadu, India, *Taiwan Water Conservancy*, 61 (4): 126– 134.

Rejesus, R.M., Palis, F.G., Rodriguez, D.G.P., Lampayan, R.M. and Bouman, B.A.M. (2011) Impact of Alternate wetting and drying (AWD) water-saving irrigation technique: Evidence from rice producers in the Philippines, *Food Policy,* 36: 280–288.

Richards, M. and Sander, B.O. (2014) Information note on alternate wetting and drying in irrigated rice. Implementation guidance for policymakers and investors, CGIAR, CCAFS, IRRI.

Sadeghi, S.M. and Danesh, R.K. (2011) Effects of water deficit role at different stages of reproductive growth on yield components of rice, *World Applied Sciences Journal*, 13 (9): 2021–2026.

Salim, M. and Shehzad, F.D. (2008) Alternate wetting and drying (AWD) of paddy fields: A water saving technology, *Science, Technology and Development*, 27 (3–4), Islamabad, Pakistan.

Sarkar, S. (2001) Effect of water stress on growth, productivity and water expense efficiency of summer rice, *Indian Journal of Agricultural Sciences*, 71 (3): 153–158.

Sharma, C., Tiwari, M.K. and Pathak, H. (2008) Estimates of emission and deposition of reactive nitrogenous species for India, *Current Science*, 94 (11) 1439-1446.

Sidhu, M.K.K. and Benbi, B.S. (2011) Methane emission from rice fields in relation to management of irrigation water, *Journal of Environmental Biology*, 32: 169–172.

Siopongco, J.D.L.C., Wassmann, R. and Sander, B.O. (2013) Alternate wetting and drying in Philippine rice production: feasibility study for a clean development mechanism, *IRRI Technical Bulletin*, No. 17, Los Baños, Philippines.

Sujono, J., Matsuo, N., Hiramatsu, K. and Mochizuki, T. (2011) Improving the water productivity of paddy rice (*Oryza sativa* L.) cultivation through water saving irrigation treatments, *Agricultural Sciences*, 2 (4): 511–517.

Tabbal, D.F., Bouman, B.A.M., Bhuiyan, S.I., Sibayan, E.B. and Sattar, M.A. (2002) On-farm strategies for reducing water input in irrigated rice; case studies in the Philippines, *Agricultural Water Management*, 56: 93–112.

Thiyagarajan, T.M. and Biksham, G. (2013) *Transforming Rice Production with SRI (System of Rice Intensification) Knowledge and Practice*, National Consortium on SRI (NCS), Hyderabad, India.

Toung, T.P. (2007) Alternate wetting and drying irrigation (AWD): A technology for water saving in rice production. Paper presented at crop cutting ceremony, BADC Farm, Modhupur, Tangil http://knowledgebank-brri.org/AWD/AWD_technology.pdf, (accessed 28 March 2015).

Towprayoon, S., Smakgahn, K. and Poonkaew, S. (2005) Mitigation of methane and nitrous oxide emissions from drained irrigated rice fields, *Chemosphere*, 59: 1547–1556.

Uphoff, N. (2006) *The System of Rice Intensification (SRI) as a Methodology for Reducing Water Requirements in Irrigated Rice Production: Exploring Options for Food Security and Sustainable Environments*, IRRI, Los Baños.

Uprety, D.C., Dhar, S., Hongmin, D., Kimball, B.A., Garg, A. and Upadhyay, J. (2012) Technologies for climate change mitigation: agriculture sector, Department of Management Engineering, Technical University of Denmark.

Van der Hoek, W., Sakthivadivel, R., Renshaw, M., Silver, J.B., Birley, M.H. and Konradsen, F. (2001) *Alternate Wet/Dry Irrigation in Rice Cultivation: A Practical Way to Save Water and Control Malaria and Japanese Encephalitis?* Research Report 47, International Water Management Institute, Colombo.

Veeraputhiran, R., Balasubramanian, R., Pandian, B.J., Chelladurai, M., Tamilselvi, R. and Renganathan, V.G. (2012) Influence of system of rice intensification on yield, water use and economics through farmers' participatory approach, *Madras Agricultural Journal*, 99: (4–6): 251–254.

Wassmann, R., Hosen, Y. and Sumfleth, K. (2009) Agriculture and climate change: an agenda for negotiation in copenhagen for food, agriculture, and the environment reducing methane emissions from irrigated rice, *International Food Policy Research Institute Focus 16*, Brief 3, Washington DC, USA.

Yagi, K. H., Tsuruta, K. K. and Minami, K. (1996) Effect of water management on methane emission from a Japanese rice paddy field: Automated methane monitoring, *Global Biogeochemical Cycles*, 10: 255–267.

Yao, F., Huang, J., Cui, K., Nie, L., Xiang, J., Liu, X., Wu, W., Chen, M. and Peng, S. (2012) Agronomic performance of high-yielding rice variety grown under alternate wetting and drying irrigation, *Field Crops Research*, 126: 16–22.

Zou, J.W., Huang, Y., Lu, Y.Y., Zheng, X.H. and Wang, Y.S. (2005) Direct emission factor for N_2O from rice-winter wheat rotation systems in southeast China, *Atmospheric Environment*, 39: 4755–4763.

11 System of rice intensification

Climate-smart rice cultivation system to mitigate climate change impacts in India

Geethalakshmi V., Mehreteab Tesfai, Lakshmanan A., Andrew Borrell, Udaya Sekhar Nagothu, Arasu M. S., Senthilraja K., Manikandan N. and Sumathi S.

Introduction

Rice is one of the most important staple crops for Indian society. India has the world's largest area of rice cultivation (44 million ha) and it is the second largest rice producing country in the world after China (Thiyagarajan and Gujja, 2013). The country will need to produce at least 130 million tons of milled rice per year by 2030 in order to feed its growing population (Gujja and Thiyagarajan, 2009). However, there is not much scope to expand the current area under rice cultivation to increase rice production due to rapid urbanization on one hand, and water constraints and shortage of suitable soils on the other hand. Enhancing the productivity per unit area is another possible way to increase rice production under the existing cultivated areas. However, the challenge is not only increasing food production, but also adapting to the changing climatic conditions that impose water scarcity and mitigating Green House Gases (GHG) emissions from the rice fields.

To maximize agricultural production, mitigate GHG emissions, and enhance food security, fundamental changes are needed in rice production systems. One alternative option is to apply the System of Rice Intensification (SRI) which requires less agricultural inputs (land, seeds, fertilizers, pesticides) and less water compared to conventional rice cultivation (Stoop *et al.*, 2002; Uphoff, 2002; Thakur *et al.*, 2010; Geethalakshmi *et al.*, 2011; Jain *et al.*, 2013). Assembled in Madagascar in the 1980s by the French agronomist Father Henri de Laulanié, SRI has been adopted by rice farmers in more than 50 countries worldwide. SRI was introduced to India in 2000 in Tamil Nadu, Puduchery, and Tripura (Thiyagarajan and Gujja, 2013). SRI is a package of agronomic practices which exploits the genetic potential of rice plants, creates a better growing environment (both above and below ground), enhances soil health, reduces inputs such as seeds, fertilizers, water and labour requirements for planting (Gujja and Thiyagarajan, 2009), and reduces fuel consumption for pumping water by 30 litres/ha (Siopongco *et al.,* 2013). On the other hand, the continuous flooding system (i.e. paddy rice) consumes more water, seeds and inorganic fertilizers than SRI (Lakshmanan *et al.,* 2012). Moreover, paddy rice is one of the major contributors to GHG emissions particularly methane and, to lesser extent, nitrous oxide.

Fields with SRI could reduce CH_4 production by about 30–60 percent and lower the Global Warming Potential significantly (Jain *et al.,* 2013; Rajkishore *et al.,* 2013). This supports previous work by several researchers (Uphoff, 2003; Barah, 2009; Thakur *et al.,* 2010; Uphoff *et al.,* 2011). In addition, SRI can reduce water consumption by 22 to 25 percent, compared to paddy rice (Lakshmanan *et al.,* 2012). The water saved by practicing SRI could be used to grow other food crops or may be reallocated and used for domestic and/or industrial purposes.

Despite these favourable reports, SRI is highly criticized and is still a controversial issue among scholars, researchers, farmers and policy makers when it comes to its performance as a water-saving technology, increasing yield and reducing GHG emissions. Dobermann (2004) reported that SRI does not show any significant increase in yields. He argued that the performance of SRI largely depends on soil and other environmental conditions in the locality, efficiency in management and farmers' knowledge and experience. Further, Sheehy *et al.* (2004) evaluated SRI and conventional rice production in three locations in China, concluding that SRI has no inherent advantage over the conventional system and that the original reports of extraordinary high yields are likely to be the consequence of error. McDonald *et al.* (2008) reviewed different research publications on SRI performance against recommended management practice (RMP) and concluded that SRI yields were not higher than RMP. In fact, in 35 site-years of data compiled from nine different Asian countries, they found no evidence of a yield advantage except for one set of experiments in Madagascar, where SRI more than doubled rice grain productivity with respect to best management practices (BMPs). These publications have spurred considerable debate (Glover, 2011).

The objectives of this chapter are (i) to review and describe evidence-based SRI performance regarding climate change impacts, adaptation, mitigation of GHG emissions, and increasing agricultural productivity in comparison with the conventional rice cultivation methods practiced by small-holder farmers in Asia and Sub-Saharan Africa, and (ii) to present and discuss case study results from SRI activities carried out in Kalingarayan canal basin (KB) and Ponnaiyar reservoir basin (PB) in Tamil Nadu state of India.

Review of literature

Climate-smart agriculture is defined as an approach that 'sustainably increases productivity, enhances resilience ("adaptation"), reduces greenhouse gas emissions ("mitigation"), promotes carbon sequestration, and increases food security and development goals' (FAO, 2010). Climate-smart agriculture addresses the synergistic effects of climate change mitigation, adaptation and food security, and minimizes their potential negative trade-offs by bringing about 'win-win' opportunities. To demonstrate that SRI technology fulfils the principles of CSA as mentioned above, we have reviewed and discussed the performance of SRI (in relation to CSA principles) in comparison with the conventional practice by small-holder farmers in some selected rice growing countries in Asia and sub-Saharan Africa (SSA) in the following section. The terms paddy and conventional rice cultivation are interchangeable in this chapter.

SRI as an adaptation measure to water scarcity

The contribution of SRI in saving water in some of the rice growing countries are presented in Table 11.1. According to Zheng *et al.* (2013), irrigation water was reduced by 1,933.5 m^3/ha (a water saving of 22 percent compared with conventional flooded paddy) with SRI practices in Sichuan province in south-western China. Water productivity was 1.12 kg of grain per m^3, with an increase of 0.30 kg/m^3 (36 percent), and irrigation water use efficiency was 1.34 kg of grain/m^3, an increase of 0.37 kg/m^3 (38 percent). Reports from other countries also similarly showed that irrigation water requirements under SRI was lower compared to flooded paddy cultivation (Table 11.1).

Table 11.1 SRI performance in climate change adaptation in some rice growing countries

Country (site)	Water used by paddy (mm)	Water used by SRI (mm)	Water saved by SRI (%)	References
China (Sichuan province)	2,377	1,933	19	Zheng *et al.* (2013)
India (Andhra Pradesh)	1,012	651	36	Satyanaryana *et al.* (2007)
India (Tamil Nadu)	1,077	938	13	This study
Vietnam (Mekong region)	n.a.	n.a.	33	Dung and Minh (2008)
Indonesia (eastern region)	n.a.	n.a.	40	Sato and Uphoff (2007)
Kenya (Mwea)★	4,626	3,510	24	Ndiri *et al.* (2012); Nyamai *et al.* (2012)
Tanzania (Mkindo)	2,882	1,026	64	Katambara *et al.* (2013)
Mali (Timbuktu region)	n.a.	n.a.	10	Styger *et al.* (2011)

n.a.: not available. ★ converted into ha from 240 m^2 plot

Source: Authors' own compilation

Table 11.2 Examples of SRI's role in mitigating CH₄- C emissions in some Asian countries

Country (site)	unit	SRI	Paddy	% change	References
China (Thaihu lake basin)	kg C/ha	22.0*	39.0	44	Yang *et al.* (2012)
India (New Delhi)	kg C/ha	19.9	32.3	38	Suryavanshi *et al.* (2012)
India (New Delhi)	kg C/ha	8.2	22.5	64	Jain *et al.*, (2013)
Thailand (irrigated)★★	kg C/ha	302.7	473.0	36	Sass (1996); Kroeze (1996); Shin *et al.* (1996)
Thailand (rainfed)★★	kg C/ha	129.3	202.1	36	Sass (1996); Kroeze (1996); Shin *et al.* (1996)
Thailand (clay soil)★★	kg C/ha	301.8	471.6	36	Sass (1996); Kroeze (1996); Shin *et al.* (1996)
Thailand (sandy loam soil)★★	kg C/ha	172.4	269.4	36	Sass (1996); Kroeze (1996) Shin *et al.* (1996)

★ Partial SRI practices. ★★ CH₄-C emissions rate are total per season

Source: Authors' own compilation

SRI as a mitigation measure to GHG emissions

Very limited work has been done to quantify the GHG emissions from SRI and conventional methods of rice cultivation. The role of SRI in reducing GHG emissions from the experiments conducted in China, India and Thailand are presented in Table 11.2. For instance in a two-year field experiment in China, total N_2O emissions from intermittent irrigation (which is one of the components of SRI) during the rice-growing period was 107 mg N_2O-N/m^2 that is an increment of 10 percent from paddy cultivation (data not shown). In other words, SRI increased N_2O emissions by 10 percent. Whereas, total CH_4-C emissions were 2.5 g C/m^2 under intermittent irrigation (II) and 3.9 g C/m^2 under paddy cultivation (i.e. a reduction by 44 percent, on average, with II). Carbon dioxide equivalents of both CH_4-C and N_2O-N emissions from the paddy field during the rice-growing period under intermittent irrigation were 788 kg CO_2-C/ha, which is a reduction of 61 percent compared with those from continuous flooded irrigation (Yang *et al.,* 2012). The reduction in CH_4 emissions by SRI was, on average, 36 percent in Thailand under clay soils, sandy loam, rainfed, and irrigated systems but, in the case of India and China, the CH_4 reduction was 38 percent and 44 percent, respectively.

Table 11.3 Comparative performance of rice (average grain yield in t/ha) under SRI and paddy in some Asian and African countries

Country	SRI	Paddy	% increase	References
China	6.2	5.0	24	Zhao *et al.*, (2010)
India	8.7	6.3	38	Satyanaryana *et al.* (2007)
Philippines	7.3	3.7	97	Thakar (2005)
Vietnam	6.8	5.6	21	Dung and Minh (2008)
Indonesia	7.2	3.9	85	Sato and Uphoff (2007)
Bangladesh	6.0	4.0	50	Latif *et al.* (2005)
Myanmar	4.4*	2.1	110	Kabir and Uphoff (2007)
Kenya	6.6	5.0	32	Ndiri *et al.* (2012); Nyamai *et al.* (2012)
Tanzania	9.9	3.8	161	Katambara *et al.* (2013)
Madagascar	8.0	2.0	300	http://sri.ciifad.cornell.edu/countries
Mali	9.0	6.7	34	Styger *et al.* (2011)

* Rainfed SRI

Source: Authors' own compilation

SRI's role in increasing productivity and food security

The effectiveness of SRI in increasing crop yields in some rice growing countries are presented in Table 11.3. The average rice grain yields with SRI varied from 4.4 t/ha in Myanmar to 9.9 t/ha in Tanzania. However, the yields were lower with paddy cultivation, varying from 2.0 t/ha in Madagascar to 6.7 t/ha in Mali. In China, Zhao *et al* (2010) conducted field experiments in the northern part of Zhejiang where they compared rice production under SRI with traditional flooding. Grain yields ranged between 5.6 and 6.9 t/ha with SRI and between 4.0 and 6.1 t/ha under traditional flooded management. On average, grain yields under SRI were 24 percent higher than traditional flooding in China. However, the variability in yield increment was as low as 21 percent in Vietnam and as high as 300 percent in Madagascar (Table 11.3). The low increase in Vietnam may be partly explained by inadequate capacity of farmers to control water and infection by root-feeding nematodes (pers. comm. Norman Uphoof, 2015). It is important to acknowledge that SRI does not always enhance rice yields compared with traditional flooded production. For example, in 35 site-years of data compiled from nine different Asian countries, McDonald *et al.* (2006) found no evidence of a yield advantage of SRI over recommended management practices, except for one set of experiments in Madagascar where SRI more than doubled rice grain productivity.

Case study from India

In India, a number of studies have been carried out on SRI performance targeted at water saving, yield productivity (Abraham *et al.,* 2014; Bhuvaneswari *et al.,* 2014; Gathorne-Hardy *et al.,* 2013; Ndiiri *et al.,* 2013; Thiyagarajan and Biksham, 2013) and mitigating GHG emissions (Jain *et al.,* 2013; Rajkishore *et al.,* 2013). However, studies on the synergistic effects of SRI in building adaptation capacity of farmers to climate change and mitigating GHG gases while sustaining rice production are very few. The objective of the present case study is to assess the performance of SRI practices (various combinations of SRI components) in relation to increasing productivity, mitigating CH_4 and N_2O emissions, and adaptation to climate change impacts in two ClimaAdapt project sites namely, Kalingarayan and Ponnaiyar canal basins of Tamil Nadu (India).

Site description

Field trials were conducted in the Kalingarayan basin (11° 58'N, 77° 58'E; 163 m a.s.l.), and Ponnaiyar canal basin (10° 57'N, 78° 25'E; 259 m, a.s.l.). In Kalingarayan basin, the climate is mostly dry with an average maximum temperature of 29°C to 38°C. The average annual rainfall is 714 mm with dry weather throughout the year except during the monsoons. The Ponnaiyar basin experiences a moderate climate with average maximum temperature ranging between 29°C to 37°C. The average annual rainfall is about 850 mm with maximum rainfall (323 mm) during the North-East monsoon season (October–December) and also a moderate rainfall (213 mm) during South-West Monsoon from June to September. There is also some rainfall (154 mm) during the hot season from March to May, but no rainfall during the winter months of January and February. The major soil types in the Kalingarayan basin area are red loamy, red sandy and lateritic soils of which red-loamy soils cover a larger area in the irrigated fields. The agricultural lands in Ponnaiyar basin are dominated by red loamy soil and clay loam soils.

SRI trial 2013

Field measurements were carried out on up to six components of SRI in selected farmers' fields (n = 26) in five villages located at Kalingarayan basin and six villages (n = 22) located in Ponnaiyar basin during *rabi*[1] 2013. The SRI components were land levelling (LL); planting less than 14 days old young single seedling (YS); square planting at 25 x 25 cm spacing (SP); green manure application (GM); alternate wetting & drying (AWD); and cono weeding (CW). Cono weeding is practiced by cono weeder which are made up of plastic materials that uproot weeds and bury them in the mud as result of the push and pull operations. The application of SRI components by the farmers in those villages varied from 17 percent (where only one component of SRI was applied) to 100 percent (where six components of SRI were fully implemented). To accommodate the variability

Table 11.4 SRI components applied (in number and %) in each treatments at farmers' fields during *rabi* 2013

Treatments	LL	YS	SP	GM	AWD	CW	No. of SRI components	SRI applied (%)
T1	×	×	×	×	×	×	6	100
T2	×	–	×	×	×	×	5	83
T3	×	–	×	–	×	×	4	67
T4	–	–	×	×	–	–	2	33
T5	–	–	×	×	–	×	3	50
T6	–	–	×	–	–	×	2	33
T7	×	×	–	×	–	–	3	50
T8	–	–	×	×	×	–	3	50
T9	–	–	–	–	×	–	1	17
T10	–	–	×	–	–	–	1	17
T11	–	–	–	–	–	–	0	0
Total	4	2	8	6	5	5	-	-

Note that the symbol x indicates which SRI component has been in the corresponding treatment

Source: Based on data from ClimaAdapt project (www.climaadapt.org)

among farmers' adoption levels, 11 treatments with various components of SRI practices including the control were formulated (Table 11.4).

The conventional method of cultivation (T11) was treated as a control and all the other treatments (T1 to T10) were modifications based on the adaptation of SRI principles, either singly or in combinations. Conventional transplanting (CT: control treatment) is characterized by: a) puddling the soils to keep them anaerobic and oversaturated; b) seedlings transplanted normally from 22 days; c) spacing is 10 cm by 15 cm; d) no green manure application but inorganic fertilizer application; e) continuous flooding; and f) use of herbicides to control weeds. Only two treatments (T1 and T7) had less than 14 day old YS planted, but the rest had 25 to 30 days old seedlings planted. Square planting (at 25 x 25 cm spacing) was the most common SRI component, applied in eight of the 11 treatments. The rice variety CO 48 in the Kalingarayan basin and ADT 45 in the Ponnaiyar basin were planted during *kharif* season[2] (August 2013) and the seedlings were transplanted at a spacing of 15 x 10 cm in T7, T9 and T11. Cono weeding was practiced four times at an interval of 10 to 15 days after transplanting (DAT) in T1, T2, T3, T5, and T6. Seven replications were maintained for each of the treatments to analyse some plant growth and yield parameters. Number of plants, number of productive tillers, number of grains per panicle, filled grains (in %), 1,000 grains weight (g) and grain yields were monitored.

Irrigation water was supplied to the SRI fields by open channel. In total, 11 to 14 irrigations were delivered during the whole crop growing season. The SRI fields were irrigated using the AWD method: water was re-applied after disappearance of ponded water and formation of hairline cracks on the soil surface to a depth of 2.5 cm. T1, T2, T3, T8, T9 followed the AWD method of irrigation, but the other treatments applied the normal conventional flood irrigation system. Recommended dose fertilizers (RDF) were applied at the rate of 100: 50: 50 kg for N: P: K per ha, respectively for all treatments. The P and K fertilizers were applied basally at once, but the N fertilizer application was split into three equal dosages, i.e. first application before transplanting, second and third applications at 25 and 45 DAT, respectively.

SRI trial 2014

In 2014, SRI was up-scaled in clusters assuming that each cluster has a similar soil type and uniform management practices were applied, except for the treatments tested. Clusters were defined based on geographical location. SRI treatment composed of 6 components, namely LL, YS, SP, GM, AWD and CW were tested in all SRI clusters in KB and PB basins, in addition to the conventional method used as a control. Each farmer's field was considered as one replication. A total of 14 replications in KB, 11 replications in Ponnaiyar dam and 9 replications in Manpathai village of PB were monitored for each of the clusters (Table 11.5). The total number of farmers' fields were 56 with a total area of 60 ha in KB. In the case of PB, they were 40 farmers' fields with a total area of 13.8 ha. In PB, due to shortage of water, a farmer's field was as small as 0.2 ha under paddy rice.

Table 11.5 Field trials conducted in clusters in Kalingarayan and Ponnaiyar basins during *rabi* 2014

Site	Cluster	Treatment*	Village	No. of farmers' fields
Kalingarayan (KB)	SRI 1 & CT 1	ASD 16	B.S. Agraharam and S.S. Agraharam	14 & 14
	SRI 2 & CT 2	BPT 5204	Vairapalayam	14 & 14
	SRI 3 & CT 3		B.P.Agraharam	14 & 14
Ponnaniyar (PB)	SRI 1 & CT 1	ADT 39	Ponnaniyar dam	11 & 11
	SRI 2 & CT 2	ADT 45	Manpathai	9 & 9

* Note that the treatments names refer to rice genotypes

Source: Based on data from ClimaAdapt project (www.climaadapt.org)

Agronomic data collection

The rice genotype BPT 5204 was transplanted on 7 October 2014 in the main field in both conventional and SRI plots. The date of sowing was 15 September 2014 in the conventional plot and 24 October 2014 in the SRI plot. In the SRI plot, 14 days old seedling were planted at a spacing of 25 cm x 25 cm, and in the conventional plot, 22 days old seedlings were planted at a spacing of 15 cm x 10 cm. In the conventional plot, hand weeding was performed twice i.e. 25 and 45 DAT but in the SRI plot, cono weeding was done at 15, 30 and 45 DAT. Five plant samples were collected from a 0.2 ha plot where grain yields were measured. The grain moisture was determined by moisture meter and the grain yield was corrected for difference in moisture content. Agronomic parameters such as number of plants/m^2, number of productive tillers per plant at maturity, number of grains per panicle, filled grains (%), 1,000 grains weight (g), and grain yield (t/ha) were recorded. In addition, cost of seeds, fertilizers, herbicides, pesticides and labour cost were also registered.

Water flow measurements

Water flow measurements were carried out in farmers' fields from 0.2 ha where SRI and conventional method of cultivation were applied. The total water flow from PB canal was measured using a cut-throat flume. The water was channelled to SRI and conventional cultivation plots and measured using a water meter (WALTEX M-Horizontal Woltmann meter, Indian make) in the delivery pipe. In the SRI plots, field water tubes were also installed to observe the water level and assist in determining when to irrigate. The plot under conventional cultivation was always under water, ponded to a height of 2.5 to 5 cm. The following measurements were calculated:

Total water use (mm) = irrigation water use (mm) + rainfall (mm)

Water productivity (kg/ha mm) = grain yield (kg/ha) ÷ total water consumed (mm)

Water saving (%) =
[(water used in CT – water used in SRI) × 100] ÷ water used in CT

Soil sampling, analysis and treatments

Soil samples were drawn from the fields of all participating farmers in the top 15 cm by making a 'V' shaped cut and collecting the soil samples from the sides of the cut. At least five samples were collected from each of the farmer's fields, and a representative sample was taken by thoroughly mixing the collected soil samples. Soil samples were analysed for physical and chemical properties before planting the rice seedlings, and at crop harvest. Organic carbon was analysed by the Walkley and Black method (Gelman *et al.*, 2011). The available N was determined by the alkaline permanganate method (Subbiah and Asija, 1956), available P by 0.5 M $NaHCO_3$ (pH 8.5) (Olsen *et al.*, 1954), and available K by neutral normal ammonium acetate method (Stanford and English, 1949). Dissolved oxygen was determined by the Azide modification of iodimetric method. In SRI and CT plots, four treatments were applied to investigate the effect of organic and inorganic fertilizers on CH_4 and N_2O fluxes from the plots, and also effects on grain yield. These treatments were: T1 (recommended dose of fertilizer, RDF); T2 (RDF + azolla); T3: (RDF + blue green algae, BGA); and T4: (RDF +azolla + BGA).

CH_4 & N_2O sampling and analysis

The closed chamber technique was used to measure direct N_2O-N and CH_4-C fluxes from soils treated with SRI and conventional rice cultivation (e.g. Jain *et al.*, 2013). The chambers were fabricated as recommended by studies on trace gas measurements under field conditions (Denmead, 2008). Open-bottom perplex chambers using 4 mm acrylic sheets with a dimension of 50 cm x 50 cm x 100 cm were fabricated. A battery (12V) operated fan was fixed to the chamber for air circulation in order to avoid plant suffocation, to mix the air inside the chamber and to draw the air samples into air-sampling bags (Tedlar®). Gas sampling was carried out in farmers' fields in PB basin from two different rice cultivation methods. These cultivation methods were SRI and CT. The chamber was inserted 5 cm into the soil. Four gas samples were withdrawn from each chamber at 15 minutes intervals (0, 15, 30 and 45 minutes) using a syringe and one-way valve pump immediately after closure, from a surface area of 0.30 m². The gas samples were often collected in the morning (09:00–10:00) and in the evening (14:00–15:00). The average of morning and evening fluxes were used as the flux rate for the day (Jayadeva *et al.*, 2009). The gas samples were collected three times a week during the active tillering, panicle initiation, flowering and maturity stages. The average of the seven days was reported as the average daily CH_4 and CO_2 emission rate for the respective stage. The CH_4 and N_2O flux assessment was continued until the maturity stage to detect the trends of emissions from the rice soils. In total, 192 gas samples were collected on 12 events between 12 June to 14 September 2013 (total 70 days).

The CH_4-C and N_2O-N concentrations were analysed using a gas chromatograph (GC, Shimadzu GC-2014) equipped with a flame ionization detector. The GC was calibrated before and after each set of measurements using 1ppm, 2.3 ppm and 5 ppm of standards (Chemtron® science laboratories Pvt. Ltd., Mumbai) to develop a primary standard curve over the concentration ranges. The CH_4-C and N_2O-N concentrations were then determined by peak area and flux was calculated based on the equation proposed by Rolston (1986) to estimate CH_4 and N_2O emissions as follows:

$$F = (V/A)(\Delta C/\Delta t)$$

where F is CH_4 (mg/ m²/hr), or N_2O (µg/m²/hr) flux rate, V is volume of the chamber above the soil (m³), A is cross-section of chamber (m²), ΔC is concentration difference between zero and t times (mg/cm³), and Δt is time duration between two sampling periods (hour).

Statistical analysis

The results in this study were statistically analysed using the Analysis of Variance (ANOVA) technique (Panse and Sukhatme, 1985).The critical significance difference (CD) of the means of the treatments were worked out at 0.05 and 0.01 probability levels.

Results and discussion

SRI growth parameters and yield: rabi 2013

Figure 11.1 depicts some of the agronomic parameters of rice grown under the SRI (T1) and conventional cultivation systems at the KB and PB sites. At both sites, SRI outperformed conventional plots in the number of productive tillers and in grain number per panicle, although there was no difference in the percentage of filled grains or 1,000 grain weight. In both basins, the highest number of plants/m^2 were recorded in T11 (i.e. the conventional plot) and lowest in T1, T2, T3, T4 and T5 where more than 50% of the SRI components were applied. The wider plant spacing (25 cm ´ 25 cm) used in SRI plots compared with the closer spacing (15 cm ´ 15 cm) applied in the conventional rice cultivation system has contributed to the large difference in the number of plants/m^2 between SRI and non-SRI plots, particularly in the early stage of the rice plant (Figure 11.1).

The average number of productive tillers per plant in the conventional plot were 14 and 13 in KB and PB, respectively, which were significantly lower (p <0.05) than all the rest of SRI plots (data not shown). In the SRI plots, the number of productive tillers ranged from 27 (T10) to 43 (T1), where all the SRI components are implemented. This may be due to the roots under SRI having good air circulation and water movement in the soils, enabling them to absorb the nutrients effectively, thereby developing more productive tillers than the paddy rice plants (Krishna *et al.*, 2008). The average number of grains per panicle in the conventional plot was 109 and 90 for KB and PB, respectively, and these were significantly lower (p <0.05) than those recorded in most of the SRI plots in both basins (data not shown). However, the percentage of grain filling measured in SRI was slightly higher than the conventional plots. The rice grain yields under SRI plot (T1) in both basins were greater than the conventional plot (T11). At the PB site, the rice grain yield is lower than KB site in all the treatments.

Table 11.6 presents mean grain yield of rice and harvest index in Kalingarayan and Ponnaniyar basins during *rabi* 2013. The mean data for the growth and yield parameters for all treatments showed a statistical difference at the 5 percent level. The difference in means for SRI and conventional plots in all parameters was significant both at 5 percent and 1 percent levels (Table 11.6). There was no significant difference observed for mean yield between T7 and T9 (where three and one component of SRI was implemented, respectively) at 5 percent level in KB and PB sites. The standard deviation among the treatments was higher for

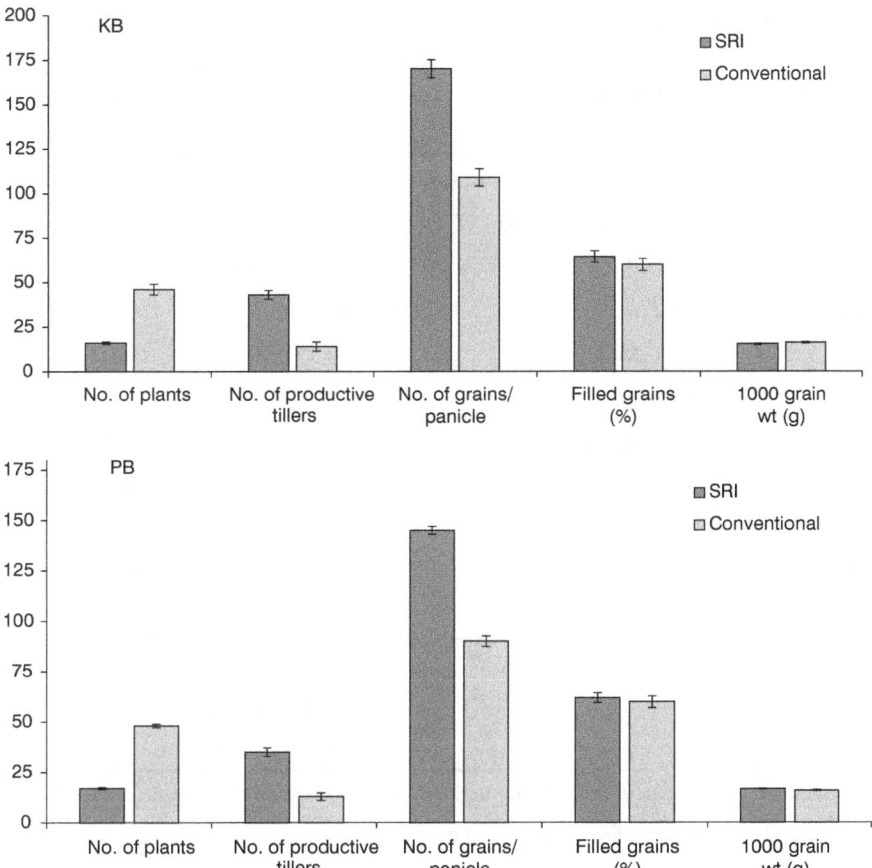

Figure 11.1 Mean values of some growth and yield parameters of rice grown under SRI and conventional in Kalingarayan basin (A) and Ponnaiyar basin (B): *Rabi* 2013. The error bars correspond to the standard errors. Note that the SRI bars represent only T1 which contains all 6 components of SRI

Source: Authors' own analysis, based on data from ClimaAdapt project (www.climaadapt.org)

the number of grains per panicle (data not shown) and lower for grain yield in KB, as well as PB sites (Table 11.6).

The highest grain yield was produced in T1 (containing all six components of SRI), averaging 8.74 t/ha across the two sites. In contrast, the conventional flooded system (T11) yielded only 67 percent of T1. Although the experiment was not a factorial design, analysis of certain combinations of components allows the value of particular components to be assessed. For example, adding the square planting (SP) component alone to the conventional system (T11) increased the grain yield (relative to T1) from 67 percent to 70 percent. Further adding cono-weeding (CW) to SP increased the relative grain yield by another 4 percent to 74 percent. Alternatively, adding green manure (GM) to SP increased the relative

Table 11.6 Mean yield parameters of rice in Kalingarayan basin and Ponnaiyar basins, *rabi* 2013

Treatments	SRI practices applied	Grain yield (t/ha)		% as of T1 (mean)
		KB	PB	
T1	LL, YS, SP, GM, AWD, CW	9.72	7.76	100.0
T2	LL, SP, GM, AWD, CW	8.69	7.06	90.1
T3	LL, SP, AWD, CW	8.04	6.31	82.1
T4	SP, GM	7.58	5.81	76.6
T5	SP, GM, CW	8.38	6.66	86.0
T6	SP, CW	7.28	5.69	74.2
T7	LL, YS, GM	7.11	5.53	72.3
T8	SP, GM, AWD	7.63	6.12	78.7
T9	AWD	7.12	5.44	71.9
T10	SP	6.87	5.33	69.8
T11	–	6.56	5.11	66.8
SEd	–	0.0895	0.0720	–
CD (0.05)	–	0.1791	0.1440	–
CD (0.01)	–	*0.2382*	*0.1915*	–

SEd: Standard error deviation (±); CD: Critical Difference at 0.05 and 0.01 significance levels

Source: Based on data from ClimaAdapt project (www.climaadapt.org)

grain yield even more to 77 percent. However, adding both CW and GM to SP resulted in a further significant increase in relative yield to 86 percent, indicating the synergistic interactions of SP, CW and GM. This demonstrates the stepwise increases resulting from particular components. Another useful comparison would be to remove particular components from T1 and see the extent to which yield is reduced. For example, T2 contains all components of SRI except YS, highlighting that the removal of this single component reduced yield by 10 percent. Removing GM in addition to YS reduced yield by another 8 percent (to 82 percent of T1).

SRI growth and yield parameters: **rabi 2014**

Figure 11.2 shows the mean values of selected agronomic parameters for the ASD 16 and BPT 5204 rice varieties grown under SRI and conventional practices in KB, and ADT 39 and ADT 45 rice varieties in PB. In general, performance of the rice varieties grown under SRI was better than under conventional flooded production at both sites, particularly in relation to the number of productive tillers and the number of grains per panicle. The rice production system (SRI vs CT) had little impact on percentage of grain filling or 1,000 grain weight at either location or in either variety. This was a similar result to the *rabi* 2013 season.

The effect of varieties, cultivation methods and interaction between variety and cultivation method and vice versa on grain yields are shown in Table 11.7 for the KB and PB sites.

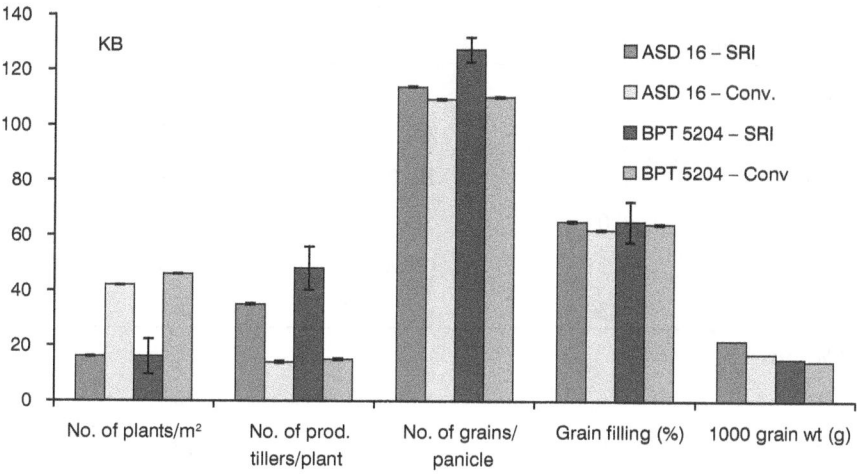

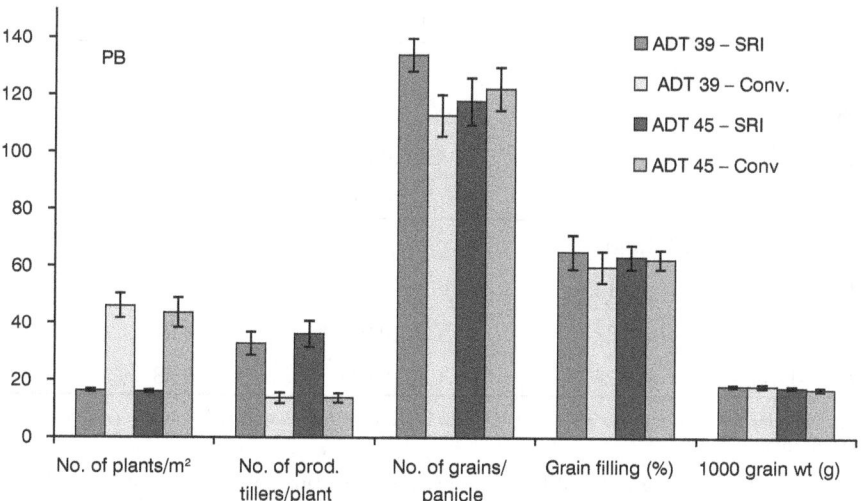

Figure 11.2 Mean values of some growth and yield parameters of rice grown under SRI (with ADT 39 and ADT 45) and CT (with ADT 39 and ADT 45) in KB (A) and PB (B) 2014 *rabi*

The error bars correspond to the standard errors

Source: Authors' own analysis, based on data from ClimaAdapt project (www.climaadapt.org)

Table 11.7 Mean grain yields of rice (t/ha) cultivated under SRI and CT in KB and PB: *rabi* 2014

KB	Grain yield (t/ha)	PB	Grain yield (t/ha)
SRI ASD 16	6.9	SRI ADT 39	7.2
CT ASD 16	5.8	CT ADT 39	5.3
SRI BPT 5204	8.4	SRI ADT 45	6.7
CT BPT 5204	5.4	CT ADT 45	5.5
Variety			
ASD 16	6.4	ASD 16 ADT 39	6.2
BPT 5204	6.9	BPT 5204 ADT 45	6.1
SEd	0.09	SEd	0.10
CD (0.05)	0.21	CD (0.05)	n.s.
CD (0.01)	0.30	CD (0.01)	n.s.
Cultivation method			
SRI	7.6	SRI	6.9
CT	5.6	CT	5.4
SEd	0.07	SEd	0.07
CD (0.05)	0.15	CD (0.05)	0.15
CD (0.01)	0.19	CD (0.01)	0.19
Variety X Cultivation method			
SEd	0.12	SEd	0.13
CD (0.05)	0.26	CD (0.05)	0.27
CD (0.01)	0.35	CD (0.01)	0.37
Cultivation method X Variety			
SEd	0.10	SEd	0.10
CD (0.05)	0.21	CD (0.05)	0.21
CD (0.01)	0.28	CD (0.01)	0.28

SEd: Standard error deviation (±); CD: Critical Difference at 5 % and at 1% levels. (n.s.: not significant)

Source: Based on data from ClimaAdapt project (www.climaadapt.org)

Higher grain yield were recorded from SRI plots than CT plots in both KB and PB sites (Table 11.7). The mean critical differences for SRI and CT cultivation methods for all growth parameters were significant both at 0.05 and 0.01 levels, except for grain filling at 0.01 level and 1,000 grain weight for PB sites only. There was a significant difference in number of plants and tillers at the 0.05 level, but no significant difference observed at 0.01 level for all growth parameters except for 1,000 grain weight (data not shown). A significant difference was detected in grain yield between the rice varieties in the KB site at 0.05 and 0.01 levels, but no significant difference was observed at the PB site. The CD between the cultivation methods was significant in grain yields at KB, as well as PB sites (Table 11.7).

Water savings by SRI

The amount of water used under SRI and CT planting methods is shown in Figure 11.3. The water consumed in SRI was lower than for CT in all crop growth stages except in the 46–60 days after transplanting (DAT) period, when water consumption was equivalent between planting methods (Figure 11.3). In total, 491 and 630 mm were used by SRI and CT, respectively. There was no additional water applied in the SRI and CT plots after 1–15 DAT because the soils were saturated after more than 200 mm water was applied during the field preparation.

Table 11.8 presents the total water used and water productivity under SRI and CT plots. The SRI plots used less water and produced higher yield compared to CT plots. SRI saved about 22 percent of total water applied. The water productivity in SRI was about 1.5 times greater than CT. In other words, the difference between the two methods was about 55 percent which corresponds to the result found by Jagannath *et al.* (2013).

Cost-benefits of SRI

A cost-benefit analysis was carried out for the crops cultivated during the 2013 and 2014 *rabi* and *kharif* seasons in both the KB and PB sites (Table 11.9). In 2013 *rabi*, the rice varieties ADT 45 in PB and CO-48 in KB site were planted in SRI and CT plots. In *rabi* 2014, however, the SRI and CT plots were planted with BPT 5204 in KB site and ADT 39 in PB site. In both production seasons and sites, the total rice production costs were slightly lower in SRI than CT treatment. The main difference between SRI and CT was that total returns were substantially higher in SRI due to higher grain yields (Table 11.9). Calculation of the benefit-cost ratio shows that SRI compared to (CT) gave a significantly higher benefit-cost ratio in the range of 1.67 (1.14) in *rabi* 2013 and 1.61 (1.16) in *kharif* 2014 at the PB site. However at the KB site, the benefit-cost ratio was even higher, ranging from 1.90 (1.38) in *rabi* 2013 to 1.71 (1.12) in *kharif* 2014.

Soil nutrients and dissolved O2 contents: rabi 2014

The organic C, available N, P, K and dissolved O_2 contents of the soils treated with SRI and CT, amended with four types of fertilizer are shown in Table 11.10. The combined application of inorganic fertilizers with organic fertilizers consisting of both azolla + BGA showed higher organic C, available N, P and

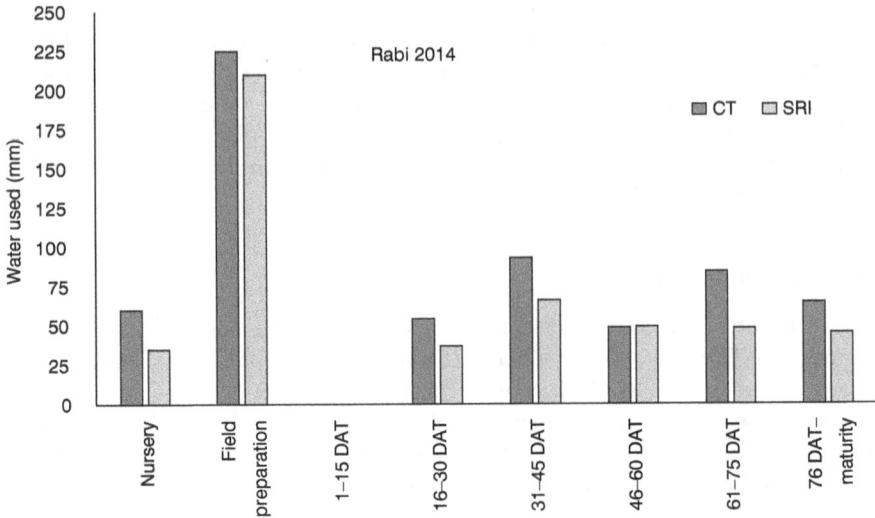

Figure 11.3 Water used by SRI and CT from nursery, field preparation and DAT until maturity stages of rice, *rabi* 2014

Source: Authors' own analysis, based on data from ClimaAdapt project (www.climaadapt.org)

Table 11.8 Water used in SRI and CT plots with grain yield and water productivity in PB during *rabi* 2014

	CT	SRI
Total irrigations (no.)	22	18
Water applied (mm)	630	491
Total rainfall (mm)	447	447
Total water applied (mm)	1,077	938
Water saving (%)	–	22
Grain yield (kg/ha)	5,146	6,951
Water productivity (kg/ha mm)	4.77	7.40

Source: Based on data from ClimaAdapt project (www.climaadapt.org)

K, and dissolved O_2 contents in both SRI and CT. The SRI plots (with RDF + azolla + BGA treatment) contained higher nutrients and dissolved oxygen than the rest of the treatments. This indicates that SRI soils were efficient in taking up the soil nutrients and were well aerated and drained. As a result, the SRI plots render higher grain yield than CT. The CT soils were deficient in oxygen and thus were poorly aerated and drained due to the continuous flooding of the rice fields. The soils under RDF contained the lowest organic C, available N, P, K, and dissolved O_2 in SRI as well as CT. The dissolved O_2 content of the CT soils

Table 11.9 Economics analysis of SRI and CT cultivation during *rabi* 2013 and 2014 in the PB and KB sites

Particulars	Rabi 2013				Rabi 2014			
	PB (ADT 45)		KB (CO-48)		PB (ADT 39)		KB (BPT 5204)	
	SRI	CT	SRI	CT	SRI	CT	SRI	CT
Land (Lease rate)	7,000	7,000	8,000	8,000	7,000	7,000	8,200	8,200
Labour (family)	3,000	3,000	3,000	3,000	3,000	3,000	3,100	3,100
Seed (variety)	150	850	180	950	120	800	160	780
Land preparation and transplanting	12,100	8,700	12,300	9,100	12,700	9,200	13,700	9,600
Herbicides	0	1,450	0	1,400	0	1,500	0	1,350
Fertilizers	2,800	3,200	3,000	3,200	2,100	3,300	2,800	3,300
Organic manures	3,400	4,100	3,500	4,300	1,500	900	1,400	1,200
Pesticides	350	800	300	600	400	985	350	780
Manual weeding	2,200	3,000	2,400	3,300	2,600	3,100	2,750	3,200
Irrigation	1,150	2,600	1,350	2,650	1,300	2,300	1,450	2,600
Harvesting	11,100	11,700	11,400	12,000	11,100	11,700	11,400	11,900
Total cost	43,250	46,400	45,430	48,500	41,820	43,785	45,310	46,010
Yields – paddy (kg/ha)	7,760	5,110	9,200	6,560	7,200	5,300	8,400	5,400
Yields – byproducts (kg/ha)	2,510	3,730	3,230	4,510	2,400	2,300	2,500	2,300
Returns on paddy	65,960	43,435	78,200	55,760	61,200	45,050	71,400	45,900
Returns on straw	6,275	9,325	8,075	11,275	6,000	5,750	6,250	5,750
Total returns	72,235	52,760	86,275	67,035	67,200	50,800	77,650	51,650
Profit	28,985	6,360	40,845	18,535	25,380	7,015	32,340	5,640
Benefit cost ratio	1.67	1.14	1.90	1.38	1.61	1.16	1.71	1.12

All costs are expressed in Indian Rupees/ha, currently, 1 Indian Rupee is equivalent to 0.0157 US $ (or 1 US $ ~63.7Rs). n.a. not available

Source: Based on data from ClimaAdapt project (www.climaadapt.org)

Table 11.10 Organic C, available N, P, K and dissolved O_2 contents under SRI and CT cultivation methods

Treatments		Org C (%)	Ava. N (kg/ha)	Ava. P (kg/ha)	Ava. K (kg/ha)	Dis. O2 (mg/L)	Grain yield (kg/ha)
SRI	RDF	0.52	221.1	34.5	434	3.23	6,827
	RDF+ azolla	0.59	231.0	36.0	442	3.42	6,902
	RDF + BGA	0.56	239.8	37.2	452	3.55	6,987
	RDF + azolla + BGA	0.60	249.3	37.6	464	3.69	7,182
CT	RDF	0.52	199.5	33.5	421	1.75	4,814
	RDF+ azolla	0.57	206.5	33.6	431	2.11	4,889
	RDF + BGA	0.55	214.3	36.7	439	2.25	5,339
	RDF + azolla + BGA	0.59	226.5	36.6	448	2.35	5,156

RDF: recommended dose fertilizer; azolla; BGA: blue green algae. Note: results are the outcome of the soil sample analysis drawn on 45 DAT. ADT 39

Source: Based on data from ClimaAdapt project (www.climaadapt.org)

amended with only RDF (1.75 mg/L), was less than the SRI plot treated with combined organic and inorganic fertilizers by over 50 percent.

CH₄ and N₂O fluxes

Among the different systems of rice cultivation, SRI recorded lower methane fluxes than conventional systems (Figure 11.4). SRI cultivation consists of components such as organic fertilizers, square planting, cono weeding, alternate wetting and drying, and enhanced aeration in the soils. This has resulted in higher dissolved oxygen contents and higher storage of carbon in the soil (Table 11.10), and lower methane emissions than CT treatments from active tillering to crop maturity stages (Figure 11.4A). Cono-weeding stirs the soil, facilitating aeration and increasing the soil's redox potential (Eh) (Rajkishore *et al.*, 2013).

Methane flux increased progressively with the advancement of crop stage, reaching the highest value at panicle initiation and slightly declining towards maturity. It is likely that more intense reducing conditions at the panicle initiation stage that might have favoured the activities of methanogenic bacteria, contributing to highest methane emissions at this stage. However, the N_2O emissions from SRI plots were higher than CT at tillering and panicle initiation stages, but similar at heading and maturity stages. Hence, there is a trade-off between enhanced N_2O emissions and reduced CH_4 emissions from rice soils

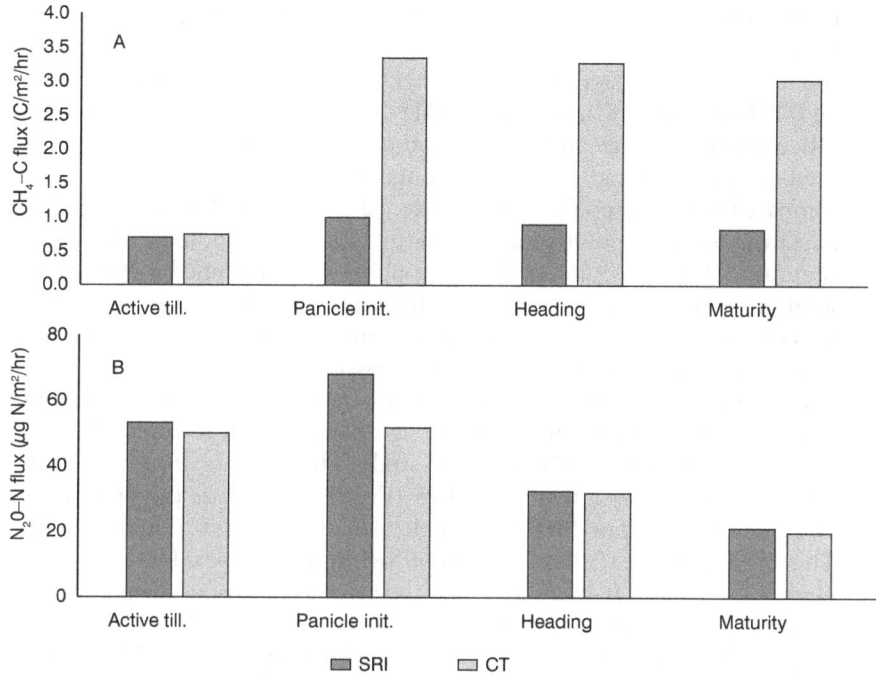

Figure 11.4 Influence of SRI and CT cultivation systems on (A) methane and (B) nitrous oxide fluxes along the growth stages of rice during *rabi* 2014

Source: Authors' own analysis, based on data from ClimaAdapt project (www.climaadapt.org)

in SRI, as opposed to the conventional flooded fields which normally favour complete denitrification of nitrates and/or ammonium to gaseous nitrogen (N_2) thereby reducing N_2O emissions (Fig. 11.4B).

Adoption and upscaling SRI: opportunities and constraints

Some of the available opportunities that could be used to disseminate and upscale SRI to many farmers on a wider area are discussed below:

Opportunities for adoption and upscaling SRI

1   *Adaptation to climate change:* The soils under SRI plots were well aerated and drained since the dissolved oxygen content (3.2 to 3.7 mg O_2/L) was greater than the soils under the conventional plots (1.8 to 2.4 mg O_2/L). This has contributed to increased water use efficiency. The SRI plots used 22 percent less irrigation water in the *rabi* 2014 season, on average, compared with traditional flooded production (see Table 11.8). Studies carried out in others countries showed that the irrigation water savings varied from 10 to 40 percent (Table 11.1). Water savings also translated into reduced

production costs in terms of labour and power. The cost-benefit analysis in Table 11.9 shows that the cost-benefit ratio (CBR) under SRI was about 1.5 times greater than CT. Other researchers found similar results on CBR (Adusumilli and Laxmi, 2011). The water savings resulting from SRI could be used to grow other food crops or could be reallocated and used for domestic and/or industrial purposes. Moreover, the time savings compared with CT could be shifted to other livelihood activities such as off-farm work, which gives additional income for the household. Hence, adopting SRI practices in rice-based cropping systems could be a part of the solution to water scarcity related to climate change. Moreover, SRI builds up farmers resilience to the changing climatic conditions by producing 'more from less' while reducing GHG emissions.

2   *Mitigation to GHGs emissions:* One of the benefits of SRI is the mitigation of net CH_4 emissions from flooded rice soils as shown in Table 11.2, and this has been confirmed by this case study. The methane emissions from SRI plots were lower than CT plots (Figure 11.4a) but the cumulative N_2O emissions from SRI were higher than CT plots (Figure 11.4b). This highlights the trade-off effects of SRI in mitigating GHG emissions. Higher losses of N_2O-N indicate that the aerobic/anaerobic changes in SRI are probably associated with less efficient use of N fertilizers. However, proper nitrogen management such as applying need-based N fertilizer applications, use of slow release urea fertilizers and deep placement of N may curtail the increments of N_2O emissions, increase N use efficiency, and contribute to reduced Global Warming Potential (GWP). Field measurements by Jain *et al.* (2013) have shown that GWP was reduced by 28 percent in SRI over flooded paddy cultivation. Moreover, mitigating the net CH_4 emissions from rice fields increases carbon storage in soils and the organic matter content which, in turn, improves soil properties. In general, mitigating GHGs through SRI could help farmers earn carbon credits if policy mechanisms are developed.

3   *Increasing productivity and food security:* SRI increased the number of productive tillers and grains per panicle compared with CT (Figures 11.1 and 11.2). Moreover, the grain yields from SRI plots plus the harvest indexes were higher than CT plots (Tables 11.6 and 11.7). Similar results have also been found for grain yield under SRI and paddy in other rice growing countries (see Table 11.3). The water productivity from SRI was 1.5 times greater than CT (Table 11.8). The higher water productivity gives more crop per drop and reduces the irrigation water requirement (Figure 11.3) with higher labour productivity. The benefit-cost ratio under SRI (1.6 to 1.90) was higher than CT (1.1–1.4) due to higher yields obtained from less production costs. The summed effects of SRI enhances the multi-dimensional aspects of food security i.e. food production, food availability, access to food, food utilization and stability.

Constraints to adoption and upscaling SRI

The aforementioned benefits of SRI fulfil the core principles of CSA. These benefits can be achieved if all the components of SRI (see Table 11.4) are implemented in a given rice field. Otherwise, partial implementation of the components of SRI will at least achieve some benefits aligned to CSA principles. Under changing climate conditions, the conventional method of rice cultivation does not offer as many opportunities for farmers as SRI. Despite these facts, the implementation and adoption of SRI by many rice farmers in different parts of the world remains either low, or the rate of adoption is slow. This situation also applies for the Indian rice farmers. For instance, the adoption rate of SRI by farmers was 5 percent in the Kalingarayan basin (n = 502) and 10 percent in the Ponnaiyar basin (n = 384) in Tamil Nadu (ClimaAdapt, 2013).

Although rice farmers in India have seen the advantages of applying SRI in the neighbouring farmers' fields or demonstration sites, they are reluctant to implement all six components of SRI listed in Table 11.4. Rather, farmers practice one, or a few of the SRI components in combination, or follow the conventional way of rice cultivation unless situations force them otherwise. One notable example is the 2012 summer and winter seasons, when the monsoon rainfall was low, forcing many farmers to apply SRI because they faced severe water shortages. This has resulted in a 30 percent increase in farmers' productivity (ClimaAdapt, 2013). There are several reasons why farmers are reluctant to apply all of the SRI components.

1   *Changing farmers' perceptions towards SRI:* Planting more than one seedling has been practiced by farmers for many decades under conventional rice cultivation. Farmers are concerned about the survival of a single seedling if they adopt the YS component of SRI, fearing that this practice will lead to crop failure. Moreover, farmers believe that the yield from conventional rice cultivation is reasonable considering the cost of production. Therefore, it remains a challenge to change farmers' mind-set to a new method of rice cultivation such as SRI. Farmers also assume that SRI is not suitable to their soils. Wider spacing is perceived as waste of land area (SP, another SRI component).

2   *Shortage of inputs:* Although farmers show a willingness to practice components of SRI such as field levelling (LL) and cono weeding, they encounter a shortage of farm implements such as cono weeders and levelling implements. These implements are also unaffordable to small-holders due to a lack of capital. Moreover, there is a shortage of good quality water for irrigation, especially for those fields that are near urban centres, as toxic substances are discharged into the irrigation canals.

3   *Supply of irrigation water and power:* As most of the irrigation schemes in the lowland rice areas are dependent on a power supply, irrigating at the right time is constrained by frequent power cuts in the irrigation areas. This deters farmers from applying AWD (a component of SRI), since they are not sure when the next power will come, and for how long it will last.

4 *High labour cost:* There is a shortage of skilled labour for transplanting at the right time. The labour costs are high for weeding and for applying green manures (an SRI component). Close follow-up and monitoring of crop growth is required while implementing SRI compared with CT, and this increases the cost of labour to farmers.

5 *Inadequate extension and CC awareness:* The local farmers lack knowledge on SRI principles, applications and climate change impacts because of weak extension services and the lack of training. For instance, about 45 percent of the farmers interviewed in KB are not aware of the change in climatic conditions (ClimaAdapt, 2013). Farmers face problems in maintaining nurseries (e.g. the rice seedlings were scorched due to intense sunlight when plastic sheets were used). Farmers' in Tamil Nadu have some difficulty in using cono weeders in clayey soils. Demonstration through in-field trials and training farmers on SRI practices should be promoted in order to enhance adoption and upscaling SRI.

6 *National policy frameworks*: There are no national policy frameworks supporting the dissemination and uptake of water-saving technologies such as SRI in the rice growing countries of SSA and Asia, including India (Reddy and Venkatanarayana, 2013). Although SRI was introduced to India 15 years ago, there are still controversial issues raised by farmers, researchers and policy makers with regard to SRI's performance in increasing adaptation to CC impacts, reducing GHG emissions, and increasing yields and food security. Recently, some of the recommendations from the Climarice I and II projects (i.e. introducing modified cultivation methods such as SRI) are now being included in the final draft of State Climate Change Action Plan for the Tamil Nadu state of India.[3]

Conclusions and recommendations

The performance of SRI in relation to the three core principles of CSA (i.e. increasing adaptation to CC, mitigation of GHG emissions, and increasing productivity and food security) have been reviewed by analysing research results in different rice growing countries of Asia and SSA. According to these publications, SRI has outperformed conventional rice cultivation with respect to increasing adaptation to CC impacts (Table 11.1), reducing GHG emissions, particularly CH_4 (Table 11.2) and increasing yield (Table 11.3). These results are also consistent with the case study from Tamil Nadu state (India) which demonstrated the superiority of SRI compared to the conventional planting method in the various plant growth and yield parameters (Tables 11.6 and 11.7), adaptation to water scarcity (Table 11.8), economic advantages in terms of the benefit-cost ratio (Table 11.9) and mitigation of CH_4 emissions (Figure 11.4a). However, potential trade-off effects have been observed by SRI (Figure 11.4b) in terms of increased N_2O emissions from soils, posing a concern to the implementation and adoption of SRI. Furthermore, some studies on SRI do not demonstrate increases in grain yield compared to conventional flooded rice production.

The beneficial effects of SRI suggest that this water-saving technology could be up-scaled. Farmers will only adopt the full components of SRI on a large scale if they actually benefit from using the technology. The interactions among researchers, policy-makers and stakeholders, including farmers, should be strengthened to increase our science-based knowledge of SRI, enabling us to develop policy guidelines targeting SRI adoption and, if appropriate, upscaling activities. The following measures could help to promote the adoption and upscaling of SRI at the local level:

- demonstrating SRI performance in relation to CSA principles to farmers and other key stakeholders;
- raising awareness campaigns and training on principles and applications of SRI and climate change impacts on rice production;
- providing low-cost levelling equipment and markers from local suppliers;
- promoting hand-held motorized cono weeders; and
- developing policy guidelines on water use for agriculture and law enforcement procedures.

Notes

1 The *rabi* cropping season (in India) is from October-March (winter).
2 The *kharif* cropping season in India is from July –October during the south-west monsoon.
3 http://www.indiaenvironmentportal.org.in/files/file/tamil%20nadu%20climate%20change%20action%20plan.pdf

References

Abraham, B., Araya, H., Berhe, T., Edwards, S., Gujja, B. Khadka, R.B. Koma, Y.S., Sen, D., Sharif, A., Styger, E., Uphoff N. and Verma, A. (2014) The system of crop intensification: reports from the field on improving agricultural production, food security, and resilience to climate change for multiple crops, *Agriculture and Food Security*, 3 (4): 1–12.

Adusumilli, R. and Laxmi, S.B. (2011) Potential of the system of rice intensification for systemic improvement in rice production and water use: the case of Andhra Pradesh, India, *Paddy Water Environment*, 9: 89–97.

Barah, B.C. (2009) Economic and ecological benefits of system of rice intensification (SRI) in Tamil Nadu, *Agricultural Economics Research Review*, 22: 209–214.

Bhuvaneswari, K., Geethalakshmi, V., Lakshmanan, A., Anbhazhagan, R. and Nagothu, U.S. (2014) Climate change impact assessment and developing adaptation strategies for rice crop in western zone of Tamil Nadu, *Journal of Agrometeorology*, 16 (1): 38–43.

Denmead, O.T. (2008) Approaches to measuring fluxes of methane and nitrous oxide between landscapes and the atmosphere, *Plant and Soil*, 309: 5–24.

Dobermann, A. (2004) A critical assessment of the system of rice intensification (SRI), *Agricultural Systems*, 79: 261–281.

Dung N.T. and Minh, L.N. (2008) System of rice intensification – advancing small farmers in mekong region, Oxfam America, East Asian Regional Office. http://sri.cals.cornell.edu/countries/vietnam/vnOxfamPPD0608.pdf [Accessed 10 March 2015].

Food and Agricultural Organization (FAO) (2010) *"Climate-Smart" Agriculture: Policies, Practices and Financing for Food Security, Adaptation and Mitigation*, Rome, Italy.

Gathorne-Hardy, A., Reddy, D.N., Venkatanarayana, M. and Harriss-White, B. (2013) A life cycle assessment (LCA) of greenhouse gas emissions from SRI and flooded rice production in SE India, *Taiwan Water Conservancy*, 61 (4): 110–125.

Geethalakshmi, V., Lakshmanan, A., Rajalakshmi, D., Jagannathan, R., Sridhar, G., Ramaraj, A.P., Bhuvaneswari, K., Gurusamy, L. and Anbhazhagan, R. (2011) Climate change impact assessment and adaptation strategies to sustain rice production in Cauvery basin of Tamil Nadu, *Current Science,* 101 (3): 342–347.

Gelman, F., Binstock, R. and Halicz, L. (2011) *Application of the Walkley-Black Titration for Organic Carbon Quantification in Organic Rich Sedimentary Rocks*, Report GSI/13/ 2011, Jerusalem, online, http://www.gsi.gov.il/_Uploads/ftp/GsiReport/2011/Gelman-Faina-GSI-13-2011.pdf [Accessed 15 February 2015].

Glover, D. (2011) Science, practice and the system of rice intensification in Indian agriculture, *Food Policy,* 36: 749–755.

Gujja, B. and Thiyagarajan, T.M. (2009) *New Hope for Indian Food Security? The System of Rice Intensification,* Gate keeper 143, International Institute for Environment and Development, London.

Jagannath, P., Pullabhotla, H. and Uphoff, N. (2013) Meta analysis evaluating water use, water saving and water productivity in irrigated production of rice with SRI vs. standard management methods, *Taiwan Water Conservancy*, 61: 14–49.

Jain, N., Dubey, R., Dubey, D.S., Singh, J., Khanna, M., Pathak, H. and Bhatia, A. (2013) Mitigation of greenhouse gas emission with system of rice intensification in the Indo-Gangetic Plains, *Paddy Water Environment*, DOI 10.1007/s10333-013-0390-2.

Jayadeva, H. M., Setty, T. K. Prabhakara, Gowda, R. C., Devendra, R., Mallikarjun, G.B. and Bandi, A.G. (2009) Methane emission patterns and their associated soil microflora with SRI and conventional systems of rice cultivation in Tamil Nadu, India, *Agricultural Science Digest*, 29: 4.

Kabir, H. and Uphoff, N. (2007) Results of disseminating the system of rice intensification with farmer field school methods in northern Myanmar, *Experimental Agriculture* 43: 463–476.

Katambara, Z., Kahimba, F.C., Mahoo, H. F., Mbungu, W.B., Mhenga, F., Reuben, P., Maugo, M. and Nyarubamba, A. (2013) Adopting the system of rice intensification (SRI) in Tanzania: A Review, *Agricultural Sciences,* 4: 369–375.

Krishna, A., Biradarpatil, N.K. and Channappagoudar, B.B. (2008) Influence of system of rice intensification (SRI) cultivation on seed yield and quality, *Karnataka Journal of Agricultural Science*, 21 (3): 369–372.

Kroeze, C., Mosier, A., Nevison, C., Oenema, O., Seitzinger, S. and Van Cleemput, O. (1996) IPCC guidelines for national greenhouse gas inventories, Reference manual, *Agriculture* 3: 53–71.

Lakshmanan, A.S., Geethalakshmi, V., Latha, P. and Nagothu, U.S. (2012) Role of photosynthethic Diazotrophs in reducing methane flux from rice soil ecosystem, Technical brief.

Latif, M.A., Islam, M.R., Ali, M.Y. and Saleque, M.A. (2005) Validation of the system of rice intensification (SRI) in Bangladesh, *Field Crops Research*, 93: 281–292.

McDonald, A.J., Hobbs, P.R. and Riha, S.J. (2006) Does the system of rice intensification outperform conventional best management? A synopsis of the empirical record, *Field Crops Research*, 96: 31–36.

McDonald, A.J., Hobbs, P.R. and Riha, S.J. (2008) Stubborn facts: still no evidence that the system of rice intensification out-yields best management practices (BMPs) beyond Madagascar, *Field Crops Research*, 108: 188–191.

Ndiri, J.A., Mati, B.M., Home, P.G., Odongo, B. and Uphoff, N. (2012) Comparison of water savings of paddy rice under system of rice intensification (SRI) growing rice in Mwea, Kenya, *International Journal of Current Research and Review*, 4: 63–73.

Ndiri, J.A., Mati, B.M., Home, P.G., Odongo, B. and Uphoff, N. (2013) Adoption, constraints and economic returns of paddy rice intensification in Mwea, Kenya, *Agricultural Water Management*, 129: 44–55.

Nyamai, M., Mati, B.M., Home, P.G., Odongo, B., Wanjugo, R. and Thuranira, E.G. (2012) Improving crop productivity and water use efficiency in basin rice cultivation in Kenya through SRI, *Agricultural Engineering International: CIGR Journal*, 14: 1–9.

Olsen, S.R., Cole, C.V., Watanabe, F.S. and Dean, L.A. (1954) *Estimation of Available Phosphorus in Soils by Extraction with Sodium Bicarbonate*, Circular No. 939. U.S. Department of Agriculture.

Panse, V.G. and Sukhatme, P.V. (1985) *Statistical Methods for Agricultural Workers*, 4th ed., ICAR, New Delhi.

Rajkishore, S.K., Doraisamy, P., Subramanian, K.S. and Maheswari, M. (2013) Methane emission patterns and their associated soil microflora with SRI and conventional systems of rice cultivation in Tamil Nadu, India, *Taiwan Water Conservancy*, 61 (4): 126–134.

Reddy, D.N. and Venkatanarayana, M. (2013) *SRI Cultivation in Andhra Pradesh: Achievements, Problems and Implications for GHGs and Work,* online, http://mpra.ub.uni-muenchen.de/52115/ [accessed 10 December 2014].

Rolston, D.E. (1986) Gas flux. In: Klute, A., (ed.), *Methods of Soil Analysis,* 2nd edn, Monograph No. 9, 1103-1109, American Society of Agronomy and Soil Science Society of America, Madison, WI.

Sass, R.L. (1996) IPCC guidelines for national greenhouse gas inventories Reference manual, *Agriculture*, 3: 399–413.

Sato, S. and Uphoff, N. (2007) A review of on-farm evaluations of system of rice intensification methods in Eastern Indonesia, *CAB Reviews: Perspectives in Agriculture, Veterinary Science, Nutrition and Natural Resources*, 2: 54.

Satyanaryana, A., Thiyagarajan, T.M. and Uphoff, N. (2007) Opportunities for water saving with higher yield from the system of rice intensification, *Irrigation Science*, 25: 99–115.

Sheehy, J.E., Peng, S., Doberman, A., Mitchell, P.L., Ferrer, A., Yang, J., Y., Zou, Y., Zhong, X. and Huang, J. (2004) Fantastic yields in the system of rice intensification: fact or fallacy? *Field Crops Research*, 88: 1–8.

Shin, Y.K., Yun, S.H., Park, M.E. and Lee, B.L. (1996) Mitigation options for methane emission from rice fields in Korea, *Springer (e-journal)*, 25 (4): 289–291.

Siopongco, J.D.L.C., Wassmann, R. and Sander, B.O. (2013) *Alternate Wetting and Drying in Philippine Rice Production: Feasibility Study for a Clean Development Mechanism*, Technical Bulletin, No.17, International Rice Research Institute, Manila, Philippines.

Stanford, D. and English, L. (1949) Use of flame photometer in rapid soil tests of K and Ca, *Agronomy Journal*, 4: 446.

Stoop, W., Uphoff, N. and Kassam, A. (2002) A review of agricultural research issues raised by the system of rice intensification (SRI) from Madagascar: Opportunities for improving farming systems for resource-poor farmers, *Agricultural Systems*, 71: 249–274.

Styger, E., Attaher, M.A., Guindo, H., Ibrahim H., Diaty, M., Abba, I. and Traore, M. (2011) Application of system of rice intensification practices in the arid environment of the Timbuktu region in Mali, *Paddy Water Environment*, 9: 137–144.

Subbiah, B.V. and Asija, G.L. (1956) A rapid procedure for estimation of available nitrogen in soils, *Current Science*, 25: 259–260.

Suryavanshi, P., Singh, Y.V., Prasanna, R., Bhatia, A. and Shivay, Y.S. (2012) Pattern of methane emission and water productivity under different methods of rice crop establishment, *Paddy and Water Environment*, 11: 321–332.

Thakur, A.K., Rath, S., Roychowdhury, S. and Uphoff, N. (2010) Comparative performance of rice with System of Rice Intensification (SRI) and conventional management using different plant spacing, *Journal of Agronomy and Crop Science*, 196 (2): 146–159.

Thiyagarajan, T.M. and Biksham, G. (2013) *Transforming Rice Production with SRI (System of Rice Intensification) Knowledge and Practice*, National Consortium on SRI (NCS), Hyderabad, India.

Thiyagarajan, T.M. and Gujja, B. (2013) *Transforming Rice Production with SRI (System of Rice Intensification) Knowledge and Practice*, National Consortium on SRI, India.

Uphoff, N. (2002) Opportunities for raising yields by changing management practices: the system of rice intensification in Madagascar. In: Uphoff, N. (ed.), *Agroecological Innovations*, London, Earthscan.

Uphoff, N. (2003) Higher yields with fewer external inputs? The system of rice intensification and potential contribution to agricultural sustainability, *International Journal of Agricultural Sustainability*, 1: 38–50.

Uphoff, N. (2015) The system of rice intensification and beyond: improving the management of many crops for more food production, climate resilience, and farmer enablement. Paper presented at Oxfam America Side-Event 2015, WFP Symposium 14 October 2015, Des Moines, IA, USA.

Uphoff, N., Amir, K. and Richard H. (2011) SRI as a methodology for raising crop and water productivity: productive adaptations in rice agronomy and irrigation water management, *Paddy and Water Environment*, 9: 3–11.

Yang, S., Peng, S., Xu, J., Luo, Y. and Li, D. (2012) Methane and nitrous oxide emissions from paddy field as affected by water-saving irrigation, *Physics and Chemistry of the Earth*, 53–54: 30–37.

Zhao, L, Wu, L., Li, Y., Lu, X. Zhu, D. and Uphoff, N. (2009) Influence of the system of rice intensification on rice yield and nitrogen and water use efficiency with different N application rates, *Experimental Agriculture*, 45: 275–286.

Zheng, J., Zhongzhi, C., Xuyi, L. and Xinlu, J. (2013) Agricultural water savings possible through sri for future water management in Sichuan, China, *Taiwan Water Conservancy*, 61: 50–85.

12 Gendered adaptation to climate change in canal irrigated agro-ecosystems

Rengalakshmi R. and Udaya Sekhar Nagothu

Introduction

Agriculture in India is the major source of livelihood to 70 percent of the rural population, but its contribution to the Gross Domestic Product is only 14 percent. Indian agriculture is characterized largely by small holdings. According to the Agricultural Census (Ministry of Agriculture, 2010–11), nearly 85 percent of the farmers are smallholders (below 2 ha) operating 45 percent of the total cultivated area. Overall, women play an important role in agriculture in India and particularly on smallholder farms. Out of the total cultivable area of 182 million ha, only 140 million ha is under net cultivation, of which 62 million ha (34 percent) is under irrigation in India (FAO, 2010a). Productivity is 130 percent higher in irrigated agro-ecosystems compared to rain-fed systems (Government of India, 2011). Hence irrigation is considered as one of the potential strategies for agricultural intensification to meet the increasing food demands and cope with climate change and variability. Climate change will significantly influence the role of women in irrigated agriculture and hence it is important to put special emphasis on mainstreaming gender while developing new adaptation strategies in the agriculture sector.

Irrigated agro-ecosystems are mostly located in semi-arid and arid regions which are highly vulnerable to increasing climate change and extreme weather (Turral *et al.*, 2011). As a consequence, we see that out migration from these areas has increased in recent years due to increased climate vulnerability risks. It is often men that migrate to cities in search of alternative livelihoods, thereby increasing the household burden on women. There is a growing attention towards climate adaptation measures in irrigated agro-ecosystems to deal with the increased climate variability (Pathak *et al.*, 2014). In order to tackle the water scarcity problems in climate vulnerable irrigated agro-ecosystems, the FAO (2010b) has been recommending the development and adoption of efficient water management approaches that are suitable for smallholders. Since 2000, studies have suggested integrating climate change perspectives into development plans (Klein *et al.*, 2005; UN-Water, 2007). Specific research studies were carried out to provide technical and policy recommendations to address the impacts of climate change on water resources (Cliche and Saravia, 2011; Shaw, 2011).

However, a majority of these studies did not put needed emphasis on gender issues, despite the significant role of women in irrigated agriculture and the increasing burden they are facing due to climate risks. From several research studies and national surveys, it is well proven and recognized that women play a significant role in ensuring household food and nutritional needs in most developing countries (FAO, 2003; Turral *et al.*, 2011). At the farm level, both men and women complement each other to ensure agriculture production and income. Still there is a need for more context specific research on differentiated roles and responsibilities of women and men, specific vulnerabilities and their capabilities for adaptation in the context of changing economic, climate and agrarian systems.

This chapter will discuss the role of gender in agriculture and the importance of gender mainstreaming in climate adaptation strategies. The chapter focuses on gender dimensions in selected adaptation initiatives in irrigated agro-ecosystems and to what extent they meet the criteria of climate smart agriculture (CSA) systems. The process of gender mainstreaming approach is presented. This is followed by results from the ongoing ClimaAdapt project (www.climaadapt. org), where the implementation of gender sensitive approaches are showcased. Finally, the chapter provides some conclusions and policy recommendations.

Gender, climate change and agriculture – a literature review

Globally, except in Europe, the proportion of women in the agriculture work force has been increasing over the past four decades (Doss, 2011). In Southeast Asia, women provide up to 90 percent of the labour for rice production, and in Sub-Saharan Africa they are responsible for 80 percent of food production (FAO, 2010b). In India, 79 percent of rural women are engaged in agriculture (as against 63 percent of men), of which 81 percent of women workers are from socially and economically deprived groups (NSSO, 2009–2010). In the context of changing agrarian systems, the roles of men and women in small farms are also undergoing rapid changes. The recent study in India by Dand (2010) shows that role of men in subsistence farming is declining rapidly as they shift to non-farm wage employment in urban and peri-urban areas. As a consequence, women are managing work in the household and on the farm. As a result of growing male off-farm migration, women are taking over the management of the farms, and in some areas they even constitute a larger share of the agricultural workforce (GOI, 2011; Rao, 2012). In this context, the existing gender gap in access to productive resources like land, credit, technology and information are further challenging women's role in farm management and increasing women's burden and workloads in ensuring household food security, enhancing production and income (FAO, 2005; Skinner, 2011; UNDP and ACCESS, 2014).

Among the several new challenges that small holders face, climate change is the most critical one, which unduly increases the risks in production (Harvey *et al.*, 2014). The Intergovernmental Panel on Climate Change (2007) indicated that the impacts of climate change on communities vary due to gender, class, locality, health and social status, ethnicity and age differences. Nevertheless, the negative consequences of differential impact of climate change on women and marginalized groups of the community are primarily due to gender inequality and women are particularly vulnerable to the negative impacts of climate change (MacGregor, 2010). The gendered impact of climate change is mainly due to the lower social position of women at household and community levels, and the unequal power relations that perpetuate the inequalities in roles and responsibilities, access, use and control of resources, and decision making at household and community levels. This results in variations in vulnerabilities, risks and adaptation (WEDO, 2003; WEDO, 2007; Aguilar, 2004a, 2004b; Alston, 2013; Arora-Jonsson, 2011; Codjoe *et al.,* 2012). The study of McOmber *et al.* (2013) revealed that 'women farmers are left out in accessing and using many forms of communication channels that are critical to their ability to adapt to a rapidly changing climate'. The dissimilarities between men and women further varies due to social, economic, demographic and political factors. To strengthen the case, Cannon *et al.* (2014) further argued that the whole process of development would be defeated if social issues including gender in climate change are not considered, because the underlying factors which keep women and the marginal groups in a vulnerable position are deeply entrenched in social and cultural practices, attitudes and traditions. Thus it is essential to understand the context specific constraints of men and women farmers, existing gendered coping strategies, and gender differentiated needs and priorities while enhancing their adaptive capacity. Moreover, women and men perceive the causes and impacts of climate change differently. As a result they also respond differently to adapt to and mitigate the impacts of climate change.

Development practitioners believe and advocate that mainstreaming the climate-smart adaptation strategies with socially inclusive and gender sensitive approaches in agricultural development sector have the potential to reduce the gender gap. However, in practice it depends upon how it is being conceptualized and facilitated. Though some environmental change-based development projects have started to integrate and mainstream gender into the processes (Ginige *et al.*, 2009; Tacoli *et al.*, 2014), the results and experiences so far are not as encouraging as expected in sustaining the gender equitable outcomes (Woodford-Berger, 2004; Porter and Sweetman, 2005; Walby, 2005; Kapoor, 2012). On the other hand, some of the projects demonstrated the new opportunities to bring gender outcomes, like women moving into non-traditional roles (Babugura, 2009), engaging in income-generating activities and attaining economic empowerment (Speranza, 2006; Bäthge, 2010; Klasen, 2013) and participating in off-farm employment sectors (Whittenbury, 2013). Still, in many contexts, these 'gender role changes may not be sufficient to bring uncontested positive changes in gender relations' (Tatlonghari and Paris, 2013). To date, there are few studies focused on the impact of adaptation strategies on gender outcomes and hence further location-specific studies are needed (Bradshaw and Linnekar, 2014).

The commitment of formal and informal institutions is important as they largely promote, govern and perpetuate gender mainstreaming across levels and scale. The very basic and foremost important point is to understand the socio-cultural contexts, and identify the main drivers of gendered vulnerabilities to climate change impacts (Bacchi and Eveline, 2010) and integrate gender as an inherent part of the project design and implementation (Wright and Chandani, 2014). According to Schipper and Langtson (2014) the main reason for failure of many projects is the lack of proper gender analysis and integration during the design phase of programmes. Ferris *et al.* (2013) added to the learning that there is inadequate monitoring and evaluation of gender within the programme, especially in the collection and analysis of gender-disaggregated data and information. In addition, Otzelberger (2011) stressed the capacity of the development practitioners and policy makers to understand and integrate gender strategies, and adopt systematic procedures for gender mainstreaming and allocation of specific budgets in programmes to make sure that all the activities are gender responsive (Wright and Chandani, 2014).

On the other hand, the importance as well as potential in reducing gender inequalities in agriculture was clearly pointed out in a FAO study (Turral *et al.*, 2011), which indicated that if women farmers had the same access to resources as men, the agricultural output in the developing world would increase by an average of 2.5 to 4 percent, and reduce poverty among an estimated 100 million people across the globe. Though the evidence-based research clearly brought out gendered impacts of climate change, so far policies and actions designed to address and implement adaptation strategies are not enough to bring the desired transformative gender outcomes (Alston, 2014; Kapoor, 2012) due to its time-bound and project-based interventions.

ClimaAdapt project: context and background

Given the importance of the links between climate change and gender in agriculture, this chapter focuses on gender dimensions of sustainable agriculture initiatives undertaken in irrigated agro-ecosystems in an ongoing project, ClimaAdapt[1], being implemented in three states of southern India (Tamil Nadu, Andhra Pradesh and Telangana). The interdisciplinary project focuses on climate change adaptation in agriculture in India. It emphasizes the links between research, innovation and capacity building, and encourages stakeholder engagement, including gender mainstreaming. In addition, capacity building, knowledge management and gender components are well integrated in all the project areas. The specific focus of gender mainstreaming[2] in the ClimaAdapt programme enabled a better understanding of the different roles of men and women in agriculture in the context of increasing climate variation and changes. The project has mapped the perceptions of men and women and the impact the perceptions have on livelihoods and coping strategies. This chapter aims to examine the process involved in gender mainstreaming in the project using the following research questions:

- How do the roles of men and women (or gender) change with climate variability?

- What are the outcomes of gender mainstreaming in implementing CSA initiatives?
- How do the outcomes from field level learning influence policy development?

The following section provides a description of the background of the study areas, the process adopted in gender mainstreaming and the methodology adopted to study the research questions.

Methodology

Description of the Study area: Tamil Nadu

Tamil Nadu state, located on the southeast part of India, is a leading producer of rice and horticultural crops. The ClimaAdapt project was introduced simultaneously in June 2012 in fourteen hamlets (sub-villages) in the state. These hamlets were located in two different areas, one along the Kalingarayan canal in Erode district and the other in the Ponanaiyar basin in Trichirapalli district (refer to Figure 9.1b in Chapter 9). The annual average rainfall in Kalingarayan is 980 mm, distributed equally in both the southeast and northwest monsoon season. Rice is the predominant crop followed by turmeric, sugarcane, vegetables and groundnut. Agriculture is the primary livelihood source of nearly 70 percent of the households, supplemented by livestock production in 45 percent of the households. In addition, agricultural wage labour is the secondary occupation for smallholders, especially for women. The region is predominantly inhabited by *Goundars*, a farming community, besides the landless agricultural labourers. The region is experiencing a shift in the cropping system from monocropping of rice paddies to mixed cropping systems with banana, turmeric and vegetables. Women are extensively involved in farming operations in different cropping systems.

The second site in Tamil Nadu is the *Ponnaniyar* reservoir (an earthen dam), in operation since 1974, with a total cultivated area of 2,100 ha. It is important to note here that in seven out of the last 23 years, the reservoir was not filled up or empty, because every third year the area experienced a drought. The area receives an annual average rainfall of 850 mm, mostly during the northeast monsoon. The farming system in the region is more diverse compared to the Kalingarayan basin, and more than 90 percent of the local population depends upon agriculture for livelihood. Rice, sorghum, vegetables and sesame are the major crops grown, and more recently flowers and mushrooms are becoming popular among women farmers due to market demand. This is followed by small scale livestock and poultry.

Description of the study area: Andhra Pradesh and Telangana states

In 2012, the ClimaAdapt project was introduced in the two districts namely Nalgonda and Guntur in Andhra Pradesh state (Refer to Figure 9.1a in Chapter

9). The state was bifurcated into two in May 2014, Andhra Pradesh and Telangana. The district of Guntur falls within Andhra Pradesh, whereas Nalgonda was allocated to Telangana. In both the field sites, *Reddiyars* and *Naidus* are the predominant farming communities followed by the landless agricultural wage labourers. Rice and vegetables are the major crops in the project areas. Agriculture along with dairy farming is the primary livelihood source, supplemented with income from wage labour in smallholder farmers. This is applicable to both men and women who share the work in smallholder farms. Water is the main constraint in all the four project areas. Prior to the construction of the Nagarjuna Sagar irrigation project in the area, rainfall and open wells were the two major source of irrigation. Farmers cultivate two crop seasons in a year; rice in the wet lands and vegetables, millets and other crops in the dry land areas.

Major constraints

Water scarcity is a major constraint in all the three states. Poor quality of water for irrigation is becoming an increasing concern due to industrial pollution in the Kalingarayan basin. Increasing salinity due to overexploitation of ground water is a major concern in the *Ponanaiyar* region. In the case of Andhra Pradesh, though the canal irrigation water supply is meant to grow two crops of rice in a year, erratic rainfall patterns allows farmers only a single crop in some years. This is evident from the frequent droughts that Guntur region has experienced during the last decade. In the past, millets were cultivated in the region and it has gradually been replaced by rice due to water availability. Besides, commercial crops such as chillies, tobacco and cotton are also dominant in the Guntur and Nalgonda districts. However, with the reduction in canal irrigation water supply, the dependency on the village level tanks and bore wells has increased, and the cropping intensity reduced. Women play a major role in all the farming activities in the region.

Gender mainstreaming – a practical approach

The ClimaAdapt project aimed at mainstreaming gender by adopting a project cycle approach. In this regard, the first consultative workshop for the project scientists and field staff was organised in January 2014. Key gender issues in the study sites were identified during the workshop and a detailed engendered work plan was prepared. In the plan, gender issues were integrated at the activity level by identifying appropriate areas and strategies. Also a *detailed gender strategy* was prepared with the participation of scientists from all four project areas with an objective to understand why gender should be addressed at all stages of project implementation. Following this process, *three separate consultative workshops* were organized for the project staff in 2014, in line with the project cycle approach with a focus on gender mainstreaming in the planning, implementation, monitoring and evaluation stages. *Community level gender sensitization workshops* were organized for men and women farmers and field staff to strengthen their

Table 12.1 CSA interventions identified to promote adaptation at basin level

Tamil Nadu	Andhra Pradesh and Telangana
• Green manuring in paddy • System of rice intensification • Integrated nutrient management with special focus on bio-inoculants • Weather-based crop insurance	• Alternate wetting and drying • Machine transplanting for rice under SRI • Direct sowing of rice • Weather-based crop insurance

Source: Data from ClimaAdapt project (www.climaadapt.org)

awareness and develop appropriate step-by-step strategies to ensure proper integration of gender perspectives in all stages of CSA technology testing and dissemination (Table 12.1). CSA technologies already identified and tested in the project were considered for gender mainstreaming. A participatory stakeholder approach was adopted in the respective project sites in consultation with men and women farmers, and officials of the Agricultural Department and Public Works Department of the respective State Governments.

The different technologies were introduced and tested on the farmers' fields using a cluster approach. Each cluster consisted of 20 to 25 farmers, depending upon the land type and distribution, with the active participation of men and women farmers. Necessary inputs as well as institutional linkages were established with institutions like the Agricultural Department and private agencies supplying farm inputs, training and information. Detailed step-by-step protocols were evolved to integrate gender in the different work components of the project. Following this, gendered monitoring and evaluation plan was prepared for the process, output and outcome level results and staff members were orientated to observe and document the changes.

In total, eight community managed Village Knowledge Centres (VKC[3]) were established at different field sites to facilitate extension services on the climate risks and adaptation related information and knowledge to both men and women farmers. Considering the importance of the capacity building component in the project, a *detailed guideline manual on gender sensitive training programmes* was also prepared for the project staff covering the different processes in conducting the training programmes; planning and designing, implementing and monitoring and how the knowledge is transformed into action level. In addition, detailed guidelines were developed for *gender responsive communication* using multiple communication media including the local VKCs. The nature of access to information and knowledge through VKCs was important to prepare gender responsive strategies to guide social mobilization, need assessment, content preparation, communication media and dissemination.

Relevant policies and sectoral plans of the respective states were analysed through a gender lens to understand the status and gap in addressing gender issues. In this context, gender audit was used as a tool to understand, identify and analyse the hindering factors in mainstreaming gender in climate change policies both at formulation and implementation level. The important policies studied were the National Action Plan for Climate Change, the National Mission on Sustainable Agriculture, and National Mission on Water and Andhra Pradesh Farmers' Management of Irrigation Systems Act, March 1997, for Andhra Pradesh and Telangana sites. In the case of Kalingarayan as well as Ponanaiyar basin in Tamil Nadu state, three policies including the State Action Plan on Climate Change (SAPCC), the National Action Plan for Climate Change (the National Mission on Sustainable Agriculture and the National Water Mission), and the Tamil Nadu Farmers' Management of Irrigation Systems Act 2000 were analysed. The areas identified for policy making or advocacy were capacity building and adoption of CSA technologies in improving water use efficiency in the context of climate change. Apart from this, a strategy for effective gender mainstreaming in policies related to climate change as well as water management at the state level and up-scaling of the interventions were developed and discussed among partners.

Data collection and analysis

A mix of qualitative and quantitative methodologies were used to understand the gendered climate change issues, outcomes of capacity building and knowledge management as well as the technology implementation process. Gender analysis was conducted in all the four field sites through eight Focus Group Discussions (FDG) (two in each site, with 10–15 members) with the participation of men and women farmers in separate groups, from April–June 2014. The main purpose of the FGDs was to study the gendered impacts of climate change, agriculture and livelihoods. It was supplemented by key informant interviews and participant observation to triangulate the collected information and data. The detailed division of labour and roles in production and social reproduction domains, gendered participation in implementing the identified CSA technologies, access to and control over resources and services, and men and women's role in household and community level decision making was recorded. In addition, the issues associated with climate variability and change in on-farm livelihoods and coping strategies were discussed. Subsequently, the data were analysed to identify the key factors that influence the adoption of the identified CSA practices. This was followed by analysing the economic, institutional and cultural factors and trends that influence the access and control over resources and adoption of the technologies. The participation of women and men farmers in the training and capacity building programmes were assessed using the trainee attendance registers from August 2013–July 2014. At the end of each training programme a brief assessment format was used to evaluate and document the level of awareness on the climate smart technologies among men and women participants. In addition, a baseline study was done during October to December 2014 to understand the level of awareness of climate information and climate smart technologies, modern information and communication technologies amongst 500 farmers in Tamil Nadu and 400 in Andhra Pradesh and Telangana states respectively. Subsequently, during February 2015, a brief survey and discussion was carried out in all the project sites to understand the adoption pattern of men and women farmers on the CSA technologies introduced.

Results and discussion

Gendered impacts: climate change, agriculture and livelihoods

Both men and women farmers in the study sites were adapting to the changing environmental conditions in order to maximize the use of natural resources like water and soil for higher productivity. Reduced availability of water for irrigation due to below average rainfall and its erratic distribution, and time of water release from reservoir to start the preliminary planting operations including land preparation or nursery making are the key risks in crop production across all the study sites. In order to combat these risks, farmers have been adopting strategies such as: (i) changing the cropping system with less water demanding crops; (ii) ensuring minimum irrigation using ground water through digging additional bore wells in the farm, and (iii) shifting to non-farm livelihoods, in particular men migrating to the peri-urban and urban areas, and increased dependence on unskilled work. One of the key constraints in all the project areas is the lack of adequate opportunities for alternate livelihoods in off-farm and non-farm sectors, except for small-scale livestock production.

The new adaptation technologies have different implications for men and women farmers. The latter are more affected by the shift to new technologies as compared to male farmers. In the Kalingarayan basin, medium and large farmers, mostly men, lease out their land in order to diversify their livelihood options, often by entering into non-farm businesses in the nearby urban areas. Women in the middle-income families are largely confined to social reproduction and household work. This situation undermines women's position in the households, especially in decision making, and reiterates the social position and identity of men as the primary income earners and breadwinners of the household. Whereas, in the case of smallholders and agricultural labourer households, womenfolk are mostly engaged in farming in leased lands and also work as labourers in the ongoing governmental social welfare programs for unskilled labour work, while men take up semi-skilled off-farm work in nearby urban and peri-urban factories and shops. This leads to an increase in women's workload in the households as men shift to non-farm employment in the nearby areas or temporarily migrate to urban areas. In normal years, though men and women equally participate in the agricultural work, women's labour contribution and gender roles vary based on the type of cropping system, as well as the size of the land holding.

In the Andhra Pradesh and Telangana project areas, women on average carried out 80 percent of the work in paddy, and 72 percent work in vegetable and cotton crops respectively. Whereas, in the Kalingarayan basin, 90 percent of the agricultural operations in paddy and turmeric crops and 80 percent of work in paddy, millets and vegetable crops in the Ponanaiyar basins were managed by women. In general, migration to urban areas in search of alternative income is increasing in recent years as observed in all the project areas. Out-migration of male members resulted in increased burden on women in the household. Seasonal migration of men from small and marginal farming households is increasing from the villages in the Ponnaniyar basin. In the Andhra Pradesh and Telangana project areas labour migration is mainly short term for harvesting operations in the neighbouring districts. This was mainly attributed to the changing cropping system from rice to maize, cotton and vegetables as well as mechanization of farming operations.

Lack of access to productive resources to women farmers, especially water, limits the potential for increasing the productivity of the crops and adoption of climate smart technologies. Women do not own land, and this restricts the women farmers' membership, participation and representation in Water Users Associations (WUA). Dasthagir (2012) has done a separate analysis of women farmers and their membership in WUA in Tamil Nadu, and argued that women water users do not constitute a homogeneous category, though the Tamil Nadu Farmers' Management of Irrigation Systems Act recognizes all women holding title to land as women farmers and eligible to be a member in WUAs. The percentage of women holding membership in WUA was found to vary from 8–12 percent in the Kalingarayan basin, and less than 6 percent in the Ponanaiyar basin, whereas women's membership was limited in Andhra Pradesh and Telangana project sites. In addition to the membership, the quality of the participation in WUA activities as well as decision-making power needs further study. Apart from access to water and land, gendered differences were also observed in the access to credit, technologies, institutional linkages and information. The baseline study of climate information and technologies in the field sites revealed the large differences between men and women farmers.

The continuous engagement and participation of both men and women farmers in addressing the environmental risks and challenges led to the development of locale-specific traditional knowledge. However, there is a wide gap between men and women farmers in accessing new information, knowledge and skills like weather forecast information, weather-based crop insurance and climate smart agricultural technologies (Table 12.2). Irrespective of the gender, the individual farmer's ability to take decisions is largely determined by the information and knowledge he or she possesses. Hence access to knowledge and capacity to use it in their livelihoods plays an important role in ascertaining the individual's position in the household. In addition, there is a wider variation among men and women farmers in the use of communication channels as well as the purpose for which it is used. Men predominantly use mobile phones, peer groups and Agricultural Department and Farmer Training Centres to receive weather and market information, inputs (seeds and fertilizers) as well as advisory services for crop and pest management. Women receive information on micro credit, livestock management practices and health information mostly through men in the household, neighbours and Self Help Group members. The differences were attributed to prevailing social norms in the division of labour

Table 12.2 Number of men and women farmers who were aware about climate change prior to CSA interventions

Field sites	Number of men farmers	Number of women farmers	Total
Ponaniyar	44	14	58
Kalinagarayan	66	18	84
Nalgonda	48	12	60
Guntur	56	8	64

Source: Data from ClimaAdapt project (www.climaadapt.org)

and decision making, women's work burden in the social reproduction role and limited awareness on the climate change related issues and adaptation measures. As a result women farmers faced constraints in their capacity to diversify their livelihoods and are thus more vulnerable to the changing resource availability, especially lack of water, and to the extreme weather events.

Overall, it was found that though women's role in agricultural operations are increasing in the context of climate change and changing socio-economic context, the main gender concerns for adopting CSA initiatives for women farmers compared to men were: (i) limited access and skills to use the information and knowledge necessary to cope with increasing climate variability and change, and risk management due to inadequate access to modern communication modes; and (ii) reduced access to productive resources, especially technology and decision-making power to implement and manage the risks that arise due to climatic factors, and limited opportunities and resources to take up alternate livelihoods.

Implementing gender sensitive interventions

Knowledge management

In the study, gender sensitive knowledge management strategies were adopted in: (i) capacity-building activities to improve the awareness, knowledge and skills on CSA initiatives through training and exposure visits; and (ii) access to climate-risk related locale specific information through VKCs to reiterate the learning and necessary support for adoption at the field level.

Capacity building

Based on the gender sensitive training guidelines, specific emphasis was given to mobilize women farmers and necessary context specific steps were taken to ensure 50 percent participation of women farmers in all the capacity building programmes in the ClimaAdapt project. In addition, community level gender sensitization programmes for both men and women leaders were organized which helped them to understand the concept of 'gender' in their local context, and the importance and need to promote the participation of women in the training programmes and enhancing their awareness, knowledge and skills linked to climate information and services. Prior to the sensitization programme, women farmers' participation in the local as well as external training was less than 50 percent in the Andhra Pradesh and Telangana project areas, and 62 percent and 48 percent in the Ponanaiyar and Kalingarayan basin, respectively. However, the sensitization of both the community as well as the project team members enabled a change in their approach and planning mode that led to a large increase in the participation of women in capacity-building programmes (more than 95 percent) on CSA initiatives, as well as seeking climate risk information from VKCs or government agencies.

The pre- and post-training survey results were encouraging in terms of increasing awareness and knowledge on CSA technologies among men (67 percent) and women (72 percent). From the FGDs it was observed that, as an outcome of the increased knowledge and skills, women farmers have started to take part in the discussions at the household level on the adaptation initiatives and risk management measures to cope with climate change.

Access to climate information

VKCs established at the village level were playing a crucial role in facilitating the access to information for both men and women. Gender mainstreaming helped to provide appropriate climate related information. This was done through separate and periodic needs assessment of men and women farmers and agricultural labourers and designed delivery through locally accepted modes of communication (mobile phones, video conference, bulletin boards, help lines, public addressal system, webinars, face to face interactions in village level plant clinic programmes etc). VKCs enabled a significant increase in access to awareness (41 percent women), schemes and entitlements, new information and knowledge and new skills (35 percent women) related to adaptation and mitigation practices in agriculture and other livelihoods. Men and women farmers regularly received medium-range weather forecasts and advisories on a bi-weekly basis which helped them to take targeted decisions to cope with the changing weather events. Trainings contents were simplified and suitable delivery methods such as public address systems, voicemail or notice boards were adapted to reach farmers. Special attention was paid to women farmers and women-headed households. In all the project areas, such facilities were made available and facilitated by women staff (88 percent are women).

At the output level, the participation of women is increasing in the VKC management committees (an increase from 12 percent to 65 percent), serving as platforms for voicing the issues of women farmers and agricultural labourers and also sharing information horizontally. VKCs could serve as key nodes for communication at the village level. The 'plant clinic' programmes organized at the village level helped to reduce the gap between the 'lab to land' mode of technology transfer. Among the participants, 38 percent were women farmers.

As a result we see that new knowledge is paving the way for a new role for women in plant health management. At the outcome level, we also see a change in men and women farmers' approach to adopt new agronomic practices to address climate change. This could be attributed to timely access to information and new technologies. It was observed that during the 2014 and 2015 cropping seasons, 80 percent of the farmers (57 percent men and 23 percent women) shifted to the alternate wetting and drying method of irrigation, observing weather forecast information to plan and take up farming activities, using machineries such as mechanized paddy transplanters, cono weeders, sprayer and seed drill, practices like planting short duration varieties, soil test-based application of fertilizers and economic threshold level-based pesticide application.

Access to technologies and its implications

Six technologies were introduced to improve the farmer's adaptive capacity. There were clear gender differences with regard to access and adoption of the technologies. Though the technologies are gender neutral, the prevailing socioeconomic conditions including gender roles on the farms determined the preferences for the adoption of the technologies (see Tables 12.3 and 12.4).

The results from ClimaAdapt project showed that the rate of adoption and the choices of technologies varied among the men and women farmers. Although there is a clear division of labour among men and women while implementing different agronomic practices, their roles overlapped on smallholder farms.

In the current study, the technologies such as machine transplanting in System of Rice Intensification (SRI) and direct sowing of rice had different outcomes for male and female farmers. The male farmers expressed that both the technologies were helpful in optimizing water use and increasing water use efficiency in the context of decreasing water resources and reduced their dependency on manual labour, whereas for women farmers, the new technologies reduced the women's labour opportunity in transplanting, the primary labour demanding task in paddy cultivation. On the other hand, the alternate wetting and drying generated an additional job for women in weeding operations. Some of the adaptation interventions involved a trade-off in terms of women's livelihoods. In some cases, it resulted in loss of employment and wage income from labour for women. The review of past studies also indicated that 'adaptation' may not benefit the entire household or society and women may have to bear the cost of 'coping' with climate change (Elson, 1989; Rossi and Lambrou, 2008; Tandon, 2009). Hence while selecting technologies to address climate change, adequate care should be taken to look into the potential trade-offs and impacts on women and their livelihoods. We see from the experiences of ClimaAdapt project in Tamil Nadu that a gender-sensitive development strategy should support diverse economic options. So far, 150 women were trained in the Ponnaniar basin and support was provided to set up small-scale income generating enterprises in mushroom cultivation, vermi-compost and livestock production. They completed one cycle of production during 2014 and

Table 12.3 The adoption rate of technologies in the selected sites among men and women farmers in Tamil Nadu sites in 2014

Field sites	Green manure (%)		SRI (%)		Bioinputs (%)	
	Men	Women	Men	Women	Men	Women
Ponaniyar	26	74	66	34	22	88
Kalinagarayan	38	62	72	28	26	84

Source: Data from ClimaAdapt project (www.climaadapt.org)

Table 12.4 The adoption rate of technologies in the selected sites among men and women farmers in Andhra Pradesh and Telangana states in 2014

Field sites	Machine transplanting of rice in SRI		Alternate wetting and drying		Direct sowing	
	Men	Women	Men	Women	Men	Women
Nalgonda	92	8	80	20	88	12
Guntur	94	6	96	4	90	10

Source: Data from ClimaAdapt project (www.climaadapt.org)

2015 and efforts are on to expand the rural entrepreneurial skills and appropriate institutional contacts for scaling out, depending on the support from the Agricultural Department and other agencies for inputs and technologies, banks for credit and loan facilities, private agencies and individuals for marketing, and further developing and strengthening the backward and forward linkages.

Linking lessons learned from field experiences to policy development

Creating timely access to new knowledge, information and skills on CSA technologies to women farmers helps to improve their status, and their active participation in the household discussions. Though it is not impacting women's position in decision making or changing gender roles immediately, it helps in the long-run with the shifting roles and risks due to climate and economic change. Hence, the need to include women farmers as participants in the capacity-building programmes should be emphasized in climate plans and policy, so that adequate resources can be allocated and strategies specified for their inclusion. Currently, the state climate plans or national plan for that matter do not provide such room for involving women.

Similarly, in the case of knowledge management, gender responsive processes in need assessment, content creation and dissemination has demonstrated an increase in access and use of CSA related information by women farmers. Though both the state and national climate plans are indicating the need for capacity building, a targeted gendered approach is still missing. Also such an approach can be mainstreamed across government programmes, for example in the Common Service Centres which are promoted by national government across the country through appropriate changes in the current policy and actions.

In addition, while advocating CSA technologies it is important to critically analyse the pros and cons considering the local socio-economic context, especially the implications for women and men's livelihoods. Where negative implications for women's employment opportunities are likely, suitable alternate economic opportunities and actions need to be provided in the plan, so that implementing agencies may prepare comprehensive plans by anticipating the outcomes. Also, it is important to regularly revisit the policies and make changes according to the needs and demands of the people and rural settings.

Earlier irrigation was commonly managed by men except in women-headed households, irrespective of the cropping system. In all the four project areas it was common that the WUA's were male dominated and do not encourage the participation of women. As men are moving out of agriculture to off-farm work and migration to urban areas, women are taking a larger role in agriculture. However, government policies are not recognizing their position and legalizing their access to land and water rights. In case of the Andhra Pradesh and Telangana project areas it was explicit that women were not entitled to get the membership in the WUA due to lack of land rights. Whereas in case of Tamil Nadu, women are legally entitled to get the membership due to the Participatory Irrigation Management Act, however, their participation in the institution is highly limited. Hence, considering the significant amount of work carried out by farm women, state policy level changes are necessary to consider both men and women of the households to be eligible as members in the WUAs. Adequate awareness programmes and an enabling environment needs to be created to promote the participation of women in WUA's in Andra Pradesh and Telangana states. In case of Tamil Nadu, awareness has to be created at the village level among men and women as well as the Public Works Department officials to reiterate and encourage the participation of women in water management and related institutions.

Concluding remarks

From the study, it was evident that irrespective of the geographical location, the role of gender, challenges and opportunities were more or less similar in all the three states with some minor variations. This was mainly due to the differences in socio-economic and policy dimensions between the states. It is clearly evident from the field experiences of the ClimaAdapt project that the gender-sensitive knowledge management strategy and investment plan to engage more women and maintain a gender balance offers a useful approach to reduce the gender gap, and thereby help improve access to new information, technologies and institutional linkage. The specific strategies adapted include gendered vulnerability analysis and periodical need assessment, developing appropriate and gender sensitive contents, adapting multiple gender friendly dissemination or communication mode to access the resources and opportunities essential for adoption of climate smart technologies by both men and women, and building the capacity of women to perform non-traditional gender roles.

It is clear from the project results that gender-sensitive capacity-building methods adopted in the project enabled equal and active participation of women. This was possible as adequate care was taken and context specific strategies were introduced to ensure the participation of women farmers in both local training and exposure visits. This is a crucial approach for mainstreaming gender.

Adequate attention is necessary to strengthen the knowledge and skills of men and women farmers to improve their participation in adaptation interventions. It is essential to consider the socio-economic context while introducing CSA technologies. Nevertheless, the policies on climate change are largely gender neutral and as a result addressing gender disparities and operationalizing the gender equity issues are challenging in the field through development programs. We see the deliberate need for integration of gender issues in the policy documents and plans, and ensure adequate allocation of sufficient resources in the budget for implementation of the policies.

It is clearly evident in the ClimaAdapt project that gender was emphasized from the planning phase and subsequently adequate resources were made available in the project for implementation of the activities linked to gender. The project adopted a comprehensive approach, not only to build the capacity of the scientific staff to have a shared understanding on gender mainstreaming in the project, but also to increase the awareness amongst policy makers and other stakeholders.

Notes

1 www.climaadapt.org
2 Gender mainstreaming is described as 'the process of assessing the implications for women and men of any planned action, including legislation, policies or programmes, in all areas and at all levels. It is a strategy for making women's as well as men's concerns and experiences an integral dimension of the design, implementation, monitoring and evaluation of policies and programmes in all political, economic and societal spheres so that women and men benefit equally and inequality is not perpetuated' (http://www.un.org/womenwatch/osagi/pdf/ECOSOCAC1997.2.PDF).
3 The VKCs are computer-based knowledge centres with Internet connection and provide need based static and dynamic information. A set of VKCs operating in a region are connected with a 'hub' in the Village Resource Centre as nodal point, which receives the generic information and adds value by converting it to locale-specific information. The local community manages the VKCs; access is ensured to all irrespective of caste, class, gender and or age. Need-based content creation is regularly done on the basis of the need assessment and feedback from the local women and men who visit the centre. The local village people are trained in the management of modern ICT, like operating the computers and hardware maintenance.

References

Aguilar, L. (2004a) *Gender and Biodiversity*, IUCN report, online, https://www.cbd.int/gender/doc/fs_uicn_biodiversity.pdf (Accessed 18 November 2015).

Aguilar, L. (2004b) *Climate Change and Disaster Mitigation* (IUCN), online, http://www.fire.uni-freiburg.de/Manag/gender%20docs/DRR-Climate-Change-Gender-IUCN-2009.pdf (Accessed 21 October 2015).

Alston, M. (2013) Women and adaptation. *WIREs Climate Change*, 4: 351–358.

Alston, M. (2014) Gender mainstreaming and climate change, *Women Studies International Forum*, 47: 287–294.

Arora-Jonsson, S. (2011) Virtue and vulnerability: discourses on women, gender and climate change, *Global Environmental Change*, 21(2): 744–1326.

Babugura, A. (2009) *Gender and Climate Change: South Africa Case Study*, Heinrich Böll Foundation, Southern Africa, online, https://za.boell.org/sites/default/files/downloads/GCC_south_africa.pdf (Accessed 11 May 2015).

Bacchi, C. and Eveline, J. (2010) Approaches to gender mainstreaming: What is the problem presented to be? In: *Mainstreaming Politics: Gendering Practices and Feminist Theory*. South Australia: University of Adelaide Press. 111–138.

Bäthge, S. (2010) *Climate Change and Gender: Economic Empowerment of Women through Climate Mitigation and Adaptation?* GTZ Working Papers, GTZ, Eschborn, online, http://www.oecd.org/dac/gender-development/46975138.pdf. (Accessed 17 April 2015).

Bradshaw, S. and Linneker, B. (2014) Gender and environmental change in the developing world. IIED Working Paper, IIED, London, online, http://pubs.iied.org/10716IIED,

Cannon, T., Schipper, L., Bankoff, G. and Kruger, F. (2014) *World Disasters Report: Focus on Culture and Risk*, International Federation of Red Cross and Red Crescent Societies.

Cliche, G. and Saravia, M. (2011) Information and communication technologies, water and climate change: issues and research priorities in Latin America and the Caribbean, paper presented at the Association for Progressive Communications (APC) ICTs, Climate Change and Water Workshop, Johannesburg, South Africa, 7–10 July, online, https://www.apc.org/en/system/files/ICTs_Climate_Change_Water.pdf. (Accessed 08 April 2015).

Codjoe, S.N., Atidoh, A., Kafui, L. and Burkett, V. (2012) Gender and occupational perspectives on adaptation to climate extremes in the Afram Plains of Ghana, *Climatic Change*, 110 (1–2): 431–454.

Dand, S. (2010) Women in subsistence agriculture: vital producers still food insecure, in PWESCR, women's productive resources: realizing the right to food for all. discussion paper in a Roundtable with the UN Special Rapporteur on the Right to Food, Chennai, India.

Dasthagir, K.G. (2012) Fencing women in water user associations: an appraisal of gender strategy for participatory irrigation management in Tamil Nadu, IWMI-Tata Water Policy Research Highlight, 2012, online, http://www.iwmi.cgiar.org/iwmi-tata/PDFs/2012_Highlight-24.pdf (Accessed 08 October 2015).

Doss, C. (2011) If women hold up half the sky, how much of the world's food do they produce? ESA Working Paper No. 11-04, Rome, Agricultural Development Economic Division Food and Agricultural Organisation of the United Nations.

Elson, D. (1989) The impact of structural adjustment on women: Concepts and issues. In: Bade, O., (ed.) *The IMF, the World Bank and the African Debt*. Vol. 2, London, Zed Book Ltd, pp. 19–23.

Ferris, E., Petz, D. and Stark, C. (2013) Disaster risk management: a gender-sensitive approach is a smart approach. In: *The Year of Recurring Disasters: A Review of Natural Disasters in 2012,* Brookings Institution, online, http://www.brookings.edu/~/media/Research/Files/Reports/2013/03/natural%20disasters%20review/ND%20Review%20Chapter%204.pdf. (Accessed 12 October 2015).

Food and Agriculture Organization of the United Nations (FAO) (2003) *World Agriculture towards 2015/2030: An FAO Perspective.* Earthscan Publishers, Rome, London.

Food and Agriculture Organization of the United Nations (FAO) (2005) *Gender: The Missing Component in the Response to Climate Change.* Food and Agriculture Organization, Rome.

Food and Agriculture Organization of the United Nations (FAO) (2010a) *"Climate Smart" Agriculture: Policies, Practices and Financing for Food Security, Adaptation and Mitigation,* Food and Agriculture Organization, Rome.

Food and Agriculture Organization of the United Nations (FAO) (2010b) AQUASTAT database, online, http://www.fao.org/nr/water/aquastat/main/index.stm (Accessed 02 October 2015).

Ginige, K., Amaratunga D. and Haigh, R. (2009) Mainstreaming gender in disaster reduction: why and how? *Disaster Prevention and Management,* 18: 23–34.

Government of India (GOI) (2011) *Report of the Working Group on Disadvantaged Farmers, Including Women for the Twelfth Five Year Plan (2012–2017),* Final Report, November 2011, Planning Commission, New Delhi.

Harvey, C.A., Rakotobe, Z.L., Rao, N.S., Dave, R., Razafimahatratra, H., Rabarijohn, R.H., Rajaofara, H. and MacKinnon, J.L. (2014) Extreme vulnerability of smallholder farmers to agricultural risks and climate change in Madagascar, *Philosophical Transactions of the Royal Society B* 369: 20130089, online, http://dx.doi.org/10.1098/rstb.2013.0089.

Intergovernmental Panel on Climate Change (IPCC) (2007) *Climate Change: Impacts, Adaptation and Vulnerability, IPCC Working Group II Report,* Chapter 19, IPCC, online, http://www.ipcc.ch/pdf/assessment-report/ar4/wg2/ar4-wg2-chapter19.pdf (Accessed 09 April 2015).

Kapoor, A. (2011) *Engendering the Climate for Change: Policies and Practices for Gender-just Adaptation,* Alternative Futures Research Report. Heinrich Boll Foundation, Christian Aid, New Delhi, online, www.alternativefutures.org.in.

Kapoor, A. (2012) A gender policy framework for state action plans for climate change, Policy brief, Alternative Futures, New Delhi.

Klasen, S. (2013) Gender, growth and adaptation to climate change. In: Cela, B. *et al.* (eds.) *Powerful Synergies: Gender Equality, Economic Development and Environmental Sustainability,* UNDP, online, http:// www.undp.org/content/dam/undp/library/gender/Gender%20and%20Environment/Powerful-Synergies.pdf (Accessed 16 April 2015).

Klein, R.J.T., Schipper, E.L.F. and Dessai, S. (2005) Integrating mitigation and adaptation into climate and development policy: three research questions, *Environmental Science & Policy,* 8(6): 579–588.

MacGregor, S. (2010) A stranger silence still: the need for feminist social research on climate change, *Sociological Review,* 57, Carter, B and Charles, N (eds) Issue Supplement S2, Special Issue: Sociological Review Monograph Series: Nature, Society and Environmental Crisis, pp. 124–140.

McOmber, C., Panikowski, A. McKune, S., Bartels, W. and Russo, S. (2013) Investigating climate information services through a gendered lens. CCAFS Working Paper no. 42. CGIAR Research Program on Climate Change, Agriculture and Food Security

(CCAFS), Copenhagen, online https://cgspace.cgiar.org/rest/bitstreams/19665/retrieve. (Accessed 14 April 2015).

Ministry of Agriculture (2010–11) Agriculture Census, online, http://agricoop.nic.in/imagedefault/whatsnew/Agriculture%20Census.pdf. (Accessed 23 October 2015).

NSSO 2009–2010 (2011) National Sample Survey Office (NSSO) 2009–2010 (66th round) - Schedule 1.0 (Type 1) - Consumer Expenditure, online, http://catalog.ihsn.org/index.php/catalog/1903 (Accessed 06 October 2015).

Otzelberger, A. (2011) *Gender-Responsive Strategies on Climate Change: Recent Progress and Ways Forward for Donors*, BRIDGE, Institute of Development Studies.

Pathak, H., Pramanik, P., Khanna, M. and Kumar, A. (2014) Climate change and water availability in Indian agriculture: Impacts and adaptation, *Indian Journal of Agricultural Science*, 84 (6): 671–679.

Porter, F. and Sweetman, C. (2005) *Mainstreaming Gender in Development: A Critical Review. Has Gender Mainstreaming Made a Difference after 10 Years?* Oxford: Oxfam.

Rao, N. (2012) Male 'providers' and female 'housewives': A gendered co-performance in rural north India, *Development and Change*, 43 (5): 1025–1048.

Rossi, A. and Lambrou, Y. (2008) *Gender and Equity Issues in Liquid Biofuels Production Minimizing the Risks to Maximize the Opportunities*, Food and Agriculture Organization of the United Nations.

Schipper, L. and Langston, L. (2014) Literature review of the project on "gender and climate compatible development: Drivers and challenges to people's empowerment", Climate and Development Knowledge Network, online, http://cdkn.org/wp-content/uploads/2014/12/Gender-literature-review_publication.pdf. (Accessed 10 April 2015).

Shaw, R. (2011) Information and communications technologies (ICTs) and water in the face of climate change, paper presented at the Association for Progressive Communications (APC) ICTs, Climate Change and Water Workshop, Johannesburg, South Africa, 7–10 July 2011. https://www.apc.org/en/system/files/ICTs_Climate_Change_Water.pdf (Accessed 16 October 2015).

Skinner, E. (2011) *Gender and Climate Change Overview Report*, BRIDGE Cutting Edge Pack on Gender and Climate Change. Institute of Development Studies, Brighton.

Speranza, I.C. (2006) Gender-based analysis of vulnerability to drought among agro-pastoral households in semi-arid Makueni district, Kenya. In: Premchander, S. and Müller, C (eds.) *Gender and Sustainable Development,* NCCR-North-South, pp. 119–146.

Tacoli, C., Polack, E., Nhantumbo, I. and Tenzing, J. (2014) *Building Resilience to Environmental Change by Transforming Gender Relations*, International Institute of Environment and Development, London, UK.

Tandon, N. (2009) The Bio-Fuel Frenzy: What options for rural women? A case of rural development schizophrenia. In: Terry, G. (ed.) *Climate Change and Gender Justice*, Practical Action Publishing and Oxfam GB, Oxford, pp. 127–142.

Tatlonghari G.T. and Paris T.R. (2013) Gendered adaptations to climate change: a case study from the Philippines, In: Alston, M. and Whittenbury, K. (eds.) *Research, Action and Policy: Addressing the Gendered Impacts of Climate Change*, Springer, Amsterdam, pp. 237–250.

Turral, H., Burke, J. and Faurès, J.M. (2011) *Climate Change, Water and Food Security*, FAO Water Reports 36. Food and Agriculture Organization of the United Nations, Rome. Online, http://www.fao.org/docrep/014/i2096e/i2096e.pdf. (Accessed 06 October 2015).

United Nations Development Program (UNDP) and ACCESS (2014) *Empowering Women in Agriculture: Closing the Gender Gap through Mahila Kisan Sashaktikaran Pariyojana (MKSP),* Policy Paper.

UN-Water (2007) *Farms, Rivers and Markets: A Whole-of-System Approach to Doing More with Less Water,* Business case to the National Water Commission, Australia.

Walby, S. (2005) Gender mainstreaming: productive tensions in theory and practice, *Social Politics: International Studies in Gender, State and Society,* 12 (3): 32–343.

Whittenbury, K. (2013) Climate change, women's health, wellbeing and experiences of gender based violence in Australia, In: Alston, M. and Whittenbury, K. (eds.) *Research, Action and Policy: Addressing the Gendered Impacts of Climate Change,* Springer, Netherlands, pp. 207–222.

Women's Environment and Development Organization (WEDO) (2003) *Untapped Connections: Gender, Water and Poverty.* Women's Environment and Development Organization, New York.

Women's Environment and Development Organisation (WEDO) (2007) Changing the climate: why women's perspectives matter, Information Sheet, online, http://www.wedo.org/wp-content/uploads/changing-the-climate-why-womens-perspectives-matter-2008.pdf. (Accessed 10 May 2015).

Woodford-Berger, P. (2004) Gender mainstreaming: What is it (about) and should we continue doing it? *IDS Bulletin,* 35 (4): 65–72.

Wright, H. and Chandani, A. (2014) Gender in scaling up community based adaptation to climate change, In: Schipper, L. *et al.* (eds.). *Community Based Adaptation To Climate Change: Scaling It Up,* New York: Routledge.

13 Climate mitigation and adaptation

Institutional and policy support in India

Solveig Kolberg and Udaya Sekhar Nagothu

Introduction

India inhabits an intriguing dual position in global climate politics. On one hand, the country is a developing economy with low levels of historical and per capita emissions, and on the other, it is a rapidly growing economy with increasing emissions (Dubash, 2013). India is the second fastest growing economy with one-third of the world's poorest people (Olinto and Uematsu, 2013). Practically the entire country is highly or extremely vulnerable to climate change (CC) and extreme weather events, primarily floods and droughts. As much as 65 percent of the country is defined as drought prone, about 12 percent (nearly 40 million ha) as flood prone, and 8 percent is susceptible to tropical cyclones (Government of Tamil Nadu, 2013). More than two-thirds of Indian agricultural land is rainfed (Neate, 2013), and 68 percent of net sown area is drought prone. The Indian Council of Agricultural Research (ICAR) forecasts significant negative CC impacts within the medium-term (2010–2039), including a reduction in agricultural yield between 4.5 to 9 percent. This yield reduction corresponds to a cost of 1.5 percent of the Gross Domestic Product (GDP) per year, given that agriculture contributes 15 percent to India's GDP (ICAR *et al.,* 2013). In their study, Dasgupta *et al.,* (2013) project a decline of up to 18 per cent in food grain production between 2030 and 2050. On top of that, food and nutrition security is also challenged by the fact that more than half of the population in India is projected to live in urban areas, implying that 14 percent of the world's urban population will be Indians by 2050 (IPCC, 2014). Urbanization will lead to a change in food consumption habits and a higher burden on the Indian farmers to produce diverse foods including milk and meat products.

Agriculture in India is primarily dependant on monsoons, more specifically on the distribution patterns during the growing seasons. Hence any change in the intra-annual variability is likely to have an adverse effect on farmer performance and food security. Though India's average annual rainfall is abundant by global standards, much of its rainfall occurs in relatively brief deluges during the monsoon, and varies considerably across different regions (Wyrwoll, 2012). The intra-annual variability is a major challenge for farmers and the government while planning for future adaptation strategies. Groundwater is an important source of irrigation in the country that supplements the demands of 70 percent of the irrigated area. However, groundwater resources in most parts of India have been overexploited and are no longer sustainable unless serious steps are taken to replenish the water tables. This situation is compounded by land degradation, high poverty levels, poor general health and malnutrition, population pressure and a consequential strain on natural resources. Climate change has heterogeneous spatial impacts in India. Most noticeable, the north-western region of India, the so-called Green Revolution frontier states, are the worst affected. Nevertheless, these states continue to be major suppliers of food, especially wheat and rice. Whereas regions of eastern India, including the states of Odisha, Bihar and Assam have been comparatively less impacted by the Green Revolution (Jacoby *et al.*, 2011).

India has a federal governance structure, and is administered by a Central Government and the various state governments with independent institutional and legal systems. With an advanced institutional landscape, India is preparing itself to deal with the multi-faceted nature of CC (GIZ, 2014). According to Ahmed and Fajber (2009), vulnerability to natural hazards in India is due to poverty and marginalization, and not simply due to the physical effects of the hazards themselves. As CC impacts worsen, effective institutions and policies become crucial to support any adaptation or mitigation initiatives that aim to reduce the vulnerability of smallholders and the poor. Institutions in the form of organizations, rules and norms are critical building blocks for short- and long-term CC adaptation initiatives (McGray and Sokona, 2012). Globally, India stands out as one of the few developing countries that has launched a strong policy for fighting CC (Taneja *et al.*, 2014), and in 2008, the National Action Plan on CC (NAPCC) was launched under the leadership of the Prime Minister of India (GOI, 2008). The plan consists of eight national missions, and among these, the National Mission for Sustainable Agriculture that aims to transform agriculture to adapt to CC and the National Water Mission that aims to improve water use efficiency and manage water resources sustainably are the most relevant for ensuring food security. Simultaneously, various state level and sector specific initiatives were launched to put the national policy into action. Technical programs such as the National Initiative on Climate Resilient Agriculture (NICRA) was initiated in 2011 across the country. However, these scattered initiatives are not adequate for a large country like India with varied agro-ecological zones and vulnerability.

India considers the Agriculture sector out of bounds to greenhouse gas (GHG) emission reduction, given the sensitivity and the potential consequences for economic growth and livelihoods of millions of smallholders. Therefore, India's official position in global negotiations has been one in the realm of adaptation and opposed to mitigation in the agricultural sector (Swain, 2014). Moreover, the participation has been based on the principle of 'common but differentiated responsibilities and respective capabilities' (Government of Tamil Nadu, 2013). The policy documents often prefer the term Climate Resilient Agriculture (CRA) to CSA, which also to some extent reflects India's position on the global mitigation commitments. While there are some similarities in the two concepts, CSA unlike CRA aims to enhance the achievement of national food security and development goals by sustainably increasing agricultural productivity and incomes; strengthening resilience in agriculture (adaptation); and reducing agriculture's contribution to GHG emissions (broadly seen as mitigation) (see Chapter 1 of this book). CRA will pursue the third pillar of mitigation only when it is a co-benefit to the first two priorities of increasing production and building resilience to CC and other shocks. While agriculture is exceedingly vulnerable to CC impacts, agriculture is indeed a major contributor to the problem. Estimates show that agricultural sector emissions contributed to 17.6 percent of the total emissions in India in 2007 (Planning Commission, 2014).

Thus in order to meet the future challenges, agriculture in India should not only be climate resilient but also climate-smart and sustainable, thereby addressing the combined challenges of adaptation and reducing net GHG emissions if possible. It is thus important to strengthen institutions and policies at different levels in order to build the capacity and enable smallholders to adapt to climate and economic changes in a sustainable manner. This chapter examines the role and potential of the current institutional and policy support and implications for implementing climate-smart agriculture (CSA) in India. In the subsequent sections, the institutional landscape, and policy framework and its relevance for implementation of CSA is reviewed, followed by the discourse on Indian climate politics, how it has evolved, and how the national level adaptation initiatives in India address climate change. The importance of addressing gender in policy development is discussed briefly. Experiences from three states in India, namely Tamil Nadu, Andhra Pradesh (AP) and the newly formed Telangana are discussed. Finally, the emerging CSA opportunities and conclusions are presented.

Institutional landscape and policy framework

Emerging discourses on Indian climate politics

Isaksen and Stokke (2014) identify three broad CC discourses in India, namely the Third World, Win-Win and the Radical Green discourse. There has been a gradual shift from the Third World discourse to the Win-Win discourse in recent years, while the Radical Green discourse, compatible with the leading international CC discourse of ecological modernization, runs parallel, supporting alignment between international and Indian politics (Table 13.1). Since the 1980s, when multilateral concerns about CC emerged, India has been at the centre of the discussions as it consistently brought in the development and the poverty dimensions (third world discourse) of the argument (Ghosh, 2010). India's initial position was 'climate equity' or per capita-based policy formulations, due to its low levels of historical and per capita emissions, and that mitigation would be a hurdle for third world countries to reach their targeted growth and development goals. Arguments circulated around India's perception

Table 13.1 Discourses on climate politics in India

	The Third World discourse	*The Win-Win discourse*	*The Radical Green discourse*
India's national interest	• Primacy of economic growth	• Primacy of economic growth	• Social equity and ecological balance
Relevance of CC for India	• International climate politics as a threat to economic growth	• CC as a threat and possibility of development and diplomacy	• CC as a threat for vulnerable groups
Climate policy at the national scale	• Growth as a precondition for sustainability, no domestic climate action	• Greening of economic growth, national action plan on CC	• Alternative development strategies
Climate policy at the international scale	• Protecting India's space for growth, principles of common but differentiated responsibilities and per capita emission levels	• Proactive climate diplomacy in pursuit of India's international interests, emission intensity targets.	• Proactive climate diplomacy

Source: Adapted from Isaksen and Stokke (2014)

of fairness. India's cumulative contribution to global emissions since 1850 corresponds to only 2.4 percent, while that of the USA is 29 percent of the total emissions. Comparing the cumulative emissions since 1990, India has a share of only 4 per cent, while China has 15 percent and the USA about 23 percent. Moreover, India, as one of the fastest growing economies in the world, will need to keep its economic growth rate between 8–10 percent per year for the next 2–3 decades to address the population needs (Parikh, 2012). Thus, India and other developing countries have argued that while developed countries should handle reducing their emissions first, they should also provide funding for mitigation and adaptation in other countries (Thaker and Leiserowitz, 2014).

India has so far made clear that it will not sign any international commitment to reduce emissions or intensity targets. However, the country has taken some significant steps in developing the NAPCC and articulated self-imposed domestic targets that its per capita emissions will not exceed the average of the per capita carbon emission levels of developed countries. As a consequence, many other developing countries are following in the footsteps of India, whereas developed countries like the USA have so far declined to sign a legally binding treaty to reduce net GHGs. This shows the serious difficulties and differences in opinions at the global level between countries in reaching an agreement to contribute to the delivery of a global public good such as net GHG emission reductions. It will be interesting to see if India and other countries including China and the USA will change their position at the Paris Climate Change Conference 2015 in December. We hope that there will be a shift in the current positions and that the global community will reach a consensus on their mitigation efforts.

In 2009, India proclaimed its ambition to reduce emission intensity of its GDP by 20–25 percent by 2020, in comparison with the 2005 levels. Accordingly, the Expert Group on 'Low Carbon Strategies for Inclusive Growth' was set up by the Planning Commission of India. Their final report provided inputs to the Twelfth Five Year Plan draft (Planning Commission, 2014). Already, during the Cancun climate negotiations in 2010, the then Indian Minister of Environment and Forests declared that *'All countries should take binding commitments in an appropriate legal form'*, and this statement led to a heated political debate in India. From this statement, one can conclude that there have been substantial changes in India's positions, most notably the commitment to voluntary intensity targets in 2009 and India's proactive role in climate negotiations in the Copenhagen and Cancun Summits subsequently. At the national level, the NAPCC of 2008 promotes CC awareness, adaptation and mitigation, energy efficiency and natural resource conservation (GOI, 2008). At the same time, it pursues overall economic growth through measures that encourage development objectives with co-benefits for addressing CC as viewed by some states in their strategies (Government of Tamil Nadu 2013). Nevertheless, the development agenda in the country focuses on rapid economic growth, and that it is not desirable for India to implement strategies exclusively for mitigating CC. None of the agriculture-related national missions under NAPCC focus on reducing net GHG emissions.

The development of national climate policy since 2007 can be understood as an institutionalization of the Win-Win discourse (Isaksen and Stokke, 2014). The general shift from Third World discourse to a Win-Win discourse at a national level stems from factors such as increased national awareness. This is due to felt impacts and strong domestic and international pressure, in particular from non-governmental organizations. Moreover, the Indian government's CC discourses are shifting as a consequence of its recognition of the 'co-benefits' of an alignment between its development and CC mitigation objectives, and thus increasing 'flexibility' on mitigation targets (Dubash, 2013; Isaksen and Stokke, 2014; Thaker and Leiserowitz, 2014). This implies that while the Win-Win discourse has facilitated the development of national climate policy in India, the dominance of the Third World discourse in foreign policy is still persistent, and this has limited the evolvement of India's role in the international climate negotiations (Isaksen and Stokke, 2014). All along, policy development in India was under influence by the different CC discourses apart from the political commitments. This has led to ambiguity in the policy discourse and lack of clarity in its global position to some extent. In the following section, we present how the development of policy and institutions have shaped the agriculture sector response to CC in India.

Institutionalizing adaptation response to climate change

There is no global consensus yet on sharing the responsibility for CC adaptation and mitigation (Gupta and Nagothu, 2012). According to Ostrom (2009), single policies adopted globally are not capable enough to build sufficient trust among citizens and business communities for a comprehensive and transparent collective action that will successfully reduce global warming. Similarly, no single government or institution can solve global collective action problems. A polycentric approach at various levels that oversee local, regional and national stakeholders' interests could help to reduce net GHG emissions (Ostrom, 2009). As climate shocks become more extreme and unpredictable, institutions and institutional arrangements at central, state and district levels will be crucial for India's ability to respond to increasing climate risks in the agricultural sector and beyond. Institutions at the national level in India could play a critical role in providing leadership not only to the states within the country but also to its regional neighbours in South Asia. The current Indian agricultural policy and institutional framework at national and state level aims at simultaneously addressing the increasing demand for food security and growth, as well as adapting to the challenges resulting from a changing climate.

At a national level, the Central Government, mainly through the Ministry of Environment, Forests and CC, takes the lead in developing climate adaptation and mitigation strategies. As a follow up of the NAPCC and its missions, several climate initiatives were launched at national and state levels (GOI, 2011). Amongst the most important initiatives was the launch of the National Initiative on Climate Resilient Agriculture (NICRA)[1] in 2011 by the ICAR[2]. NICRA's main objective is to improve the resilience of Indian agriculture to CC and climate vulnerability through strategic research and technology demonstration (NICRA, 2015). The program focuses on enhancing agricultural productivity for national food and nutritional security for all, particularly the resource poor small and marginal farmers who would be affected most by CC. Mitigation, one of the three main pillars of CSA, is of concern, yet it is not given priority. The government has implemented several other policy initiatives to support agriculture, particularly to achieve the objectives of climate-resilient agriculture. This is evident from several action programs that were incorporated into development plans, and support adaptation by farmers. Among them the most important initiatives are (Taneja *et al.,* 2014):

- The National Project on Organic Farming;
- The National Project on Promotion of Balanced Use of Fertilizers;
- The National Food Security Mission (NFSM);
- The Accelerated Irrigation Benefit Programme;
- The Weather Based Crop Insurance Scheme (WBCIS); and
- The National Agricultural Insurance Scheme (NAIS)

One of the main constraints is that the programs are often top-down and overlapping in their objectives with the state level initiatives. The state governments make the final decision on most agricultural initiatives, since agriculture is a state subject in India as listed in the Constitution. Some of the initiatives, in particular, the NFSM will need adequate funding support and this can affect other development programs. Much debate has been generated regarding the sustainability of the program in the long-run (Kishore *et al.,* 2013). Moreover, state governments such as Tamil Nadu do not completely agree with the objectives of NFSM. Other programs such as WBCIS and NAIS need technical support at different stages, and many states are not prepared to handle these programs. Observations in Tamil Nadu, Telangana and Andhra Pradesh showed that the state governments do not have the necessary capacity to handle the large data sets generated from the weather stations and use them precisely in WBCIS and other climate services. On the other hand, a majority of the crop insurance providers in the agricultural sector are privately based, and they do not trust the data from the weather stations set up by the Indian Meteorological Department. There is a large gap in the potential and the actual use of the data in providing timely and reliable climate services to farmers across the country.

The Government of India is also developing 'a low carbon strategy for growth' consistent with its standing at the global level (Planning Commission, 2014). This is acheivable mostly by shifting to clean and renewable energy sources. So far, there are no recommendations to reduce emissions from the agricultural sector, despite its significant contribution to GHGs. Within the agricultural sector, energy usage is rapidly increasing due to the extensive use of electricity and diesel water pump sets by farmers. With the groundwater levels significantly decreasing, this can impose a greater burden on the energy use and supply (Shah *et al.,* 2007). Shifting to solar pump sets can help in both adaptation and mitigation. The government should view this as an opportunity and cooperate with the private sector to support smallholders to adopt more solar powered pump sets and other alternative renewable energy sources.

Gender and climate policy

Climate change itself and linked policies are likely to have far-reaching impacts on gender, particularly in developing countries (Terry, 2009). Gender is one of numerous key socio-cultural dimensions typically included in CC vulnerability assessments. However, it is rarely addressed in adaptation research and planning (Lambrou and Nelson, 2010). The importance of addressing gender in designing conscientious climate policies stem from the fact that: men and women contribute differently to the causes of CC; are differently affected by CC; react differently to its impacts; and, given the choice, favour different solutions of mitigating and adapting the impacts of CC (Flegel *et al.,* 2009). As a consequence, CC interventions that are not gender-responsive often result in widening the existing gender divide (Skinner, 2011). Nevertheless, transforming gender relations is not about 'adding' women to existing power structures and institutions but using an approach that address both women's and men's needs and concerns with their changing role due to climate change (Kapoor, 2011).

The NAPCC recognizes the differential impacts of CC on society and the states (GOI, 2008). This is the very first time in India that a high-level policy document has acknowledged the importance of gendered impacts of CC[3], and the need to address gender concerns in adaptation interventions (Ahmed and Fajber, 2009). Unfortunately, this standpoint has not been included in the NAPCC's assessment of the effects of CC or its outlines of mechanisms that could support people to adapt. Partly, this is because of the lack of sex-disaggregated data on climate risks and documentation of lessons learned (Ahmed and Fajber, 2009). In fact, all of NAPCC's eight associated missions have indirectly some gender relevance as they all aim for adaptation and mitigation through inclusive and sustainable development (Parikh *et al.*, 2012). Several of India's agriculture livelihood related plans, policies, programs and adaptations aim to be gender-sensitive and to work towards gender equity. However, as Kapoor (2011) argues, in the four adaptation-focused missions, especially those related to agriculture and water under NAPCC, and the up-coming State Action Plans on Climate Change, adaptation plans are largely techno-centric focused and gender-blind or at best, gender neutral. He concludes that in terms of livelihoods, the focus is more on addressing women's practical needs and not their strategic needs. The situation is no different at the state level, and gender so far is not adequately considered while preparing the adaptation strategy plans for agriculture. Chapter 12 in this book has already presented a vivid account of the importance of gender mainstreaming in climate change adaptation.

Case Studies from three Indian States

All regions of India will be adversely affected by CC, among these are the states of Andhra Pradesh, Telangana and Tamil Nadu, where CC will particularly impact the regional environment in terms of water availability (quality and quantity) and eventually the way farmers in the region will manage the soil and water resources and biodiversity. In this section, we will describe the agriculture sector's institutional strategy of the three states, and seek to examine how the intentions of the NAPCC has trickled down to the state-level planning process and initiatives. Andhra Pradesh and the Telangana States had a common CC adaptation strategy and a joint State Action Plan for Climate Change (SAPCC) draft report until it was divided into two states in 2014. The state level information on climate adaptation strategies and policy development was gathered mainly from secondary literature, experiences from the ongoing ClimaAdapt project in which the authors are closely involved (www.climaadapt.org) and to some extent supported by data from brief interviews with selected key informants. As the SAPCC process is rather new, there is limited literature available on the adaptation strategy planning process at the state level.

The impacts of CC in India were already observed on agriculture production, raising concerns about long-term food security (Geethalakshmi *et al.,* 2011;

Dasgupta *et al.*, 2013; Mishra *et al.*, 2013). The vulnerability and resilience varies between the states depending on the agro-ecological conditions, and the response from the state governments and farmers to adapt to CC. Even within the states, the degree of vulnerability varies, as observed in Andhra Pradesh and Telangana, where the upstream parts of the Krishna River Basin experienced more droughts as compared to the downstream that suffered from floods (Geethalakshmi *et al.*, 2011). This implies that adaptation strategies at the state level have to be carefully planned to address inter- and intra-state variability. Our observations show that SAPCC preparation at the state level so far was not based on sound scientific knowledge. The involvement of different stakeholder groups, especially farmers and private sector is also rather limited. In the latter part of this section, we briefly discuss the ongoing process and the gaps within the planning process in the three states. Recognizing that the impacts of CC vary across states, sectors, locations and populations, and that different approaches are needed to fit specific sub-national contexts and conditions, the NAPCC demands that all Indian States prepare their own SAPCCs in line with the NAPCC mandate. The SAPCCs are crucial for fulfilling the objectives of the NAPCC.

Tamil Nadu

Tamil Nadu has a population of more than 72 million people (GOI, 2011), and it is one of the better-off states in India, performing relatively well on human development indicators (Mehrotra, 2011). About 40 percent of the state population is dependent on agriculture for their livelihoods. Nevertheless, the income from agriculture and related sectors is declining, and contributing to only 22 percent of the state's income (Government of Tamil Nadu, 2013). Despite the decreasing contribution, the state is dependent upon the agricultural sector to ensure food security and sustain livelihoods, in particular, in the rural areas. The state is experiencing slower economic growth in recent years, due to population pressure, increased urbanization, higher food demand, and in addition more fragmentation of land holdings, less availability of cultivable land and decrease in ground water resources (Government of Tamil Nadu, 2014). Since the early 1980s, the state has launched several social reforms and welfare schemes to empower, especially, the marginalized groups in rural areas including smallholders.

Tamil Nadu has an average annual rainfall of around 920 mm. It is critically dependent on monsoon rains for recharging its water resources, especially the Northeast monsoon that provides 48 percent of the yearly rainfall. The major crop is rice, and most of the rice cultivation is in the flat delta regions of the Cauvery River. More than 80 percent of the 8.5 million farmers are small and marginal farmers. These are among the factors that make agriculture in Tamil Nadu highly vulnerable to climate change. The state is also known for its widespread network of tanks that were traditionally used as a primary source of irrigation, and formed an important source of water for other multiple uses (Palanisami, 1993). The tank irrigation system was neglected in the last 4–5 decades for various reasons. Restoring the tank irrigation system can be one of the most suitable options for climate proofing in Tamil Nadu. The Government of Tamil Nadu in 2014 had the ambition to become the agronomically best state in the country. Consequently, the government has started many new initiatives in recent years, such as the System of Rice Intensification (SRI), crop diversification, new rice and other crop varieties to enhance agriculture productivity. As of today, Tamil Nadu has the second highest paddy productivity after Punjab state, despite being one of India's driest states (Africare *et al.,* 2010). Barah (2009) and others have demonstrated higher economic returns due to increased production, and considerably lower cultivation costs in SRI compared to non-SRI rice cultivation, with high savings in water use (22 to 39 percent) and the use of seeds (92 percent). As the former president of the World Bank, Robert Zoellick (Zoellick, 2009) expressed:

> Everyone cites India's Green Revolution. But I am even more intrigued by what is known as SRI, or system of rice intensification. Using smart water management and planting practices, farmers in Tamil Nadu have increased rice yields between 30 to 80 per cent, reduced water use by 30 per cent, and now require significantly less fertilizer.

In particular, water saving in agriculture is essential, as it consumes 75 percent of the state's water (Africare *et al.,* 2010). Moreover, a majority of this water is used for irrigating rice, thus making it necessary to improve the water use efficiency in rice cultivation.

The process of preparing the SAPCC in Tamil Nadu started in July 2010 with the formation of a steering committee, headed by the Chief Secretary, Government of Tamil Nadu, and the draft SAPCC was completed in October 2013 (Government of Tamil Nadu, 2013). During the preparation process, consultations were held with representatives from different sectors and scientists. However, the involvement of different stakeholders in the consultation, in particular farmers, was limited. ClimaAdapt scientists were invited to a few of these meetings and were invited to provide policy inputs to the agricultural sector. A thorough bottom-up planning process could have given better feedback and an opportunity to prepare a more realistic plan.

In Tamil Nadu, the key informants interviewed revealed that most officials were aware of the institutional responsibilities in developing and implementing the SAPCC and that overall, adequate funding is available for the implementation of the plan. One of the key informants disagreed, saying that though strategies for each sector were prepared and proposals discussed, adequate funds were not allocated for their implementation. All key informants agreed that the capacity and training needs for implementation of the SAPCC have not been assessed properly. This is also in line with our observation that there are very few training programs targeted to improving farmers' adaptive capacity. The key informants felt the need for sensitizing all the stakeholders and organizing more focused and sector-oriented training to meet challenges from CC. One of the key informants also expressed the importance of identifying suitably resourced persons and organizations to disseminate knowledge effectively.

The SAPCC in Tamil Nadu states the relevance of gender, aims to carry out gender-focused monitoring and evaluation, conduct gender-sensitive audits of adaptation program and schemes, build capacities of women and men to implement participatory programs at the village-level, and build capacities on gender and adaptation within all governance institutions at all levels (Government of Tamil Nadu, 2013). This includes the development of participatory and gender-just local action plans on adaptation (LAPAs) at the village level (Government of Tamil Nadu, 2013). However, when asked if gender had been adequately mainstreamed in the SAPCC, one of the key informants did not think any action had been taken so far, whereas, the others felt the need for giving more emphasis by putting in practice what has already been specified in the plan. According to the key informants, three good examples where policy has already been put in practice in Tamil Nadu included: i) The investments made so far had been put into research to address the impacts of climate change; ii) Initiatives had been taken by the Water Resource Organization to enhance water use efficiency and conveyance efficiency through desilting of major reservoirs and canals; and iii) Policy and technical initiatives had been made to promote SRI that is seen as a good case of CSA practice. Finally, the key informants made some suggestions for better implementation of the SAPCC that included:

- Upgrade agricultural technology in the state continuously;
- Increase awareness among the farmers to adopt climate-resilient technologies;
- Climate-proof government schemes and other welfare programs;
- Set up a Centre of Excellence for Climate Adaptation with adequate technical resources to train various department officials;
- Integrate climate adaptation knowledge in all training programs even at the student level;
- Develop a time-bound research plan to support SAPCC implementation;
- Initiate intensive training programs to disseminate climate knowledge at all levels;
- Integrate adaptation plans with ongoing government development schemes;
- Develop specific strategies for the State in all sectors; and
- Make the adaptation projects location specific.

Andhra Pradesh and Telangana States

Around 60 to 70 percent of the approximately 83 million inhabitants of the combined states of Andhra Pradesh and Telangana rely on agriculture as their principal livelihood (Acosta-Michlik *et al.,* 2008). Therefore, despite a massive population shift to urban areas, agriculture is critical to the economy of the two states with rice, cotton, chilli and millets being the major crops. Combined, the two states of Telangana and Andhra Pradesh have the third largest drought prone area in India. The semi-arid areas are mainly fed through groundwater discharges and partly supplemented by monsoon rains (Kakumanu *et al.,* 2014). Studies show that the two states are prone to CC impacts resulting in water scarcity and the decline in food production has been triggered by increased crop failures, and an increased risk of the spread of disease and periodic incidences of drought (Geethalakshmi *et al.,* 2011). More recently, the states have witnessed erratic monsoons, and in particular delays in the onset of monsoons making it difficult for smallholders to plan their seasonal calendar for crop cultivation. Andhra Pradesh is subject to floods and salinity in the coastal areas that need to be addressed in the SAPCC. In the case of the newly created Telangana, the main problem is droughts, as the state is predominantly semi-arid and receives less rainfall compared to Andhra Pradesh. The division of the states will further aggravate the water-sharing situation from the two major rivers Krishna and the Godavari that flow in the two states. It is in the mutual interest of the two states to address this issue in an amicable manner and agree on the water sharing and management.

The SAPCC draft report of Andhra Pradesh (finalized before the division of the states) was prepared by the Environment Protection Training and Research Institute (EPTRI), and submitted to the Ministry of Environment and Forest, Government of India, New Delhi in July 2011 for approval (EPTRI, 2011). Both adaptation and mitigation strategies are listed as priorities in the SAPCC. The SAPCC states that stakeholders are central to the adaptation process and that a participatory approach was adopted while preparing the document. However, from the document it appears that the participatory process was limited to a few workshops organized across the state where over 200 stakeholders representing a cross-section of government institutions, administrators, national NGOs, professionals and academia covering a broad range of sectors were engaged (EPTR1, 2011). At these workshops, stakeholders discussed and communicated their views and opinions on sectoral CC issues. Our observation is that SAPCC is primarily a descriptive document where the strategic interventions in agriculture are limited to a few tables, with some general recommendations that are not based on sound scientific evidence. Although promoting gender equality is listed as one of the goals, it has not been discussed or included in the section related to agriculture. The document does not recognize CSA as one of the strategic approaches to achieve the adaptation and mitigation objectives. However, it realizes the importance of promoting diverse crops and development of new cropping systems, with enhanced capacity for CO_2 fixation, which in turn can result in producing high biomass and increased productivity. As agriculture is recognized as the most vulnerable sector to CC, more effort should have been devoted to the agricultural sector with detailed, farmer approved and scientifically-based recommendations. SAPCC does not discuss the funding options needed for implementation of the various measures recommended and sources of funding. The importance of training and capacity building has not been emphasized in the plan.

Telangana and Andhra Pradesh have to prepare two new plans in the current situation. At present, the two state governments are preoccupied with the reorganization of their administrative departments. However, the original plan prepared before the division of the states could serve as a basis for the new plans. While preparing the respective climate adaptation plans, the states should learn from the experiences and attempt to develop more transparent and realistic plans that are acceptable to farmers and other stakeholders, emphasize capacity building and investments, and give importance to including gender and environmental priorities. One of the main challenges for AP and Telangana would be to raise funds for implementation of the SAPCC. The involvement of private sector and farmer organizations could increase the ownership and help in better implementation of the plans.

Institutional arrangements enabling adoption of smart agricultural technologies

Institutions

Indeed, both national and state level institutions in India have considerable expertise relevant to CC and adaptation technologies. Such knowledge has to be compiled, tested through farmer-led demonstrations and shared through farmer-to-farmer exchange programs, farmer training centres and non-governmental agencies. It should be combined with targeted training and capacity building at all levels supported by proper investments. This was also confirmed in our case studies and discussions with stakeholders in the ClimaAdapt project who expressed their interest to receive training in adaptation technologies. Ensuring desired adaptation outcomes will require integrated efforts at different levels. One of the greatest challenges will be to ensure that institutional and policy initiatives of the states and central government are transparent and well-coordinated. This is because the federal structure is a challenge for the Indian Government, as actions or real implementation for mitigation and adaptation initiatives are under the state-level legislative and executive realm (Gupta and Nagothu, 2012). Another major challenge is to develop a cross-sectorial integrated approach (GIZ, 2014), as the agricultural sector must be aligned with other relevant sectors as part of a holistic response. This is not the case so far in India, and sectoral integration is limited to a few meetings and discussions. Integration has to be reflected at the program level, where different sectors should converge the funding and make joint efforts for implementation.

Adaptation and mitigation

Mitigation efforts that reduce the net GHG emissions will require international cooperation beyond the Third World discourse. Adaptation, however, does not directly require international efforts because most adaptation is local and private (Tol, 2014). However, international funding and knowledge exchange networks would be beneficial to ensure sharing of technologies and knowledge already developed elsewhere. Thus the role of projects such as ClimaAdapt and their outputs and outcomes will be crucial in supplementing evidence-based knowledge and developing scaling out frameworks for large-scale implementation of promising technologies.

India has been compelled to sign a legally binding emission mitigation treaty, but it has not yielded to the international pressure. It will be interesting to see what the 2015 Paris CC Convention will produce. Despite steps towards voluntary national mitigation targets, the focus of India's national and state-level policies and programs has and will in the future primarily be on resilience in terms of economic growth and adaptation, and not on mitigation, at least for the agricultural sector. Thus one can argue that India's agricultural sector responses still seem to rely on a Third World discourse, reaffirming its international standpoint, while the overall discourse in the country has advanced more towards a Win-Win discourse, also including mitigation. Climate policies should both reduce future risks, principally via mitigation, and address relevant remaining risks via adaptation in the multifaceted causal chain that begins and ends with individuals (Smith and Stern, 2011). Given the share of and potential to reduce emissions in the agricultural sector, mitigation must come higher on the agenda if possible, and not just as a co-benefit to increase productivity and adaptation. In India, both formal and informal institutional arrangements will continue to play a central role in sharing information, enabling innovation and facilitating investment for improving adoption of smart agricultural technologies among farmers.

Farmers primarily respond to CC to realize immediate benefits for survival and protection of their livelihoods. Since, adaptation to CC is highly local, its effectiveness depends on local and extra-local institutions through which incentives for individual and collective action are structured (Meinzen-Dick *et al.*, 2013). While local level institutions play a significant role in enabling smallholders to transform coping capacity into adaptive capacity (Berman *et al.*, 2012), the scale and complexity of the CC challenge will need institutional innovation at all scales and levels (Ostrom, 2009; Ostrom, 2010). Funders, government organizations, NGOs, and other actors will also need to take on additional roles as network brokers, facilitating access to resources and information. Much investment is required to build the capacity of all actors to fulfil these roles.

CSA technologies

CSA technologies should include both indigenous knowledge and advanced scientific methods, efficient resource use and integrated ecosystem management. CSA is about changing the agricultural management so that it is environmentally sustainable and economically profitable to farmers. Thus, successful CSA interventions will require both knowledge about existing and potential CSA practices in addition to how the biophysical world is changing. Institutional change to support and enable CSA projects and programs will need bridging between researchers, practitioners, farmers, private sector and policymakers who understand the institutional needs and are determined to make them efficient and inclusive. Even though all Indian states experience the adverse effect of CC, some areas will be more affected than others. Uncertainty and differences make it difficult to plan for future resource management, hence preparedness through strong institutions and policy framework will be necessary.

Gender mainstreaming

Even though all eight missions of the National Action Plan on CC (NAPCC) of India recognize the gender concerns, gender specific measures and proactive gender agenda is largely missing from planning and implementation at the state level. Realizing that the threats of CC are not gender-neutral, nevertheless understanding and assessing the risk involved is fundamental to enhancing resilience, especially in vulnerable rural communities. Investing in CSA may be particularly challenging for women and the poor, as the machinery and inputs needed may be beyond the reach of those with cash constraints, and no access to finance (Meinzen-Dick *et al.*, 2013). Knowledge and access to information are other barriers. For example, in Andhra Pradesh, India it was reported that only 21 percent of women compared to 47 percent of men received information on cropping strategies for coping with climate change (El-Fattal, 2012). If CSA should be effective and produce the benefits expected, then both women and men must be able to access opportunities and operate in a supportive social, political and economic environment. To make policies truly gender-responsive, gender has to be mainstreamed throughout the life cycle of a policy (design, implementation, monitoring, and evaluation), and CSA gender-sensitive indicators must be developed.

Concluding remarks

This chapter has reviewed the critical role of institutions and policies enabling adaptation to CC in general and CSA in particular in India. Though the country has come far in its efforts to develop policy and institutions at the national level, much has to be done at the state level in terms of finalizing the adaptation strategies, investing in training and capacity building of farmers, scaling out technology and gender mainstreaming. States need strategies that are transparent, implementable and supported by continuous funding and technologies that are sustainable and easily adaptable by farmers.

There is a need to design better policies and to integrate and strengthen climate concerns and solutions into institutional arrangements that can lead to desired mitigation and adaptation outcomes. To address this, all actors and stakeholders must be brought together in the search for solutions. Moreover, scientists must provide realistic and timely data on CC, and scenarios to politicians and bureaucrats, to support them to make on-time decisions (Gupta and Nagothu, 2012; Nagothu *et al.*, 2012). Institutions play a crucial role to help smallholders identify the practical steps to be taken to adopt CSA practices. Institutions must raise awareness about the importance of gender and engaging both men and women actively, as it is fundamental to the successful implementation of

SAPCC. This cannot be ignored in any context where women play a major role in agriculture and rural livelihoods.

New adaptation practices and solutions constantly emerge, nevertheless, any policy or institutional response has to take into consideration India's wide range of agro-climatic conditions across the country, as what is suitable in one state or community, may differ from the next. Solid and reliable local level institutions can help in the implementation of existing and new cutting-edge ideas that can become game changers, especially for the poor and women. Ultimately, the goal is to harness the power of viable solutions that give people opportunities to not only adapt and create resilient agriculture, but also to create a smarter agriculture that could be part of the global solution to mitigate CC for the good of India.

Notes

1   Now named National Innovations on Climate Resilient Agriculture
2   ICAR covers agriculture (including horticulture, fisheries and animal sciences) research and education across the entire country, and has 100 ICAR institutes and 71 agricultural universities across India, making it one of the largest national agricultural systems in the world (ICAR, 2015).
3   The NAPCC states:
    The impacts of CC could prove particularly severe for women. With climate change, there would be increasing scarcity of water, reduction in yields of forest biomass, and increased risks to human health with children, women and the elderly in a household becoming the most vulnerable. With the possibility of decline in availability of foodgrains, the threat of malnutrition may also increase. All these would add to deprivations that women already encounter and so in each of the Adaptation programmes, special attention should be paid to the aspects of gender.' (GOI, 2008)as is now evident from observed increases in the global average air and ocean temperatures, widespread melting of snow and ice and rising global mean sea level. In fact, according to the latest scientific report from the Intergovernmental Panel on Climate Change (IPCC)

References

Acosta-Michlik, L., Kelkar, U. and Sharma, U. (2008) 'A critical overview: Local evidence on vulnerabilities and adaptations to global environmental change in developing countries', *Global Environmental Change*, 18 (4), 539–542.

Africare, Oxfam America, WWF-ICRISAT Project (2010) *More Rice for People, More Water for the Planet.* Hyderabad, India: WWF-ICRISAT Project.

Ahmed, S. and Fajber, E. (2009) 'Engendering adaptation to climate variability in Gujarat, India'. In: Terry, G. and Sweetman, C. eds. *Climate Change and Gender Justice*. Practical Action Publishing in association with Oxfam GB.

Barah, B. (2009) 'Economic and ecological benefits of System of Rice Intensification (SRI) in Tamil Nadu', *Agricultural Economics Research Review*, 22, 209–2014. Available at: http://ageconsearch.umn.edu/bitstream/57397/2/4-DrBC-Barah.pdf

Berman, R., Quinn, C. and Paavola, J. (2012) 'The role of institutions in the transformation of coping capacity to sustainable adaptive capacity', *Environmental Development*, 2(1), 86–100. Available at: http://dx.doi.org/10.1016/j.envdev.2012.03.017.

Dasgupta, P., Bhattacharjee, D., and Kumari, A. (2013) 'Socio-economic analysis of climate change impacts on food grain production in Indian states', *Environmental Development*, 8(1), 5–21. Available at: http://dx.doi.org/10.1016/j.envdev.2013.06.002.

Dubash, N.K. (2013) 'The politics of climate change in India: Narratives of equity and co benefits', *Wiley Interdisciplinary Reviews: Climate Change*, 4 (3), 191–201.

El-Fattal, L. (2012) *Climate-Smart Agriculture is "Smarter" When Informed by a Gender Perspective*. WOCAN Policy brief.

EPTRI (2011) *State Action Plan on Climate Change for Andhra Pradesh*. Hyderabad, India: Environmental Protection Training and Research Institute (EPTRI). Available at: http://www.nicra-icar.in/nicrarevised/images/State%20Action%20Plan/AP-SAPCC. pdf [Accessed 14 September 2015].

Flegel, T., Gotelind, A., Röhr, U., Mungai, C.N., Davis, F. and Hemmati, M. (2009) *Gender into Climate Policy: Toolkit for Climate Experts and Decision-makers*. Wiesbaden: GenderCC –Women for Climate Justice.

Geethalakshmi, V., Lakshmanan, A., Rajalakshmi, D., Jagannathan, R., Sridhar, G., Ramaraj, A.P. Bhuvaneswari, K., Gurusamy, L. and Anbhazhagan, R. (2011) 'Climate change impact assessment and adaptation strategies to sustain rice production in Cauvery basin of Tamil Nadu', *Current Science*, 101 (3), 342–347. Available at: http://sa.indiaenvironmentportal.org.in/files/file/rice.pdf

Ghosh, P. (2010) 'The clima change debate: view from India'. In: Pachauri, R. K. ed. *Dealing with Climate Change: setting a global agenda for mitigation and adaptation*. New Delhi, India: The Energy Resources Institute.

GIZ, (2014) Climate change adaptation in rural areas of India (CCA-RAI). Available at: http://www.giz.de/en/worldwide/16603.html [Accessed 25 November 2015].

Government of India (GOI) (2008) *National Action Plan on Climate Change*, India: Government of India Prime Minister's Council on Climate Change. Available at: http://www.moef.nic.in/downloads/home/Pg01-52.pdf

Government of India (GOI) (2011) *Census of India*. Available at: http://www.censusindia. gov.in/

Government of Tamil Nadu (2013) *State Action Plan on Climate Change, " Towards Balanced Growth and Resilience"*, Draft – October 2013. Tamil Nadu, India: Government of Tamil Nadu. Available at: http://www.indiaenvironmentportal.org.in/files/file/ tamil%20nadu%20climate%20change%20action%20plan.pdf

Government of Tamil Nadu (2014) *Demand No. 5-Agriculture*. Policy Note, 2014–2015. Tamil Nadu, India: Agriculture department.

Gupta, S. and Nagothu, U.S. (2012) 'Climate change and adaptation in the water sector: institutional and policy challenges and options'. In: Nagothu, U.S. *et al.* eds. *Water and Climate Change. An Integrated Approach to Address Adaptation Challenges*. New Delhi: Macmillian Publishers India Ltd..

Indian Council of Agricultural Research (ICAR) (2015) About us. Available at: http://www.icar.org.in/en/aboutus.htm [Accessed 12 August 2015].

Indian Council of Agricultural Research (ICAR), National Initiative on Climate Resilient Agriculture (NICRA) and Central Research Institute for Dryland Agriculture (CRIDA) (2013) Towards climate resilient agriculture through adaptation and mitigation strategies. Enabling farmers to cope with climate variability through land, water, crop and livestock management in vulnerable districts of India. Hyderabad, India. Available at: http://www.nicra-icar.in/nicrarevised/images/Books/NICRA%20 Climate%20Resilient%20Agriculture%20Brochure.pdf

Intergovernmental Panel on Climate Change (IPCC) (2014) *Climate Change 2014: Impacts, Adaptation, and Vulnerability. Part A: Global and Sectoral Aspects. Contribution of Working Group II to the Fifth Assessment Report of the Intergovernmental Panel on Climate Change,* Field, C.B., Barros, V.R., Dokken, D.J., Mach, K.J., Mastrandrea, M.D., Bilir, T.E., Chatterjee, M., Ebi, K.L., Estrada, Y.O., Genova, R.C., Girma, B., Kissel, E.S., Levy, A.N., MacCracken, S., Mastrandrea, P.R. and White, L.L. eds. United Kingdom and New York: Cambridge University Press. Available at: http://ipcc-wg2.gov/AR5/

Isaksen, K.A., and Stokke, K. (2014) 'Changing climate discourse and politics in India. Climate change as challenge and opportunity for diplomacy and development', *Geoforum,* 57, 110–119. Available at: http://www.sciencedirect.com/science/article/pii/S0016718514001948.

Jacoby, H., Rabassa, M. and Skoufias, E. (2011) 'Distributional implications of climate change in India', Policy Research Working Paper 5623, 1–56, The World Bank.

Kakumanu, K.R., Kuppanan, P., Nagothu, U.S. and Reddy, N. (2014) *Climate Change Adaptation Initiatives in Andhra Pradesh and Telangana States : A Review and Synthesis of Field Research Work and Lessons Learnt,* India: ClimaAdapt project.

Kapoor, A. (2011) *Engendering the Climate for Change: Policies and Practices for Gender-just Adaptation,* New Delhi: Alternative Futures and Heinrich BÖll Foundation (HBF).

Kishore, A., Joshi, P.K. and Hoddinott, J. (2013) A novel approach to food security, online, http://www.ifpri.org/gfpr/2013/indias-right-to-food-act (Accessed 26 September 2015).

Lambrou, Y. and Nelson, S. (2010) *Farmers in a Changing Climate. Does Gender Matter? Food Security in Andhra Pradesh, India.* Rome: The FAO. Available at: http://www.fao.org/docrep/013/i1721e/i1721e00.pdf

McGray, H. and Sokona, Y. (2012) 'Why institutions matter for climate change adaptation in developing countries'. World Resource Institute. Available at: http://www.wri.org/blog/2012/05/why-institutions-matter-climate-change-adaptation-developing-countries [Accessed 1 July 2015].

Mehrotra, S. (2011) *India Human Development Report 2011: Towards Social Inclusion.* New Delhi: Planning Commission Report, Government of India. Available at: , http://econpapers.repec.org/bookchap/oxpobooks/9780198077589.htm [Accessed 21 October 2015].

Meinzen-Dick, R., Bernier, Q. and Haglund, E. 2013. 'The six "ins" of climate-smart agriculture: Inclusive institutions for information, innovation, investment, and insurance'. CAPRi Working Paper No. 114. Washington, D.C.: International Food Policy Research Institute. Available at: http://ebrary.ifpri.org/utils/getfile/collection/p15738coll2/id/128110/filename/128321.pdf.

Mishra, A., Singh, R., Raghuwanshi, N.S., Chatterjee, C and Froebrich, J. (2013) 'Spatial variability of climate change impacts on yield of rice and wheat in the Indian Ganga Basin', *Science of the Total Environment,* (468–469), 132–138. Available at: http://dx.doi.org/10.1016/j.scitotenv.2013.05.080.

Nagothu, U.S., Gosain, A.K., Barton, D.N., Palanisami, K., Tirupathaiah, K., Reddy, K.K., Stålnacke, P., Deelstra, J. and Gupta, S. (2012) 'Climate change and impacts on water resources: guidelines for adaptation in India', Hyderabad, India: International Water Management Institute (IWMI); Aas, Norway: Norwegian Institute for Agricultural and Environmental Research (Bioforsk); Delhi, India: Indian Institute of Technology; Hyderabad, India: Water and Land Management Training and Research Institute (WALAMTARI) . Climawater Project. p. 14. Available at: http://hdl.handle.net/10568/39078

National Initiative on Climate Resilient Agriculture (NICRA) (2015) About NICRA. Available at: http://www.nicra-icar.in/nicrarevised/index.php/home1 [Accessed 2 June 2015].

Neate, P. (2013) *Climate Smart Success Stories from Farming Communities Around the World*, Wageningen: CGIAR Research Program on Climate Change, Agriculture and Food Security (CCAFS) and the Technical Centre for Agricultural and Rural Cooperation (CTA). Available at: https://cgspace.cgiar.org/rest/bitstreams/24750/retrieve [Accessed 30 October 2015].

Olinto, P. and Uematsu, H. (2013) The state of the poor: where are the poor and where are they poorest? Draft. The World Bank and Prem (Poverty reduction and equity). Available at: http://www.worldbank.org/content/dam/Worldbank/document/State_of_the_poor_paper_April17.pdf

Ostrom, E. (2009) A polycentric approach for coping with climate change, Policy Research Working Paper, 5095, The World Bank, http://www10.iadb.org/intal/intalcdi/pe/2009/04268.pdf [Accessed 24 October 2015].

Ostrom, E. (2010) 'Nested externalities and polycentric institutions: must we wait for global solutions to climate change before taking actions at other scales?', *Economic Theory*, 49 (2), 353–369.

Palanisami, K. (1993). 'Optimization of cropping patterns in tank irrigation systems in Tamil Nadu, India'. In: Penning de Vries, F. *et al.* eds. *Systems Approaches for Agricultural Development*. The Netherlands: Springer pp. 413–425.

Parikh, K. (2012) 'Sustainable development and low carbon growth strategy for India', *Energy*, 40 (1), 31–38. Available at: http://www.sciencedirect.com/science/article/pii/S0360544212000187

Parikh, J., Upadhyay, D.K.D. and Singh, T. (2012) 'Gender perspectives on climate change and human security in India: An analysis of national missions on climate change'. *Cadmus*, 1 (4): 180–185. Available at: http://www.cadmusjournal.org/node/151 [Accessed 19 March 2015].

Planning Commission (2014) *The Final Report of the Expert Group on Low Carbon Strategies for Inclusive Growth*. New Delhi: Planning commission. Available at: http://planningcommission.nic.in/reports/genrep/rep_carbon2005.pdf.

Shah, M., Rao, R. and Vijay Shankar, P.S. (2007) 'Rural credit in 20th century India: An overview of history and perspectives'. Available at: http://www1.ximb.ac.in/users/fac/Shambu/Sprasad.nsf/dd5cab6801f1723585256474005327c8/e78490ff090249d06525730c0030abf9/$FILE/Mihir%20Shah_rural_credit_April_2007_epw.pdf [Accessed 26 September 2015].

Skinner, E. (2011) *Gender and Climate Change. Overview Report*. UK: Institute of Development Studies. Available at: http://www.seachangecop.org/seachange/files/documents/2011_10_BRIDGE_Gender_and_climate_change.pdf Institute of Development Studies.

Smith, L.A. and Stern, N. (2011) 'Uncertainty in science and its role in climate policy', *Philosophical Transactions of the Royal Society A: Mathematical, Physical and Engineering Sciences*, 369 (1956), 4818–4841.

Swain, A.K. (2014) 'Can India reform its agriculture'. *The Diplomat*. Available at: http://thediplomat.com/2014/06/can-india-reform-its-agriculture/.

Taneja, G., Pal, B.D., Joshi, P.K., Aggarwal, P.K., and Tyagi, N.K. (2014) 'Farmers' preferences for climate-smart agriculture: an assessment in the Indo-Gangetic Plain', IFPRI Discussion Paper 01337. New Delhi: The International Food Policy Research Institute (IFPRI). Available at: http://cdm15738.contentdm.oclc.org/utils/getfile/collection/p15738coll2/id/128116/filename/128327.pdf.

Terry, G. (2009) 'No climate justice without gender justice: an overview of the issues', *Special Issue: Climate Changes and Climate Justice*, 17 (1), 5–18.

Thaker, J. and Leiserowitz, A. (2014) 'Shifting discourses of climate change in India', *Climatic Change*, 123 (2), 107–119.

Tol, R. (2014) *Climate Economics. Economic Analysis of Climate, Climate Change and Climate Policy*. Cheltenham, UK: Edward Elgar Publishing.

Wyrwoll, P. (2012) 'India's groundwater crisis. Global Water Forum'. Available at: http://www.globalwaterforum.org/2012/07/30/indias-groundwater-crisis/

Zoellick, R. (2009) 'India could be a new pole of global growth: WB President'. *Hindustan Times*, 2 December, 2009.

14 Summary and recommendations

Udaya Sekhar Nagothu

Introduction

Climate Change is a global concern affecting millions of people worldwide, particularly the poor and vulnerable. Studies have shown that the linkages between climate change, agriculture and food security are incredibly complicated with much wider implications than perceived by scientists (Crane *et al.,* 2011). The dependency of agriculture on monsoons and the poor understanding of the erratic monsoon behaviour, in particular the intra-annual variability is something that is worrying government agencies, farmers and scientific community (IPCC, 2013). Smallholders and other communities who live in rural areas, isolated locations and low-income countries are particularly vulnerable to climate change and extreme weather. In addition, poor infrastructure, lack of education and skills, and weak institutions to support the use of technology are important factors in explaining the lack of adaptation, especially in low income developing countries (AGRA, 2013). The complexity of climate change problems needs to be seen through a multidisciplinary lens that encourages the emergence of new and innovative ways to tackle the serious challenges humanity is facing. In this book, deliberate attempts were thus made to analyse the prevailing complexity through a multidisciplinary approach, and explore the divergent views in place to address climate change.

According to the IAASTD report (2008), social and economic inequities, political uncertainties and changing environmental conditions demand a new approach to sustainable agricultural production and consumption at all levels. Agroecology and Climate-Smart Agriculture (CSA) are some approaches that have emerged in recent years drawing the attention of the scientific communities, civil society and policy makers at different levels (FAO, 2013). The scientific community in support of CSA is confident about the outcomes it would generate, benefitting the agriculture sector, farming community and greenhouse gas (GHG) mitigation (FAO, 2013), whereas critics see CSA as being much more limited by the 'climate change only' and productivity focus. They are also worried about the lack of transparency and clarity of the CSA concept, and that it only serves the interests of a certain group of stakeholders, including international agencies and the corporate sector. One of their main concerns is that CSA undermines the social and economic interests of the smallholders (COP 21, 2015). However, the proponents of CSA claim that several techniques they promote are in line with agroecology. There is certainly some degree of overlap on the technical level. Nevertheless, it does not mean that CSA and agroecology are complementary. The two concepts, CSA and agroecology are, when one goes beyond the superficiality of some technical similarities and some overlap, different 'paradigms' related to the future of agricultural development.

The cases studies presented in this book reveal both similarities and differences between countries and regions and the way CSA or other approaches are understood, prioritized and supported at different levels. The different case studies also show that the degree of vulnerability and the capacity to respond varies across countries and regions. As observed so far, the different agencies perceive CSA differently, and hence the difference of opinion is likely to polarize the scientific community and civil society. Broadly, we see that within the civil society organizations and farmer movements, ecosystem-based adaptation approaches such as agroecology are preferred. They also argue that CSA does not address the wider range of environmental issues of how the current agricultural and food systems affect biodiversity, soils, ecosystem services and pollution due to effects of agrochemicals (COP 21, 2015). CSA comes from the agricultural constituency, with a focus on sustainable production for food security and income generation. Agroecology comes from the environmental constituency, based on the premise that food or crop production are ecosystem services that people harvest or benefit from. In this book, we acknowledge that no single approach can address the complexity of the problems we face today. Hence we need to look at various approaches that are available, and their suitability to address different problems and needs of the communities and regions. At the end of the day, we need to make sure that these methods or approaches are sustainable and benefit farmers and the environment at the same time.

The book opened with an introductory chapter providing a brief analysis of the CSA concept, the elements that constitute CSA as per the definition, the lack of clarity and current discourse, and diverse views and opinions about the potential and constraints, and other approaches such as agroecology that are developing in parallel. Chapter 1 presented a modified framework emphasizing that if the CSA approach has to be successful it should adapt itself to be environmental-friendly, farmer-driven, gender sensitive and address the socio-economic concerns of smallholders. It acknowledges that technology, institutional and policy support, and investments are essential for the success of any mitigation and adaptation initiatives to address climate change. This is in line with the study by Campbell *et al.* (2011), who state that the relative importance of adaptation, mitigation and productivity objectives of CSA should depend on the local context and stakeholder preferences to be considered while identifying potential synergies and trade-offs between priorities.

The chapter also emphasized that technologies imposed from the top-down cannot be successful even though they are promising. It pointed out at the same time that technologies are important, need to be simple, and have to be accessible and easy for smallholders to implement. Overall, there is a need for a holistic approach to adapt to climate change. The social and economic inequalities, lack of access to land, resources and knowledge and the constraints that smallholders face in many countries need to be acknowledged while devising new adaptation strategies (Aggarwal *et al.*, 2013). The changes happening in the current context of tenuous global economic security, urbanization, changing demographic patterns and global warming will be ever more challenging to address in a business as usual environment. Essentially, if our goal is to create or sustain something closer to an optimal balance of the social, environmental and economic aspects of agriculture and food security, the current course will simply not suffice. According to CCAFS (2014), increasing food production sustainably in the midst of climate change is achievable through new paradigms of agricultural development that can address adaptation, mitigation and productivity goals simultaneously.

As the book's summary, this final chapter revisits and assesses the variety of ways that different regions were dealing with CSA and other approaches, including agroecology and conservation agriculture (CA), in the hope of finding ways to establish the contours of a future sustainable agriculture development and food production systems. The chapter synthesizes findings from the disparate chapters and summarizes the main challenges and opportunities to address climate change impacts through various sustainable agricultural technologies. Towards the end of this chapter, some conclusions are provided specifying the way forward to the ongoing debate.

A brief summary of the chapters

Chapter 2 focused on defining *climate variability* and *climate extremes* that we experience now, and how they may change in the next four decades. One of the most immediate and obvious impacts of climate change is on the weather-sensitive agriculture sector. Both local and global impacts of climate on production of food will have a negative effect on the ability of humanity to meet its growing food demands. This chapter mainly looked at the rainfed agricultural areas and the frequency of the climate extremes in these areas. The author presented a review of the literature on projected changes in climate extremes in the coming decades, and provided some recommendations for potential shifts in crop suitability in these regions. The chapter concluded with a review of the adaptive capacity of the agriculture sector to climate extremes through CSA and other similar initiatives. The recommendations included the importance of improved seeds and cultivars for smallholder agriculture, policy instruments in support of agricultural insurance that will improve smallholder access to these tools, and value chain interventions to ensure new markets and improved efficiencies that ensure farmers better income even in the face of a changing climate.

Chapter 3 focused on South East Asia (SEA) where climate change is occurring and adding to a highly dynamic and uncertain agricultural context. According to the authors, improved water management is the key to significant transformation that agriculture in the region must undergo in order to meet the inter-linked challenges of increasing food demand and climate change. Managing current climate variability is the best indicator of the ability to manage future variability and, they highlight that effective technologies and practices – some of which are highly innovative – already exist. The case studies presented demonstrate how *water-smart interventions* can reduce the risks associated with climate variability, improve yields, and increase incomes, in both irrigated and rainfed systems, across SEA. However, there are no simple generic solutions. Multi-faceted approaches can improve carbon sinks and significantly reduce energy requirements with knock-on implications for GHG emissions.

Boosting *resilience* and productivity across SEA is about transforming landscapes and rural livelihoods. Agricultural extension services are the main conduit for disseminating information required to make such choices, but throughout SEA they are under resourced and in decline. The imperative of climate change requires that governments significantly bolster investment in extension and ensure that adequate resources are available to provide services to farmers. The chapter emphasized that there is a need for more research to understand the implications of changes made in water management at local, national, regional and global levels. The first step in unleashing the potential for growth in the smallholder agricultural sector and improving household food and nutrition security should be to privatize land tenure and allow farmers to choose the crops they want to grow (e.g. relaxing 'rice first' policies). Regional cooperation in the future is essential to combat climate change in the region.

Chapter 4 on China explored the major threats to agriculture and food production due to climate change. The authors concluded that given the importance and special characteristics of China's agriculture, varied agroecological conditions and increasing demand for food, CSA is one of the promising options to consider in the future. As pointed out earlier by Zhou (2015), China must focus on *sustainable food production systems*, reduce environmental impacts, and at the same time build resilience to climate change and reduce GHG emissions from agriculture. This will require both technological and management innovation. The *innovation-driven development strategy* launched by the Chinese government in 2013 supports sustainable intensification and production pathways. The chapter provides specific recommendations to address climate change that include rational crop zonation and cropping systems, strengthening rural infrastructure, breeding drought resistant crop varieties, enhancing water saving and water productivity to combat water scarcity, soil and water management, disaster mitigation, post-harvest protection and processing.

The 2013 strategy also emphasizes the redevelopment of intensive farming systems that improve farmer's income and climate resilience. This implies redeveloping the cereal and cash crop systems, restructuring crop- or pasture-based livestock systems and developing agroforestry systems depending on the local natural resources and socioeconomic conditions. China aims at reduction of GHG emissions from agriculture, through: improved management and recycling of livestock manures and agricultural waste; development and popularization of advanced technologies for water saving; energy-efficient farm machinery; regulated and if possible reduced use of fertilizers and pesticides; and improved soil organic carbon sequestration that will reduce GHG emissions from agriculture.

Chapter 5 analysed the main impacts of climate change on agriculture and food security in the Sub-Saharan Africa (SSA) region. Climate change affects mostly the people who are comparatively resource disadvantaged, such as smallholder farmers in SSA. These smallholders who produce 90% of the total food in SSA are at greater risk of food insecurity themselves, and the impact of climate change and extreme weather will further make them risk prone. Therefore, a rapid transformation and reorientation of present day traditional agriculture is necessary to ensure food security in the context of climate change. The authors recommended that promoting several approaches including CSA and agroecology are critical to address the multiple challenges in the SSA region. The authors recommended that integration of local knowledge and resources wherever possible is crucial to enhance sustainability of agricultural production systems by reducing overreliance on external production inputs, and input intensive crop varieties and animal breeds. Several CSA and agroecology technologies have been developed and successfully implemented in SSA and other parts of the world. These proven technologies have to be made accessible to smallholders.

Two major alliances, the Global Alliance for Climate-Smart Agriculture (GACSA) and Africa Climate-Smart Agriculture Alliance (ACSAA) were established in 2014 to promote CSA. These alliances are to be instrumental in streamlining the CSA approach, advocating for policy support, and garnering supports of donor agencies to develop and up-scale CSA technologies. However, fostering sustainable agricultural development requires coordinated efforts of various stakeholders: farmers, researchers, the private sector, civil society and policy makers towards building evidence, increasing local institutional effectiveness, strengthening coherence between climate and agricultural policies and connecting climate and agricultural financing. How much the global alliances can achieve in terms of truly integrating the interests of smallholders needs to be seen in the coming years.

Chapter 6 summarized empirical findings from a broader participatory action research based in Malawi. It illustrated how local knowledge and participatory farmer-led agroecological research could be used to enhance the resilience of smallholder agriculture in Africa. The authors argue that although climate change is imposing significant constraints on African smallholder agriculture, it is possible to use agroecological farming methods to manage these impacts and build resilience into traditional farming systems. The chapter demonstrated that under climate variability, smallholders who use a diverse range of agroecological farming practices can actually manage drought stress situations if supported with necessary inputs and knowledge (Bezner Kerr *et. al.*, 2007).

The findings also show that, when presented with climate change adaptation options, farmers tend to select practices that increase the diversity of the farming systems. Through an agroecological approach, the authors have also shown not only how communities and farming systems could be made resilient, but also how to address some of the inequalities in todays' food systems. To sum up, the chapter showed that agroecological practices such as crop diversity, cereal-legume intercropping systems, and the use of organic material to build soil health, are very effective strategies to improve food security and address social inequalities in the context of environmental change.

If the transformational challenges presented by climate change are to be addressed, resilience thinking needs to be combined with other concepts that address distributive power issues and the unsustainable nature of current food systems. Food sovereignty and agroecological approaches can help to build community resilience in environmentally and socially marginal areas, if done along with co-learning from farmers, and modification and adaptation of existing innovations. In using food sovereignty and agroecological approaches to build resilience, however, these should be implemented as a 'basket of options,' but not a 'one-size-fits-all' approaches. The authors conclude that there is a need to recognize and address existing inequalities, and also value social, environmental and political factors influencing agricultural systems. A more resilient rural community will be one that can address these political and social inequalities alongside the environmental challenges that climate change now poses.

In chapter 7, the authors provided a comprehensive analysis of technologies and practices related to *conservation agriculture* that are relevant for South Asia. CA is seen as a management strategy for saving production resources, improving farm profitability and soil health on a sustainable basis. All components of CA, i.e. no tillage, crop residue retention and diverse crop rotations, substantially influence nutrient dynamics in the soil, thereby requiring a paradigm shift in nutrient management. Efficient nutrient management under CA systems involves maintaining the nutrient levels in soil considering the plant extraction and losses, and additions resulting from various soil processes induced by CA systems. Because the CA system involves diverse crop mixtures including legumes, nutrients and their cycles need management at cropping system level than at single crop level.

The chapter recommended that yield and profitability of CA-based cereal production can be increased by adopting precision nutrient management systems. This is highly relevant for south Asia to reduce the excessive use of fertilizers and reduce the environmental pollution. The objective is to capture the spatial and temporal variability in soil fertility and to provide recommendations to feed the crops with all the required nutrients based on crop needs. Various tools, techniques and decision support systems are available to develop site-specific precision nutrient management plans for each field and dynamically fine-tune in-season nutrient management. Application of such tools in a way are *climate-smart* as they help not only to increase crop yields, to use nutrients efficiently and profitability but also to reduce the environmental footprint. However, there is a need to fill the large knowledge gap, improve capacity, and fine-tune precision nutrient management in CA systems under different agroecological conditions. Enabling policy and institutional frameworks are necessary for wide-scale adoption of CA as a sustainable agricultural practice.

Chapter 8 reviewed the potential of six soil management technologies: need-based inorganic fertilizer management; organic or green manure, bio-fertilizers, biochar; soil and water conservation measures (SWC) and CA, and the extent to which they contribute to adaptation, mitigation and productivity. Proper management of agricultural soils and effective policies have the potential to mitigate and adapt to climate change, and at the same time enhance agricultural productivity. In this chapter, the authors presented research results from an ongoing project in India (project *ClimaAdapt*) and experiences from a project in Burundi (project *Fanning the Spark*) to show how farmers build resilience to climate change. The chapter concluded that options such as need-based fertilizer management; green manuring; biofertilizers; biochar; SWC measures (in this case, stone bunds and *zai* planting); and CA, are *'climate-smart soil management technologies'*. Combined applications of the soil management technologies is likely to give more CSA co-benefits than applying a single soil management technology.

Chapter 9 focused on measuring water used in irrigated agriculture and the importance of improving *water productivity* (WP) in water scarce areas impacted by climate change. Improving WP not only helps in adaptation to droughts and water scarcity, but also mitigation indirectly. The authors presented results from field measurements done under controlled conditions in the *ClimaAdapt* project to estimate *water productivity* (WP) under different rice growing practices such as traditional paddy rice, Direct Seeded Rice (DSR), System of Rice Intensification (SRI), Modified SRI and Alternate Wetting Drying (AWD) in the Krishna and Cauvery River basins of India. The results showed an increase in WP for alternative rice growing practices compared to conventional paddy cultivation. The authors conclude that there is a need to replicate measurements on water use and water losses in other areas to obtain robust results. Introducing new practices of irrigation also demands a change from a supply-driven system to a demand-based water supply system. Such changes need policy and technical support, and increasing awareness about the importance of concepts such as WP and water use efficiency among stakeholders. This is highly relevant in the future when climate change induced droughts and water scarcity become more frequent. Stakeholder groups need to understand the immediate need to improve water management at different levels before the conflicts escalate.

Chapters 10 and 11 discussed the merits and demerits of two alternative rice cropping systems, Alternate wetting and drying (AWD) and System of Rice Intensification (SRI). The chapters presented results from farmer-led field trials in Andhra Pradesh, Telangana and Tamil Nadu states in India conducted during 2012–2015. Results showed that AWD yields were similar to the conventional paddy, but AWD used 30–45% less water and it also recorded lower costs of production and higher farmer's income than the later. The authors reported that farmers have now started to adopt AWD in the region as an alternative to the conventional paddy. Scaling up this practice needs policy support, capacity building of farmers and changes in system-level water management to meet the alternating irrigating cycles in AWD system of rice.

Data from field trials in Tamil Nadu state tested the system of rice intensification (SRI) and demonstrated that SRI outperformed the conventional paddy cultivation method in various plant growth and yield parameters. SRI used less water and recorded lower CH_4 emissions, and the benefit-cost ratio was higher than the conventional method. However, N_2O emissions from SRI were higher at certain stages of crop growth, suggesting that there are some trade-offs associated with the SRI method. Input support and building farmers' competence is necessary to scale up SRI. Government policy in Tamil Nadu supports scaling up of SRI in the state, and farmers receive incentives for adopting SRI. If managed properly, SRI can be a good alternative to the conventional paddy system of cultivation. There is scope for supplementing chemical fertilizers with organic manure and crop rotations with rice-legumes to improve soil fertility. The *ClimaAdapt* project has demonstrated that farmers are willing to take up alternative systems provided they have access to the technology, inputs and assured market for their products. The project has provided policy recommendations to the state government on scaling out SRI.

Chapter 12 provided a review of literature on the importance of gender mainstreaming in climate change adaptation, and pointed out how essential it is to understand the context specific constraints of men and women farmers, existing gendered coping strategies and gender differentiated needs and priorities, while enhancing their adaptive capacity. Moreover, women and men perceive the causes and impacts of climate change differently. As a result, they also respond differently to adapt to the impacts of climate change. Women play an important role in agriculture in the developing world and hence the element of gender should be given priority while designing the new strategies to address climate change impacts. The chapter examined the process involved in mainstreaming gender in an ongoing climate adaptation project (*ClimaAdapt*) implemented in four field sites in southern India. It showed how the role of men and women would influence the outcomes of any new adaptation and mitigation strategies. The results showed that a strong focus on gender mainstreaming and socio-economic vulnerability across scales, and especially the inclusion of women farmers, could provide scope to reduce the overall vulnerability and strengthen the resilience of canal-irrigated agroecosystem communities.

Chapter 13 reviewed the important role of institutions and policies enabling mitigation and adaptation to climate change using the case of India. All along, policy development in India was under influence by the different climate change discourses apart from the political commitments. This has led to ambiguity in the policy discourse and lack of clarity in its global position to some extent. Though the country has come far in its efforts to develop policy and institutions at the national level, much has to be done at the state level in terms of implementing the adaptation strategies, investing in training, technology upscaling and gender mainstreaming. Given the federal political structure, the different states in India need strategies that are transparent, implementable and supported by continous funding and sustainable agriculture technologies that are easy to adopt by smallholders.

The main challenge is how to design better policies, and how to integrate and strengthen climate concerns and solutions in institutional arrangements that can lead to desired adaptation and mitigation outcomes. Scientists must provide realistic and timely data on CC, and scenarios to politicians and bureaucrats, to support them to make on-time decisions (Gupta and Nagothu, 2011). Institutions and policy frameworks have to be instrumental in helping smallholders to identify what practical steps are to be taken to adopt sustainable agriculture technologies, that ultimately ensure food security and livelihoods and in the process help in decreasing net GHG emissions. Strong and reliable local level institutions can help in the implementation of existing and new cutting-edge ideas that can become game changers, especially for the poor and women.

Major types of actions necessary to promote smart agricultural technologies

The various chapters in this book analysed the current thinking and development of a range of approaches to address climate change adaptation and mitigation. The chapters were drafted by authors from different disciplinary backgrounds and perspectives, which influenced the discussions in this book. The diverse views and extensive experiences of the authors have thus provided some originality to the book, supported by an extensive literature survey in the respective chapters. As discussed earlier in the book, it is not easy to point to one single approach that will address the challenges posed by climate change. Though the book focused on CSA, other approaches such as agroecology and CA need equal consideration while devising new strategies in the agricultural sector. To summarize, five major issues or factors were identified that will largely determine the outcomes in different regions attempting to promote sustainable agricultural practices to combat climate change. These are not presented in any order of priority, and the relevance of these factors is contextual.

The first key issue to be discussed here is *the role of science and technology* and its importance in strengthening new strategies to address impacts due to climate change, and enabling policies in different regions focused in this book (South and SE Asia, China and SSA). These regions, except SSA, have invested extensively in agricultural research and extension since the 1960s. Unfortunately, a majority of the research focused on major cereal crops (rice, wheat, maize) that led to monocultures – to the extent that they occupy nearly 80% of the agricultural landscape today. This has been the mainstream food production paradigm advocated throughout the Green Revolution, supported by heavily subsidized agrochemicals, water and power (Nagothu, 2015; Pingali, 2012). Although yields of major cereal crops have increased during the decades that monocultures have proliferated, and addressed the food security on many fronts, they were environmentally unsustainable, leading to deforestation, land degradation and loss of biodiversity. As opposed to mixed cropping systems, monocultures are likely to add to the vulnerability caused by climate change and extreme weather (Akanda, 2010).

Table 14.1 Summary of recommendations

Region	Main Challenges	Main recommendations
South Asia region	Erratic monsoon, intra-annual variability in monsoons, climate extremes, droughts and floods	CA, crop diversification and crop rotations, community management of soil and water resources, reforms in land ownership rights, rainwater storage at farm and micro-water shed level, policies to encourage contract and community farming, credit and extension services, reliable weather forecasts, index-based insurance, mainstreaming gender equality
South-East Asian region	Climate change and climate extremes/variability, highly dynamic and uncertain agricultural context	Water-smart interventions; multi-faceted approaches to improve carbon sinks; privatize land tenure and allow farmers to choose crops to be grown
China	Climate change and extreme weather, high temporal and spatial variability in the monsoon climate, droughts, flooding, low night temperatures, varied agroecological conditions and vulnerability	Sustainable food production systems, rational crop zonation and cropping systems, strengthening rural infrastructure, breeding drought resistant crop varieties, enhancing water saving and water productivity to combat water scarcity, soil and water management, disaster mitigation, post-harvest protection and processing.
Sub-Saharan Africa	Climate change and extreme weather, droughts and floods, low adaptive capacity and poor infrastructure	Agroecology-based farming systems, CA, climate-smart fodder systems, millets and legume-based crop rotations, strengthening rural infrastructure, farm-level micro irrigation systems, improving smallholder adaptive capacity, mainstream gender in adaptation and mitigation

Source: Author's own analysis

New technologies and interventions are needed to increase overall food production to meet increasing global demand for food in the coming decades, while at the same time improving farmers' resilience to extreme events. There is growing realization of the need to look for alternatives, as is evident from recommendations provided in several chapters including chapters 3, 5 and 6. These alternatives include agroecology, crop diversification, crop rotations and water-smart technologies to adapt to climate change. Chapters 7 and 8 recommend CA as one of the options to combat climate change impacts. Chapters 9, 10 and 11 advocate alternative climate-smart rice cropping systems such as SRI and AWD, that not only help in improving adaptation and mitigation, but also give higher yields. Since rice is the major crop in South and SE Asia, any alternative technology that reduces water and fertilizer use without compromising on yield, and at the same reduces net CH_4 emissions will be desirable. Government policies in India, China and SE Asia are supporting the scaling out of these measures on a larger scale. They still have a long way to go before the policy is put into practice and benefits are realized. For this to happen, farmers' capacity and access to inputs to adopt the new technologies should be improved. The private sector can play a significant role in scaling out the technologies. In the process, care should be taken to ensure that smallholders are not marginalized.

As discussed in the introductory chapter, the success of adaptation and mitigation initiatives would largely depend upon the involvement of smallholders who constitute a majority of the farming community in developing countries. Scientists must also ensure that adaptation tools and technologies are simple, locally adaptable and easy for smallholders to implement. This implies the development of agricultural technologies, including their identification, testing, and scaling out through farmer-led research and farmer-to-farmer exchange of information as emphasized in chapter 6. As discussed in chapters 2 and 6, local knowledge and agricultural practices adopted by farmers will complement the scientific knowledge.

Agroecological systems are also seen as a viable option for small holders where integration of soil conservation strategies using low cost local inputs, crop rotation and other cropping systems is possible, which ultimately helps to build long-term community resilience and also provides the agriculture-nutrition connect (Bezner Kerr *et al.,* 2007; Giovannucci, *et al.,*2012). Agroecological systems are recognized as promising approaches in reducing production costs and raising product added value, while also protecting soil, water, biodiversity and the health of both producers and consumers. Reallocation of resources, capacity building and knowledge development within each country may be needed to boost agroecology at the farm level (De Schutter, 2011). Chapter 6 in this book advocates agroecology as the best option for resource poor farmers in Malawi.

Science and technology will continue to play a key role in addressing problems of climate change and food insecurity. Research and development programs should target crops and systems that can address risks from climate change, reduce pressure on land and water resources, and at the same time target crops that address the nutritional needs of the poor (Vermeulen *et al.,* 2012). The link between the scientific community, extension agencies and farmers should be strengthened to develop and implement demand-driven research agendas (Garvey, 2012). Government should invest in long-term agricultural research programs and help to build the knowledge base. Necessary policy support has to follow to support the long-term nature of specific scientific developments involving breeding of crop varieties to adapt to different climate, food and nutrition conditions (Swaminathan, 2013).

The second major issue is *land tenure, ownership rights and reforms* that played a key role in shaping agriculture development and food security in many countries. Chapters 3, 4, 5 and 12 discussed the importance of secure land ownership rights and the impacts on agricultural development and food production. Vietnam is a good example of how a change in land rights has led to dramatic increase in food production during the 1980s (Hiebert and Nguyen, 2012). Land tenure is a key factor that determines a household's investment in the land and food production. Entitlements to land and other assets also determine household's access to other inputs including credit and social status in the society (Sen, 1981). Lack of tenure and property rights to cultivate land has been a disincentive for farmers to invest in land in many countries. This problem is exacerbated by the lack of land tenure guarantees for smallholders, and, in recent years, the expansion of agribusiness and associated land investments or 'land grabbing' by both domestic and foreign enterprises (FSWG, 2011). Land grabbing in the name of land leasing is happening on a large scale in countries across Africa and other regions across the globe. Long-term land leases neither provide livelihoods to the local communities nor contribute positively to food security of the country. On the contrary, it alienates local communities from lands on which they had traditional rights for grazing, water and other resources. Giving clear title deeds over land improves the possibility for smallholders access to credit, extension services and rights to use water resources and thereby food security. One of the strong recommendations to governments is that current and future policies should aim at providing clear rights and entitlements to land and other resources to the rural poor, women and smallholders where needed.

The third key issue is *institutional and policy frameworks* that enable farmer management of climate risks and adoption of context-suitable agricultural practices, technologies and systems. Lele *et al.* (2013) emphasize that governance issues affect the choice of policies, institutions and outcomes for addressing the daunting challenges due to climate change. Hence, good governance is a basic tenet for countries to build strong institutions. There are large variations between and even within the countries and the status of the institutional frameworks that determine the potential of the farmers. South and SE Asia are better off to combat climate risks as compared to SSA. The former have invested in agriculture research, infrastructure and development. As recommended in chapter 5, it is essential to develop rural infrastructure, provide secure land tenure, and improve market links to smallholders. Short-sighted government policies have led to unsustainable farming systems in many countries in the SSA (Tesfai *et al.*, 2015). Chapter 6 discussed how the government policies to support major cereals such as maize production through subsidies has literally replaced millet and other traditional crops that have the potential to adapt to climate change and provide a diverse food basket to the rural poor.

Chapters 2, 3, 4, 5, 9 and 13 describe the long-term policy interventions that are essential to promote sustainable agricultural development through approaches such as CSA and agroecology. These include:

- institutional and policy support to increase water conservation and water availability (farm and micro-water shed level) and reduce drought risk;
- long-term programs for breeding crop varieties that are tolerant to high temperatures, drought and salinity;
- policy implementation ensuring secure land tenure to smallholders and women in particular;
- government investments in smallholder adaptive capacity programs;
- developing financial and environmental risk management tools at the household, community and regional scales to ensure timely response to extreme events; and
- adapting agricultural management practices to be more climate-smart, such as no-till agriculture, crop diversification and crop rotations, and alternative rice cropping systems.

Climate change cuts across administrative and political boundaries, sectors and agencies and hence policies have to be looking beyond these boundaries. Policy formulation needs to be supported by sound scientific data and information. That is often not the case in many countries (Nagothu, 2015). Chapter 2 concluded that information required by policy makers will need to be specific and flexible enough to quantify climate impacts specifically enough to be integrated into multiple policy and decision-making tools (Dunford *et al.*, 2014; Krishnamurthy *et al.*, 2014). These tools are necessary for developing good decision support systems. Some countries in Asia, including India, China and Vietnam, are investing heavily to build a wide network of weather stations that could provide precise data for developing weather forecasting services with accuracy. Government policy to invest in climate related data collection and services, and proper use of the data significantly reduces vulnerability (Haggblade *et al.*, 2013). The SSA region is lagging behind, and governments therefore should realize the importance of building such infrastructure that will ensure food security in the end.

Governments and communities need to prepare for the worst, and develop adaptation strategies that are low cost and not too risky for small farmers. Chapter 13 concluded that national policies and investments, as observed in India, are important to provide guidance to states or provinces to further develop climate change adaptation and mitigation strategies that are locally specific. Promoting sustainable agricultural systems that can deal with resource scarcities and help mitigate and adapt to climate change must remain the overarching goal in the regions vulnerable to climate risks (FAO, 2011). Several low cost and easily adaptable measures, such as farm ponds for rainwater harvesting, retention of crop residues, soil mulching, integration of livestock into farm units, need-based water application and alternative production systems, are all examples of building blocks to climate resilient farming systems (FAO, 2013).

Public-private partnership (PPP) models can play an important role to support climate change adaptation and mitigation initiatives. Policy should encourage the development of PPP models involving farmer organizations. With the development of ICT tools, several private agencies are investing in providing climate services to farmers. It is necessary to make sure that farmers are not exploited in the process, and are provided services at reasonable prices. All the support institutions also require restructuring to replace command and control systems with more participatory farmer-centred approaches. This includes: i) improving links between farmers, researchers and extension staff so that practical problems are addressed; and ii) investing in human capital to ensure the necessary skills are available to meet the needs of a new era of participation and engagement (Haggblade *et al.,* 2013). The formulation and effective implementation of adaptation measures requires coordination and support from public, private and civil society stakeholders at international to local levels in building evidence and assessment tools, strengthening national and local institutions, developing coordinated and evidence-based policies and increasing financing and its effectiveness (Lipper *et al.,* 2014).

The fourth key issue is *gender and overall social inclusion* that was not adequately addressed in the past. Gender has to be given importance while promoting new agricultural development strategies to address climate change. Overall, broader social inclusion will be necessary to reduce the socio-economic inequalities and vulnerability (Revenga and Shetty, 2012). It is often the poor, smallholders and women that are most affected by climate change and climate variability. Rural areas that are remote are more vulnerable and hence the risk is high. As a result, we see an increasing outmigration of men and youth from rural areas in developing countries; i.e. fewer people left to farm in the rural areas. Often the outmigration is linked to economic reasons, climate risks and the search for better livelihood options. Outmigration creates a greater burden on smallholders and women left behind to produce for the expanding urban middle class in the developing economies (Nagothu, 2015). Chapter 12 described the social and gender inequalities. Despite the central role they play, women are often marginalized when it comes to land ownership, benefits of credit programs, and food distribution or nutrition schemes.

Land reforms are needed to provide women secure land tenure, besides providing access to credit and markets. At the same time, special efforts are required to support women farmers to lead CSA innovations (CCAFS, 2014). While gender inequality is gradually reducing in the urban middle class, the situation has not changed much amongst Africa's rural poor and parts of south Asia. The plethora of gender specific hindrances women face in regard to the basic resources of production are often interrelated, and demand comprehensive strategy approaches. In future, the role of gender in agriculture and food security should be duly recognized by policy (Jones, 2012). A toolbox developed by CCAFS in 2014 enables development practitioners to mainstream gender and social equality approaches (Jost *et al.,* 2014).

The fifth key issue is *securing investments and integrating relevant stakeholder interests* to promote adaptation and mitigation programs as we move towards a new paradigm shift in agriculture and combating climate change impacts. Although short-term investments are important, long-term funding support is needed to make agriculture an engine of growth and poverty reduction. According to CIAT (2015) the three things that matter for scaling out adaptation or mitigation initiatives would be, developing a strong knowledge base (formal and informal knowledge) and filling the knowledge gaps, integrating stakeholder interests and adequate investments to scale up promising CSA practices. Despite knowing the need, governments and international agencies are not yet investing sufficiently in the agriculture sector. A major part of the investments made in the agriculture sector is diverted to the farmer subsidies to purchase fertilizers.

Agricultural subsidy programs have put a tremendous burden on each country's national expenditure and diverted the funds from improving rural infrastructure. Environmental consequences of the subsidies have been far-reaching, leading to serious pollution of the land and fresh water resources. This is particularly evident in south Asia, where farmers relied heavily on subsidized fertilizer programs since the Green Revolution. It is likely that the governments will slowly phase out subsidy programs due to financial reasons (as we are already witnessing in India and other countries), thus transferring the entire burden over to smallholders. Subsidies could be used to improve soil condition and build long-term fertility rather than fund agrochemicals (Tilman *et al.,* 2002).

Investments from the private sector are becoming important in the future, but there should also be strong policy support to engage smallholders and other stakeholders in the process. Several chapters in this book emphasized the need for stakeholder integration and securing investments to support new climate change adaptation initiatives. Private sector investments need to be encouraged, and should complement government funding to support farming communities. It is critical to integrate smallholders more fully into national and global food systems – including health systems, value chains and markets. In order to facilitate this engagement, institutions and policies must be designed for the benefit of smallholders in order to facilitate flows of necessary resources, and targeted training and capacity building.

Conclusions and way forward

Climate change leading to climate extremes will influence farming communities in the form of higher temperatures and heat waves, droughts, intensive rainfall and floods. With a warming climate and erratic monsoon behaviour, we may need to make a significant shift in the way agriculture is practised today in order to promote sustainable and profitable agricultural systems in the future. There will be regional variations in the foregoing solutions due to variability in the regional resources and climate patterns. There are at present no suitable models or approaches that can simply be replicated to address such a variability, although much knowledge already exists and lessons can be learned from previous experiences.

Approaches such as agroecology, CA and CSA have emerged in recent years to deal with multiple challenges, by focusing on integrating institutional responses. The different schools provide divergent views when we go beyond the superficiality of some technical similarities, and some overlap in the different 'paradigms' related to the future of agricultural development and for betterment of the farming communities. It will be interesting to see how long the scientific community will seek to straddle between these different paradigms, and which concept or approach offers sustainable solutions.

In this book, we have attempted to discuss the different approaches and provide examples displaying how they function in reality, and the barriers they face. We argue that in the end it is the welfare of the farmers, their livelihoods and the need for social inclusion, and protection of the environment that will enable sustainable agricultural development and food and nutrition security in the long-term. Hence scientists have to put the priorities of the farmers and the environment first, and put aside promoting their own interests.

All this requires cooperation between the governments, farmer organizations, donors, international agencies and the private sector to engage with one another through multi-stakeholder platforms, and work collectively in order to ensure long-term sustainable agricultural development and food security. This has to be combined with a real focus on infrastructure development, both financial and physical, and market access in order to provide smallholders with the necessary support. At the scientific level, scientists from different disciplines have to work together, strengthening the multidisciplinary approach to address the complexity of climate change. Climate change-related policies should be based on good scientific results that also use local knowledge and farmer experiences in developing such knowledge. Efforts should be made to facilitate sharing of climate data, climate services and agricultural technologies through knowledge platforms that cut across country barriers and regions to stop duplication of efforts. Policy and institutional support is fundamental to support any new initiatives to promote sustainable agricultural development and address climate change impacts. The way forward is genuine stakeholder cooperation at different levels, more targeted climate research that is participatory, capacity building and securing investments to support future initiatives to address climate change.

References

Aggarwal, P., Zougmore, R. and Kinyangi, J. (2013) *Climate-smart Villages: A Community Approach to Sustainable Agricultural Development*, Climate Change Agriculture and Food Security (CCAFS), Copenhagen, Denmark.

Akanda, A.I. (2010) Rethinking crop diversification under changing climate, hydrology and food habit in Bangladesh, *Journal of Agriculture and Environment for International Development*, 104 (1–2): 3–23.

Alliance for a Green Revolution in Africa (AGRA) (2013) *Africa Agriculture Status Report: Focus on Staple Crops*. Nairobi, Kenya, online, http://www.agra.org/silo/files/agra-africa-agriculture-status-report-2014.pdf [Accessed October 04, 2015].

Bezner Kerr, R., Snapp, S., Chirwa, M., Shumba, L. and Msachi, R. (2007) Participatory research on legume diversification with Malawian smallholder farmers for improved human nutrition and soil fertility, *Experimental Agriculture*, 43: 437–453.

Campbell, B. *et al.*, 2011. *Agriculture and Climate Change: A Scoping Report*. pp. 9–15. Available at: https://cgspace.cgiar.org/handle/10568/10306 [Accessed February 23, 2015].

Climate Change Agriculture and Food Security (CCAFS) (2014) *Climate-smart Agriculture. Acting Locally, Informing Globally.* The CCAFS 2014 annual report, Available at: https://cgspace.cgiar.org/bitstream/handle/10568/65717/CCAFS_2014_Annual_Report.pdf. [Accessed October 06, 2015].

International Center for Tropical Agriculture (CIAT) (2015) World Bank: Smart investments in agriculture, online, http://ciatblogs.cgiar.org/support/world-bank-smart-investments-in-triple-win-agriculture/ [Accessed September 23, 2015].

CIAT (2015) Sustainable Food Futures: Getting the Fundamentals Right, online, http://www.gfar.net/news/ciat-releases-annual-report-2015-2016 [Accessed September 14, 2015].

COP 21 (2015) Climate smart agriculture concerns, online, http://www.climatesmartagconcerns.info/cop21-statement.html [Accessed October 09, 2015].

Crane, T.A., Roncoli, C. and Hoogenboom, G. (2011) Adaptation to climate change and climate variability: The importance of understanding agriculture as performance, *Wageningen Journal of Life Sciences*, 57: 179–185.

De Schutter, O. (2011) *Agro Ecology and the Right to Food*. Report submitted by the Special Rapporteur on the right to food to UN Human Rights Council, December 20, 2010, online, http://www.srfood.org/images/stories/pdf/officialreports/20110308_a-hrc-16-49_agroecology_en.pdf [Accessed September 3, 2013].

Dunford, R., Harrison, P.A., Jäger, J., Rounsevell, M.D.A. and Tinch, R. (2014) Exploring climate change vulnerability across sectors and scenarios using indicators of impacts and coping capacity, *Climatic change*, 128: 1–16.

Food and Agricultural Organization (FAO) (2011) *Good Food Security Governance: The Crucial Premise to the Twin-Track Approach*, online, http://www.fao.org/fileadmin/templates/righttofood/documents/project_f/fsgovernance/workshop_report.pdf [Accessed May 05, 2014].

Food and Agricultural Organization (FAO) (2013) *Climate-Smart Agriculture, Sourcebook*, online, http://www.fao.org/docrep/018/i3325e/i3325e.pdf [Accessed August 15, 2013].

FSWG (Food Security Working Group) (2011) Upland land tenure security in Myanmar: an overview, online, http://www.lift-fund.org/implementing-partner/235 [Accessed September 24, 2013].

Garvey, K. (2012) Global food security: the role of science and technology, 17–19 October 2012, Wilton Park Conference report, WP1189, UK, online, https://www.wiltonpark.org.uk/wp-content/uploads/WP1189-Report.pdf [Accessed May 05, 2014].

Giovannucci, D., Scherr, S.J., Nierenberg, D., Hebebrand, C., Shapiro, J., Milder, J. and Wheeler, K. (2012) Food and Agriculture: The Future of Sustainability (March 1, 2012). *The Sustainable Development in the 21st century (SD21) Report for Rio+20*, New York: United Nations, online, http://ssrn.com/abstract=2054838 or http://dx.doi.org/10.2139/ssrn.2054838 [Accessed April 30, 2014].

Gupta, S. and Nagothu, U.S. (2011) Institutional and policy measures to address climate change in India. In: Nagothu, U.S., Gosain, A.K., and Palanisami, K, (eds), Water and Climate Change: *An integrated Approach to Address Adaptation Challenges* Macmillan Publishers, New Delhi.

Haggblade, S., Boughton, D., Denning, G., Kloeppinger-Todd, R., Cho, K.M., Wilson, S., Wong, L.C.Y., Oo, Z., Than, T.M., Wai, N.E.M.A., Win, N.W. and Sandar, T.M. (2013) A strategic agriculture and food security diagnostic for Myanmar, online, http://

fsg.afre.msu.edu/Myanmar/myanmar_agricultural_sector_diagnostic_july_2013.pdf [Accessed October 08, 2015].

Hiebert, M. and Nguyen, P. (2012) Land reform: a critical test for Myanmar's Government. *George Washington University Centre for Strategic and International Studies* 3(2): 85–98, online, http://csis.org/publication/land-reform-critical-test-myanmars-government [Accessed October 10, 2015].

Intergovernmental Panel for Climate Change (IPCC) (2013) *The Physical Science Basis. Summary for Policymakers. Contribution of the Working Group I to the Fifth Assessment Report of the Intergovernmental Panel on Climate Change*, Cambridge University Press, Cambridge.

International Assessment of Agricultural Knowledge, Science and Technology for Development (IAASTD) (2008) *Synthesis Report with Executive Summary: A Synthesis of the Global and Sub-global IAASTD Reports.* Edited by McIntyre, B.D., Herren, H.R., Wakhungu, J. and Watson, R.T. Island Press, Washington D.C.

Jones, M. (2012) First Global Conference on Women in Agriculture (GCWA): Empowering women in agriculture: rethinking the agricultural needs and actions through the eyes of women, *Food Security*, 4 (2): 305–306.

Jost, C., Ferdous, N. and Spicer, T.D. (2014) *Gender and Inclusion Toolbox: Participatory Research in Climate Change and Agriculture*, Copenhagen, Denmark.

Krishnamurthy, P.K., Lewis, K. and Choularton, R.J. (2014) A methodological framework for rapidly assessing the impacts of climate risk on national-level food security through a vulnerability index, *Global Environment Change*, 25: 121–132.

Lele, U., Klousia-Marquis, M. and Goshwami, S. (2013) Good governance for food, water and energy security, *Aquatic Procedia*, 1: 44–63.

Lipper, L. *et al.* (2014) Climate-smart agriculture for food security, *Nature Climate Change*, (4): 1068–1072.

Nagothu, U.S. (2015) The future of food security: summary and recommendations. In: Nagothu, U.S. (ed). *Food Security and Development: Country Case Studies,* Earthscan/ Routledge, London.

Pingali, P.L. (2012) Green revolution: Impacts, limits, and the path ahead, online, http://www.ncbi.nlm.nih.gov/pmc/articles/PMC3411969/ [Accessed April 27, 2014].

Revenga, A. and Shetty, S. (2012) Empowering women is smart economics, *Finance and Development*, 49 (1): 40–43.

Sen, A. (1981) *Poverty and Famines. An Essay on Entitlements and Deprivations.* Oxford University Press, Oxford.

Swaminathan. M.S. (2013) 'From Bengal famine to right to food', *The Hindu*, February 13, 2013, online, http://www.thehindu.com/todays-paper/tp-opinion/from-bengal-famine-to-right-to-food/article4409557.ece [Accessed September 13, 2013].

Tesfai, M., Adugna, A. and Nagothu, U.S. (2015) Status and trends of food security in Ethiopia. In: Nagothu, U.S. (ed). *Food Security and Development: Country Case Studies,* Earthscan/Routledge, London.

Tilman, D., Cassman, K.G., Matson, P.A. and Naylor, R.L. (2002) Agricultural sustainability and intensive production practices, *Nature*, 418: 671–77.

Vermeulen, S.J., Campbell, B. and Ingram, J.S.I. (2012) Climate change and food systems, *The Annual Review of Environment and Resources*, 37: 5.1–5.28.

Zhou, Z.Y. (2015) Food security in China: past, present and the future. In: Nagothu, U.S. (ed). *Food Security and Development: Country Case Studies,* Earthscan/Routledge, London.

Index

Taylor & Francis eBooks

Helping you to choose the right eBooks for your Library

Add Routledge titles to your library's digital collection today. Taylor and Francis ebooks contains over 50,000 titles in the Humanities, Social Sciences, Behavioural Sciences, Built Environment and Law.

Choose from a range of subject packages or create your own!

Benefits for you

» Free MARC records
» COUNTER-compliant usage statistics
» Flexible purchase and pricing options
» All titles DRM-free.

Benefits for your user

» Off-site, anytime access via Athens or referring URL
» Print or copy pages or chapters
» Full content search
» Bookmark, highlight and annotate text
» Access to thousands of pages of quality research at the click of a button.

REQUEST YOUR **FREE** INSTITUTIONAL TRIAL TODAY

Free Trials Available
We offer free trials to qualifying academic, corporate and government customers.

eCollections – Choose from over 30 subject eCollections, including:

Archaeology	Language Learning
Architecture	Law
Asian Studies	Literature
Business & Management	Media & Communication
Classical Studies	Middle East Studies
Construction	Music
Creative & Media Arts	Philosophy
Criminology & Criminal Justice	Planning
Economics	Politics
Education	Psychology & Mental Health
Energy	Religion
Engineering	Security
English Language & Linguistics	Social Work
Environment & Sustainability	Sociology
Geography	Sport
Health Studies	Theatre & Performance
History	Tourism, Hospitality & Events

For more information, pricing enquiries or to order a free trial, please contact your local sales team:
www.tandfebooks.com/page/sales

 **Routledge**
Taylor & Francis Group

The home of
Routledge books

www.tandfebooks.com